T0174773

From AI to Robotics
Mobile, Social, and Sentient Robots

From AI to Robotics
Mobile, Social, and Sentient Robots

ARKAPRAVO BHAUMIK

CRC Press
Taylor & Francis Group
Boca Raton London New York

CRC Press is an imprint of the
Taylor & Francis Group, an **informa** business

A CHAPMAN & HALL BOOK

MATLAB® is a trademark of The MathWorks, Inc. and is used with permission. The MathWorks does not warrant the accuracy of the text or exercises in this book. This book's use or discussion of MATLAB® software or related products does not constitute endorsement or sponsorship by The MathWorks of a particular pedagogical approach or particular use of the MATLAB® software.

CRC Press
Taylor & Francis Group
6000 Broken Sound Parkway NW, Suite 300
Boca Raton, FL 33487-2742

First issued in paperback 2020

© 2018 by Taylor & Francis Group, LLC
CRC Press is an imprint of Taylor & Francis Group, an Informa business

No claim to original U.S. Government works

Version Date: 20180205

ISBN-13: 978-0-367-57209-9 (pbk)
ISBN-13: 978-1-4822-5147-0 (hbk)

This book contains information obtained from authentic and highly regarded sources. Reasonable efforts have been made to publish reliable data and information, but the author and publisher cannot assume responsibility for the validity of all materials or the consequences of their use. The authors and publishers have attempted to trace the copyright holders of all material reproduced in this publication and apologize to copyright holders if permission to publish in this form has not been obtained. If any copyright material has not been acknowledged please write and let us know so we may rectify in any future reprint.

Except as permitted under U.S. Copyright Law, no part of this book may be reprinted, reproduced, transmitted, or utilized in any form by any electronic, mechanical, or other means, now known or hereafter invented, including photocopying, microfilming, and recording, or in any information storage or retrieval system, without written permission from the publishers.

For permission to photocopy or use material electronically from this work, please access www.copyright.com (http://www.copyright.com/) or contact the Copyright Clearance Center, Inc. (CCC), 222 Rosewood Drive, Danvers, MA 01923, 978-750-8400. CCC is a not-for-profit organization that provides licenses and registration for a variety of users. For organizations that have been granted a photocopy license by the CCC, a separate system of payment has been arranged.

Trademark Notice: Product or corporate names may be trademarks or registered trademarks, and are used only for identification and explanation without intent to infringe.

Visit the Taylor & Francis Web site at
http://www.taylorandfrancis.com

and the CRC Press Web site at
http://www.crcpress.com

To all the wonderful robots
that are not mentioned in this book.

Contents

CHAPTER 10 ▪ Super intelligent robots and other predictions 331

APPENDIX A ▪ Running The Examples 361

Preface

Since Walter's Turtles, technology has come a long way and we can now boast of state-of-the-art robots, such as ASIMO, PR2, NaO and Pepper. My interest in this field of study, which has taken on obsessive proportions, is due to my academic background, my participation in various open source robotic communities (Player/Stage, ROS, MORSE etc.), my teaching assignments and projects with my students and above all, a child-like desire to make robots. Not withstanding my personal desire to put together nearly all that I have learnt over the last ten years, robotics and AI truly stand to change the world as we know it. In this text spanning ten chapters, I have looked to various researchers; Braitenberg, Dennett, Brooks, Arkin, Murphy, Winfield, Vaughan, Dudek, Dorigo, Sahin, Bekey, Abney, Wendell, Takeno, Bringsjord and the Andersons, as those who have had a lasting impression on me and have shown me the proverbial light to the correct path. Other than these academic influences and motivation, science fiction, particularly Isaac Asimov, Philip K. Dick, Arthur C. Clarke, Cory Doctorow, Peter Watts etc., led me to engaging queries on various aspects of AI and robots and their influence on human society and helped me to illustrate ideas in the later chapters. Furthermore, I have been influenced by and have enjoyed a long list of movies, among which are iconic robot movies such as, Wall-E, IRobot and cult classics, such as The Metropolis, Blade Runner and West World, and the more recent ones, such as Interstellar, Real Steel, Robot and Frank, Big Hero 6 and Ex-Machina.

This text is meant for undergraduate students and may also serve as a reference for graduate students. It attempts to introduce the reader to what has been achieved in agent-based robotics over the last four decades, the ways we can implement such concepts with easily available electronics and open source tools, multirobot teams, swarm robotics and human robot interaction, as well as the efforts to develop artificial consciousness in robots. The last chapter is on the future of AI and robotics in the foreshadow of the prophecies of super intelligent AI and technological singularity.

The journey starts with mythical lore from ancient Greece and ends with a debate on a futuristic prophecy where human beings and technology merge to create superior beings as heir to our evolutionary pathways. From Haphaestus to Philon to Da Vinci to Tesla to Walter to Toda to Moravec to Braitenberg to Brooks to Kurzweil, this text is livened up with examples, pictures, one-on-one chatter with experts, schematics and cartoons.

The seven broad themes addressed across the ten chapters are;

1. AI vs. Applied AI

There is always some contrast in the theory and the practice for any discipline, and the appeal for classroom teaching versus hands-on application in the laboratory, workshop or in industry has had its contrasts. It is many folds more for AI. It has been my experience that the question often boils down to either 'learn AI' or 'make robots'. Outstanding texts in AI as Russell and Norvig and Rich and Knight don't really help much towards building a robot, and one has to dig out online tutorials, YouTube videos etc. The same is true for machine

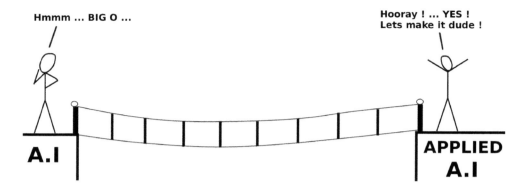

FIGURE 1 **The bridge from AI to applied AI.** For new initiates the two domains appeal as two different disciplines. An undergraduate student will find it difficult to see the convergence of the doctrines of AI; for example, planning, search and knowledge representation are the philosophical basis and analytical tools for making robots, but at a beginners level it reduces to simple programming scripts, interfacing of sensors with motors and ease of design. Therefore, the correlation between AI and robotics may not be very apparent to a young enthusiast.

learning; a text which introduces a student to neural network, such as Bishop is probably the best treatment that one can offer on the subject, but to make a neural network, the student will have to find suitable C or Python libraries. The bridge, as shown in Figure 1, between theory and application is not only distant but requires different faculties. It is also to be added that AI, by itself is not enough to make robots; the effort calls for electronics, mechanical design, sensor design and other overlapping disciplines, depending on the specifics of the robot. The book makes an effort to shorten this bridge and hopefully, make the crossing easier, by correlating concepts to their applications.

The applications cover designing simple behaviours in Chapter 2 and 3, navigation in Chapter 4, and multirobot systems and swarming, many of these are supplemented with examples from practical scenarios, algorithms and introduction to the software. Navigation is unique as it is a basic behaviour and also a design tool. Chapter 5 has a few examples with hardware which are meant as appetizers to whet the imagination and creativity of the budding robot enthusiast.

2. Deliberative vs. Reactive and the Question of Relevance

This debate between the two ways of approaching agent-based robotics was triggered by the pioneering work of Rodney Brooks in the mid 1980s. However, the thread of thought can be traced back to behaviourism and phenomenology. Behaviourism was established as a sub-discipline in psychology in late 1890s through the works of Pavlov, Thorndike and by Skinner in 1950s. Philosophical development in phenomenology by Husserl, Heidegger and Merleau-Ponty helped to cement the concepts of situatedness and embodiment. Assimilation of these ideas into robotics was done independently by three daring scientists, who had the wish to create artificial creatures: Walter in late 1940s, Toda in mid 1960s and Braitenberg in early 1980s. Walter explicitly demonstrated design of behaviour in his turtles, Toda suggested the first models of autonomous agency and Braitenberg's gedanken experiments showed that simple agency can lead to very advanced performance. These independent conclusions led to the basic principles of behaviour-based robotics.

The frame problem and the question of relevance is an inherent burden in designing a robot and forms the crux of the argument. It cannot be eradicated, and has to be either completely avoided as in behaviour-based paradigm, or consider addressing it to a minimal as in PENGI. In sheer contrast, we human beings can 'solve' the frame problem 'on the run'. Nowadays, most robots have deliberative as well as reactive modules and are supplemented with machine learning, thus overcoming the issue of relevance at least in low-level repetitive behaviours contained in a more or less known environment. However, this ugly monster raises its head time and again. The utilitarianism vs. deontology debate for autonomous vehicles is a direct extension of the deliberative vs. reactive quarrel, where all deontic approaches are pegged to a set 'frame' of pre-programmed rules and utilitarian principles are based on immediacy. Also, most attempts to develop conscious robots such as those discussed in Chapter 9 are defacto inquiries into the question of relevance: Is the dynamics of a tambourine relevant for its beat? Will looking into the mirror make the robot find its own image? Can a robot appropriate causality, and relate actions to conclusion? I would wish to believe that a robot which can 'solve' the frame problem across five senses as human beings do, would be conscious, devoid of any shade of doubt.

Fodor claimed that due to the frame problem, AI is dead. For what the last two decades of research has provided, the debate of the frame problem has added to the richness of AI.

3. Engineering Robot Behaviour

My best reference for understanding behaviour is, Ronald Arkin's tome, *Behaviour Based Robotics*, and the works of Richard Vaughan and Alan Winfield, and a good part of the book documents the engineering of desired behaviour in a robot.

Reactive systems are faster and easier to design than the deliberative or hybrid ones, and they do not need sleek hardware such as a camera or a laser scanner. The unwitting contrast is that since behaviour is emergent it is difficult to design the architecture for a desired performance. Very simple behaviours, such as 'track a light source' or 'follow a line', hardly demonstrate the inherent relation between the agent and the environment. Swarm robotics and Braitenberg vehicles help to illustrate such intimate relations and contrast of the ease of design to the unpredictable nature of performance. Gerardo Beni, very cleverly summarises this phenomenon, that [robot] behaviour stands as a convergence of two radically different ideas, unpredictability and the wish to find order. The definition of robot behaviour has been modified over the last three decades, and may be further edited in the years to come.

Walter's Turtles was the beginning of agent based robotics, but it was left to Simon and later Toda, to envisage a self-sustaining mobile agency with local landmarks and multiple goals which can negotiate tradeoff between its energy supply and the task at hand. The fungus eaters mined uranium on a far off planet and gathered fungus which when digested provided them with energy, Toda's model was also a blueprint for early path planning models. The Mars rovers can be said to be the modern-day avatars of the fungus eater, though they harness their energy from solar panels and not Martian fungus.

In the 1980s, Wilson's model of the ANIMATs, Pfeifer's principles for design and development of agent based robotics and Lumelsky and Stepanov's Bug Algorithms were motivated from mother nature and streamlined the principles for designing of automated self-sustaining mobile agency.

Braitenberg's vehicles are designs of more involved functionalities, as emotions, value systems, correspondence of information, collating aspects of reality to memory, learning and evolution, all of these from simple sensory-motor principles. Braitenberg's vehicles have served as illustrative examples in a number of chapters and is a recurring theme across the book.

AI has wilfully plagiarised from the natural world and anthropomorphic motivation in design and behaviour models derived from ethological studies has been the trend. Best examples are: Fukuda's brachiatron was motivated from the swinging motion of apes; ECOBOT series of robots modelled on the process of digestion; swarm behaviour designed on trophallaxis in insects and birds; INSBOT project which introduced robot decoys in a cockroach colony etc.

The contrast between robot groups and robot swarms further helps to appreciate the facets for engineering a behaviour. Reactive methods are the basis of multirobot swarming, this cohesion among the robots is a boon as it is flexible to dynamical changes in the local environment, and easier to scale up without hampering the performance. However, robot swarms are notorious and it is difficult to warrant their reliability and tag its performance to a factor of fault tolerance. Therefore, engineering a desired swarm behaviour is more a subject for future research than a cutting-edge technology. On the other hand, robot groups with designed with deliberative paradigm working over a local wifi network lend more safety and predictability. However, they are not very flexible, and they lack versatility and are difficult to scale up.

4. Replicating Human-Like Intelligence

Nearly all human activities can be classified as motivated either by survival or curiosity. The former is essential for our homeostasis, the latter encourages growth of culture, society, intellect, technology etc. Robot designs have exploited both these facets, both in low-level behaviour and also in sophisticated ones. Self preservation is enshrined into Asimov's laws and is also coded into simple robots, viz. Roomba, and curiosity is closely tagged to the phenomenon of emergence and is the driver of the behaviour-based paradigm.

Behaviour-based approaches are insufficient to provide for ethical agency and artificial consciousness, and the ethical question finds the best answer in developmental approaches, while the question on artificial consciousness for robots has probed into various avenues, and there is yet to be human-like conscious behaviour in robots. To fully address the question on ethics and consciousness leads to queries regarding the human brain, and robotics still awaits a breakthrough in medical science, neurology and psychology.

5. Open Source Robotics

Robotic suites Player/Stage/Gazebo and later ROS were designed for UBUNTU installations and helped unify the domains of simulation and running of the real robot. These software suites function in a client server fashion and encourage crowd sourcing in code pieces, designs and tweaks to hardware, thereby establishing a benchmark for hardware and software in robotics. ROS also integrates open source hardware such as Arduino, Raspberry Pi, ODROID, etc. making the design and development of a new robot more modular and therefore open to crowdsourcing.

6. Robots as Conscious Beings — More Than a Tool, a Machine or a Slave

As a robot enthusiast, I cannot justify a book without the mention of Isaac Asimov. His vision for a human robot society is the backdrop of Chapter 8. However, his three laws are found to be mutually conflicting. They have implementation issues and lack practical applications. Similarly, the Turing test is more a concept than a real tool to distinguish between the natural and the artificial. Such failures of Asimov's laws and the Turing test make us reflect as to what robots are and will they ever measure up to being humans, and what future they may have. Instead of labeling robots as tools or slaves, the greatest triumph of AI would be if robots can acquire free will and demonstrate human-like sentient behaviour and such ideas are discussed in Chapter 9.

Robots imbued with deontic ethics have been a success in the laboratory and may soon be employed as nurse and carer robots. However, consciousness in robots as yet has been in either a narrow concern or it is short lived — for a few seconds.

The unfortunate consequence of this triumphant robot revolution is the loss of human jobs, and this has fueled a parley which is raging across newspapers, blogs, online forums and election manifestos. If the trend continues, we may soon have a modern-day Luddite revolution which might change the rules of our social and economic systems and call for a limitation in use of automation and AI. The irony of AI's greatest achievement doesn't end here and an apocalyptic future where robots annihilate the human race or alternatively dominate our world and reduce human beings to an inferior second-rate species, starts to seem all too real. Experts in the field foresee this as a neo-Darwinian 'jump' in our evolutionary track which will be discussed in Chapter 10.

Will artificial consciousness lead us to super intelligent beings which may spell doom for humanity and eventual extinction of the human race ? The query reminds me of a Hindu mythological tale[1] that my parents told me as a kid. The story is about four disciples who have learned physiology, anatomy and medicine and have set out to hone their skills on real-world problems. These four happen upon the carcass of a dead tiger, and in the hope of testing their skills the first disciple assembles the bones of the dead tiger in an upright position, the second disciple gives muscles to this structure of bones and, not to be undone, the third disciple adds the organs. The fourth one, who was supposedly the smartest of the lot, was unabashed of his skills and despite warnings from the other three, gives life to the tiger. As a tragic end, the tiger leaps up with a growl and gobbles up the four disciples. Metaphorically, we are at the dare of the fourth disciple and AI may well pounce upon us — sooner rather than later, if we are not careful enough.

7. AI as the New Definition of Science

Science is the study of the underlying principles governing the working of the universe, and is largely based on experimentation and observation. This consistent and unified body of knowledge is as perceived by our sensory organs and inferred by the processing power of our brain. As we move towards a future dominated by automation, self enriching AI agents and virtual reality, the process of experimentation and observation will be dictated by technology, AI in particular. Therefore, AI will be the tool and discipline to lead us to science. Further, it is believed that close to technological singularity, technological growth will be off the charts, way too fast for biological intelligence such as ourselves to grasp at. With such a crescendo of processing power and plethora of information in the artificial realm, our definition of science as a consistent body of information and principles drawn from logic, and cause relating to effect will start to be strongly tied to AI.

Super intelligent robots, which are foreseen to have incredible processing power and have access to a mind boggling amount of information, are at the crossroads of romance and horror and found a number of nom de guerre. Skynet, AGI/ASI, Good's ultra intelligent self replicating machines and of course the moniker of technological singularity, which is discussed in detail in the last chapter. Kurzweil contends that by the 2040s human activities will be more in the virtual domain than the real world, and it will be accompanied with brain uploading which will eliminate death, at least as we known it, and thereby lead us towards a seamless merger between the biological and the artificial. Therefore, AI will be an overarching discipline encompassing all beings across known realms of knowledge and information, be it symbolic or sensory, and therefore indicate towards a paradigm shift and therefore a new definition of science.

Writing the book made me discover the part-time artist and the rookie cartoonist in

---

[1]This is one of the 25 tales which are conversations between King Vikramaditya and a ghost named Betal.

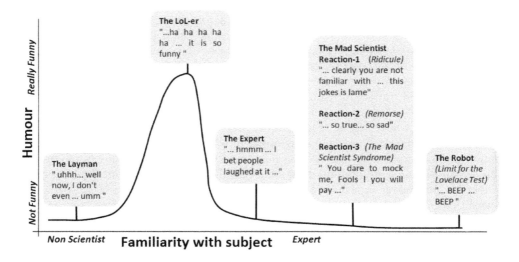

FIGURE 2 **McDermott's uncanny mountain.** Humour underscores human social interactions [223] and has defied attempts to recreate it in the artificial domain. This can be seen as an equivalence of Masahiro Mori's uncanny valley [214]. Humour is a unique attribute of human beings, with most acceptance in the average human being (IQ: 125-140). It thins out with increasing IQ and is negligible for a robot with access to all sorts of information across the world.

me and a great number of the cartoons and schematics have been sketched by yours truly. I found both laughter and wisdom in various webcomics, including the very popular ones, such as xkcd, Far Side, Dilbert, Oatmeal, smbc, Zen Pencils to the not so well known, such as Widder's Geek and Poke, McDermott's Red pen Black pen and Iyad Rahwan's cartoons on his webpage.

The cartoons has been helped by the inherent humour expressed by AI scientists and roboticists in their research papers or marked by the idiosyncrasy or opinion of a person. For example, Asimov's whiskers, Brook's elephants, Kirsh's earwig, PENGI's model on penguin, Dennett's illustration of the frame problem with variants of R2D2, the trolley problems, Hawking's opinions on a moratorium for AI, Knowledge game, Sparrow's test, Philosophical Zombie, Nagel's Bat etc. have found their way into my cartoons. Another glowing reason is that the text is written for human beings and may lack a similar appeal to robots and other super intelligent entities who may dare to read the book in the years to come, as is shown in Figure 2. Humour, particularly cartoons helps to illustrate a concept, and the satire uniquely blends it into human memory, such hardly happens for rote information, and more traditional means to teach a concept.

A mention needs to be made for the cartoon featuring Stephen Hawking and a growling Roomba robot which has a placard with ones and zeroes scribbled on it. This is now on the title page and it tells the story of how we are contemplating a moratorium on AI.

Other than my cartoons, there are breathtaking contributions from renowned animator and emmy prize winner, Maciek Albrecht with reproductions of 3 of his sketches[2] in Chapter 3 and a cheeky cartoon by Mark Shrivers in Chapter 2. There are also 4 one-on-one interviews, (1) STDR team from Aristotle University in Greece and (2) Daniel and Marcus at Yujin Robot, Seoul (3) Elizabeth Broadbent at University of Auckland, New Zealand, and (4) Junichi Takeno at Meji University, Japan. While interesting anecdotal references have been graciously provided by the Cool Farms team, Antony Beavers (Genesis at Twin

---

[2]The captioning for his 3 sketches have been done by me.

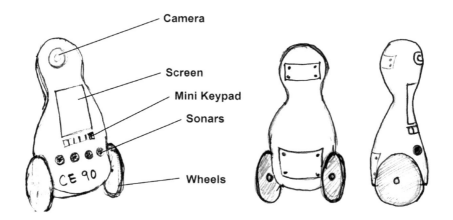

FIGURE 3 **Toby — first sketches.** These are the earliest sketches of the comic strip which I drew in the summer of 2015. The four-part comic strip illustrates various salient features of agent based robotics, and speculates on the future of human-robot interaction. A comic strip may seem out of place in a textbook, but the light humour helps to bring forth various themes, such as the semi-sentient paradigm, ethics in artificial agents, current issues in human-robot interaction, open source robotics, a future with robots with near-human mannerisms and social values, etc. The four-part Toby comic strip has been designed, developed and sketched by me, CC-by-SA 4.0 license.

Earth), Alan Winfield (The Walterians), Owen Barder's blog (Trolley Problems) and Kevin Kelly's blog (Maes-Garreau Law). For the software, Matt Zucker at Swarthmore College allowed me to use his web based Braitenberg vehicle simulator and my good friend and collaborator Luke Dunn let me use his Wall-E/Eva python chat script.

Two tools that I have used to convey ideas and information to the reader are:

1. Toby & the Walkers

Toby is a personal robot who is a mascot of sorts for the book and appears across four short comic strips. My early sketches for the robot are shown in Figure 3. This is the 2015 version of Toby and it is derived from the anthropomorphic motivations of a penguin without the beak, with a differential drive, interactive touchscreen, a mini keyboard, a camera and an array of sonars. It is priced at $600, and while the 2036 version is not discussed at length, it is left more to the reader's imagination and the promise of the future. In the first of the comic strips in the first Chapter, the Walkers are amused at the news about Philae, a robot which is exploring a comet. In the second comic strip, the Walkers purchase a C3-90 robot as a Christmas family present, and the robot connects with the little kid over bonds of perceived and apparent congeniality, while in contrast the father and the older kid approach it as a technological curio. In the third comic strip, Toby is instrumental in attending to a medical emergency of the mother. In the last comic strip, in 2036, the robot is obsolete and is given a new design and new age hardware, with a humanoid exoskeleton and the ability to communicate in near human mannerisms. This gives Toby a new avatar — more like a human being, than a mere piece of machinery. Toby's story spreads across 21 years, is a reflection of the endearing relationship a robot may have with human beings and also illustrates various facets of modern day and futuristic robotics, and hints towards the seven themes discussed previously.

2. Nomenclature for Boxed and Shaded Information

To illustrate an important piece of information, I have used 4 types of boxes, a nomenclature followed throughout the book is as follows:

1. Boxed Braces — this is used to mention a definition or a rule. They contain very important information and are closely related to the text.

> **What is schema ?**
>
> Schema is traditional to psychology and neurology, anthropomorphic motivations led Arkin to develop schema for mobile robots. Neisser defines schema as *"a pattern of action as well as a pattern for action"*, while Arbib explains schema as *"an adaptive controller that uses an identification procedure to update its representation of the object being controlled"*. In mobile robot concerns schema relates motor behaviour in terms of various concurrent stimuli and it also determines on how to react and how that reaction can be accomplished. Schemas are implemented using potential fields, individual schemas are often depicted as needle diagrams. Concurrently acting behaviours are coordinated using vector summations of the active schemas.

2. Short Box — this is meant to illustrate a piece of information that may be interesting, but cannot be expressed in great detail in the current scope of the chapter. Makes for appealing reading, but is not very closely tied with the text.

> **Moravec's Paradox**
>
> Moravec observed, that to design and develop elementary sensorimotor skills required more extensive resources than getting it to acquire high-level logical reasoning. It was seen that, making an artificial agency play a game of checkers or respond to an IQ test is much easier than replicating the perception and motor skills of a one-year-old baby.

3. Roof & Floor Box - this is used for examples and illustrations.

> *An Example In Blending Of Behaviours*
>
> This is an illustration of blending in 2 behaviours, we use stage simulator in ROS. Stage is a 2.5D simulator, I will discuss more about Stage in chapter-5. Here, I use a model of the erratic robot, which is more of a box fitted with a single laser sensor. We try to blend in 2 behaviours, (1) laser obstacle avoidance and (2) keyboard based manual changing of the heading angle.
>
> One can argue that here, changing of the heading angle is merely a fail safe arrangement and not a true behaviour, however it ought to be appreciated that this example does illustrate how a second behaviour at runtime can blend in with a lower level behaviour yielding a solution for the navigation.

4. Shaded Portions — contain very interesting information, many of which I have had the privilege of personally acquiring such. Content covers interviews and information acquired over personal emails.

> **A tête-à-tête with Daniel Stonier and Marcus Liebhardt from Yujin Robot, Seoul**
>
> Yujin Robot is a Seoul based company which developed the Turtlebot-2 with the Kobuki base. Daniel and Marcus from the company's innovation team share with us the experience of developing these robots and future prospects. This interview took place on 24 Nov 2014.
>
> **AB:** How did Yujin Robot come into being ?
>
> **DS & ML:** More than 20 years ago Yujin Robot started in automation and is now comprised of several small business groups, not all of them directly related to robotics and has also seen a couple of internal teams successfully spin off their own businesses. The R&D group entered service robotics more than a decade ago and despite being a relative adventurous proposition at the time, the driving force then, and now has been the CEOs passion for building robots.

Our greatest reach in the known universe has been to comet 67P helped by a robot.

Newer frontiers of medicine boasts of state-of-the-art robots, the Da Vinci and Zeus system. The nurse robot and Paro are the latest cutting-edge applications in the care industry and therapeutics, respectively. In warfare DARPA has provided various advanced options for the use of robots in the US military and the matter has fueled an ethical debate. Automated vehicles and delivery robots such as Starship etc. stand as the latest innovation in transportation. It is obvious that robots are getting woven into the fabric of our lives and the personal robot industry is probably one of the most potent for the next decade.

In conclusion, robots are a joy to behold. Not only are they a reflection of the convergence of human endeavour and mental faculties of design, aesthetics, logic and social interaction but they are the best examples of the handshake between technology and mother nature, and the following ten chapters reinforce these ideas.

Acknowledgements

I owe a debt of gratitude to my editor Randi Cohen and my project manager Karen Simon and I thank all those at CRC Press who supported me throughout the labourious process of writing spread across three years. A particular mention goes to Veronica Rodriguez who worked along with Randi, Shashi Kumar at the LaTeX helpdesk, Sarah Chow, who was my previous editor, Laurie Oknowski, who was the project manager for about eighteen months and Sunil Nair who offered me this opportunity. There are also those whom I did not really get to know, including Jonathan Pennell who designed the cover and other people from the artwork team, Gail Renard who was the copy editor and of course the inputs from my reviewers. Randi's continued support and encouragement stands out, and without her help this book would not have been possible. She helped with my writing, my organisation of the book, and my artwork and cartoons. She was also my 'go to' person for obtaining permissions for using images from other publishers, etc., and her eight-year-old son, and my little friend, Ari, helped me name Toby.

My understanding of the subject matter has been acquired over the last ten years. I owe it to my teachers at King's College London, a particular mention is due for Prof. Samjid Hassan Mannan, who taught me dynamical systems and has subsequently been a well wisher and a benefactor since my graduation. I owe my basic understand of robotics to Prof. Kaspar Althoefer who taught me the subject and was also my project superviser.

Help and support was readily found from the open source citadels of robotics, the player project mailing list and later the ROS mailing list and forum. I am particularly thankful to insightful emails and guidance from Prof. Richard Vaughan and help with Player/Stage/Gazebo from Rich Mattes. I remain grateful to Brian Gerkey, Tully Foote and David Lu at ROS for the continued support I have received from this community and a word of gratitude to Carlo Pinciroli and others who were a part of the ARGoS project at IRIDIA, Belgium. I remain grateful to Prof. Ronald Arkin and Prof. Alan Winfield for helping me with various topics of agent based robotics. I am thankful to Ali Emre Turgut for a very helpful Skype chat and later emails regarding swarm robotics. I am indebted to Graham Mann and Graeme Bell for inviting me to Murdoch, Australia, in the summer of 2011 and I treasure the wonderful memories of the mascot robot and the AR Drones from that trip.

A very special thanks goes to Prof. Takeno, Dr. Broadbent, the STDR team at Aristotle University, Daniel and Marcus at Yujin Robot at Seoul and the Cool Farms team at Coimbra, for the one-on-one interactions which led to four interviews and a special section in the book. I thank Etienne Li for providing useful information for their dance drama 'Robot', which included live performances with the Nao robot. I remain grateful to Prof. Beavers for providing his robotic parody on the Book of Genesis, and to Dr. Matt Zucker for the Braitenberg simulator. I also thanks all those who helped me procure the wide assortment of images for the book.

I acknowledge the input from my project students, Koushik Kabiraj, Sugandha Sangal and Kanishka Ganguly who have laboured with me and seem to have found some tangibility in my passion. I thank my pals, Goncalo Cabrita, Eliseo Ferrante, Boubacar Barry, Luke Dunn and Vasileios Lahanas all of whom have been instrumental in the making of this book.

I thank my cousin who is also my lawyer, Joydeep Banerjee. Finally, thanks to my mother, my father and my younger brother. Without these three people I would not be who I am.

About the Author

Arkapravo Bhaumik is from New Delhi, India, and has an advanced degree in mechatronics from King's College, London. His research interests are, mobile robotics, robot swarms and human-robot interaction. He is also enthusiastic about the open source philosophy and Linux. He prefers to spend time designing or writing about AI and robotics. His recent robots and simulations have employed ROS, and his research on machine learning has been with SVM to detect emotions in human faces. His excursions into natural language processing have been with Python NLTK. He is currently employed at the Birla Institute of Technology which is located at Ranchi, in the eastern part of India. Arkapravo has long been bitten by the traveler's bug. When he is not writing, making robots or busy with his students, chances are he is journeying to new destinations, doodling away with a pencil, or listening to the music of Strauss or Tchaikovsky — or doing all of these three at the same time. The best way to contact him is by his email: *arkapravobhaumik [at] gmail [dot] com* or you can say a hi to him on Twitter, *[at] asimovslegacy*.

I

Theory

" Phew ! ..... and all this to get my robot to follow a strip of line ......"

...and then there were mobile robots

" In the 1920s science fiction was becoming a popular art form for the first time ..... and one of the stock plots .... was that of the invention of a robot .... Under the influence of the well-known deeds and ultimate fate of Frankenstein and Rossum, there seemed only one change to be rung on this plot — robots were created and destroyed their creator ... I quickly grew tired of this dull hundred-times-told tale .... Knowledge has its dangers, yes, but is the response to be a retreat from knowledge? .... I began in 1940, to write robot stories of my own — but robot stories of a new variety ...... My robots were machines designed by engineers, not pseudo-men created by blasphemers."

– Isaac Asimov [367]

1.1 EARLY PIONEERS AND THE STORY TILL SHAKEY

THE wish for automated machines doing chores on human command goes back to the lore of Greek mythology. Homer writes about artificial servants with the ability of thinking, speech, and locomotion for the Gods at Mt. Olympus, which were built by the blacksmith cum craftsman Haphaestus [25].

It is some coincidence that the first evidence of an automaton is also found in ancient Greece at Alexandria, around 250 BC when Philon of Byzantium build a human-like automaton [197] which looked like a maid, holding a jug of wine in her right hand. When a cup was placed in the automaton's left palm it poured some wine, followed by water to mix in and served a drink. Philon designed his automaton with smart mechanisms and pneumatics and named it *Automatiopoeica*, which historically is said to be the first robot.

Lion-like automatons are found in the accounts of Byzantine emperor Theophilos around 820, while around medieval times, building automatons in the Muslim world is well documented in the works of Al-Jazari and Banu Musa. From the 11th to 15th century automatons such as horsemen, birds, doors etc. as parts of mechanical clocks were found both in Europe and the Arab world. Near the European renaissance, Da Vinci [283] helped by royal patronage in Italy, developed mechanically maneuverable automatons, shown in Figure 1.1.

At about the same time across the globe Japan had made its own automaton, the Karakuri Ningyo puppets. These puppets had using mechanical arrangements, where a wound up spring was employed to store energy, This was followed up with a coordination of cams, levers and gears to actuate the puppet. Meant for recreational purposes, as shown in

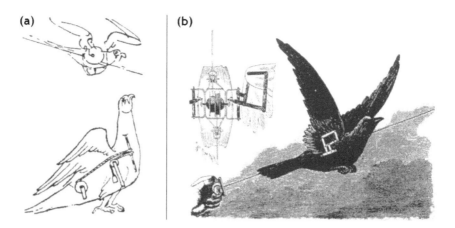

FIGURE 1.1 **Da Vinci's Bird**. These are sketches for a bird-like automaton; (a) is from Da Vinci's Codex Atlanticus and (b) is from Codex Huygens. From Rosheim [283], with kind permission from Springer Science and Business Media.

Figure 1.2, it was popularly used as a tea-serving automaton, the puppet can move forward with a cup of tea in its hands and then bow its head to offer tea to the guest, after drinking when the cup is returned by the guest to the puppet, it raises its head in a show of gratitude, turns around and returns.

More sophisticated automatons were to follow in the coming century. Vaucanson made his automated duck [368] in 1738 with mechanical arrangements as shown in Figure 1.3. His duck could flap its wings, make quacking noise, drink water, eat grain and seeds with a near realistic gulping action. The duck was designed to be the same size as a living duck, made out of gold-plated copper. Vaucanson also developed other such automatons: the flute player and the tambourine player which were life-size human figures which created an imitation of playing musical instruments and android waiters which would serve food and clear tables.

Around the late 18th century a similar mechanical arrangement was employed in Mysore, India, to build a mechanical wooden tiger which responded to turning a handle with a growl and an animated attack. Various historians have suggested that this may have been an outcome of Tipu Sultan's collaboration with the French.

All of these developments took place without any electricity, and were powered either by a smart mechanical arrangement which most of time was a coiled up spring, or by pneumatics and hydraulics. It was not until around 1890 that Tesla used electricity to build miniaturised boats which were controlled remotely using radio signals, as shown in Figure 1.4.

The coining of the word 'robot', had to wait until 1920. This etymology is derived from the Czech word 'robota' meaning committed labour. It was first used in Karel Čapek's play, R.U.R — Rossum's Universal Robots[1]. The evolution in the usage of the word 'robot' is shown in Figure 1.5 and the word cloud from Wikipedia associating it with other words and ideas is shown in Figure 1.6.

The first humanoid robot, Eric was made by W.H. Richards in 1928 and put on display in London. The Westinghouse Electric Company developed another humanoid, Electro around the 1940s which was controlled by electrical relays and could play recorded speech, respond

---

[1]Čapek's play 'Rosumovi Univerzální Roboti' was written in Czech, published in 1920, and first screened at Prague, Czechoslovakia, on 25 January 1921.

FIGURE 1.2 **Karakuri Puppets.** These puppets, made in Japan from the 17th to the 19th century, were triggered by a wound-up spring and could serve tea. Image courtesy *en.wikipedia.org*.

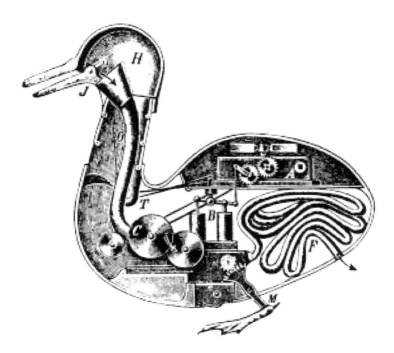

FIGURE 1.3 **Vaucanson's Duck**, made in 1738, is an example of early automaton. Image courtesy *en.wikipedia.org*.

FIGURE 1.4 **Tesla's radio-controlled boat.** Nikola Tesla patented a radio-controlled robot-boat in the winter of 1898, and named it 'teleautomaton'. Tesla demonstrated his invention at Madison Square Garden, New York City, in the Electrical Exhibition that year. Image courtesy *commons.wikimedia.org*, CC license.

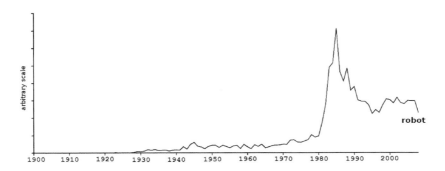

FIGURE 1.5 **Use of the word 'robot'.** Considering 1950 as a base year, around 1985 the usage of the word 'robot' saw an increase of about 12-fold, after the 1990s there has been a decline in usage due to development of nomenclature such as 'roomba', which has become a tag name for all cleaning robots, 'PUMA' and its variants which have ruled the arm robots domain, 'drones' to connote all flying robots, and newer science fiction is more about androids and cyborgs, than the simplistic R2D2 and C3P0 of the 1980s. Made using Google Books Ngram Viewer [230]

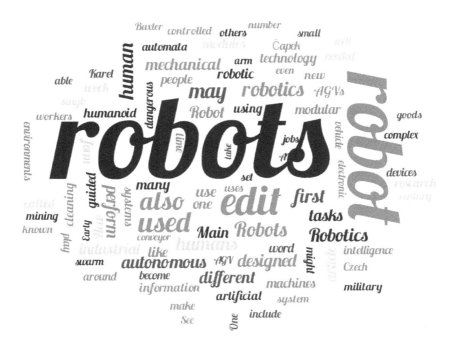

FIGURE 1.6 **Wordcloud of the word 'robot' from Wikipedia.** Nearly all the words seen here are used in the chapters to follow. Made using the wordsalad app, http://wordsaladapp.com/

FIGURE 1.7 **LUNOKHOD-1**, from a Russian lunar mission, is one of the earliest examples of teleoperation and telemanipulation. From Spenneman [308], reprinted by permission of the publisher Taylor & Francis Ltd, (http://www.tandfonline.com.)

to voice commands and moved on wheels attached to its toes. Claude Shannon in 1938 made the electromechanical mouse, which could find its way to the goal point by trial and error — a humble ancestor of the modern day micromouse.

Despite Tesla's boat and Shannon's mouse, in the first half of the last century, robots were still difficult to build because computer technology such as ENIAC relied on bulky and cumbersome electronics of gas valves which were just too big to be integrated into a mobile unit. It was not until 1949 that Walter's robots [344] heralded in the concept of a mobile computational unit readily interacting with the environment. These robots had likeness to turtles and could perform simples behaviours as following the light and avoid obstacles. Walter had achieved the first bonding of electronics and mechanical modules which could respond to an external stimuli. In the 1950s, Devol designed the Unimate robot arm which was the first programmable industrial arm manipulator.

Later, in the early 1970s, Shakey and the Stanford Cart were early robots developed from principles of autonomous agents and AI. Around the same time, researchers at Waseda University developed a full-scale humanoid, Wabot-1, which had two legs and two arms and could communicate with human beings in Japanese. From a cognitive point of view, the robot had the mental faculty of a 2-year-old kid.

The Russian moon missions led to the development of the Lunokhod programme. The Lunokhod-1, shown in Figure 1.7, was the first robot to land on another heavenly body. The Lunokhod-1 in 1970 and the Lunokhod-2 three years later were remote-controlled robotic vehicles designed to chart the lunar surface. The robot vehicles were equipped with solar panels for power by day and a radioisotope heater unit for power by night. Explorer robots,

FIGURE 1.8 **Philae (Toby, 1 of 4.)** The robotic lander on Comet 67P came alive [134] with the falling sunlight on its solar panels and reestablished contact with Earth on 13 June 2015. Cartoon made as a part of the Toby series (1 of 4) by the author, CC-by-SA 4.0 license.

since Lunokhod have come of age and gone on to explore other planets and heavenly bodies such as comets as shown in Figure 1.8.

With the microelectronics revolution in the 1970s and computer revolution in the 1980s, robotics took a leap with readily available miniaturised components and smaller processing units and user friendly software interfaces.

1.1.1 Walter's turtles

Modern day mobile robotics was born in 1949, when W.G Walter built 2 three-wheeled automated turtle-like robots [154]. These automated turtles had a light sensor, a touch sensor, a propulsion motor for a front-wheel drive, a steering motor to move around and a vacuum tube analog computer. Walter named his project *Machina speculatrix*, and he named the 2 turtles after characters in 'Alice in Wonderland', Elmer and Elsie, which also fit the suitable acronyms, **EL**ectro **ME**chanical **R**obots, **L**ight **S**ensitive, with **I**nternal and **E**xternal stability. This was the first electromechanical autonomous mobile robot. These turtles had three characteristics: attraction to moderate light and repulsion by bright light and obstacles. In various experiments, the turtles responded to a number of external stimuli:

1. **Goal seeking** behaviour. When started in the dark, the turtle made its way towards a beam of light and tracked it to reach its feeding hutch.
2. **Free will** as in the ability to to choose between alternatives, choosing to reach the nearer of two light sources when executing light tracking behaviour, staying close to it for some time and then moving on to the next light source and then returning back to the former. This cyclic behaviour continues until power runs out.
3. **Pertinacity**. The turtles showed dogged determination, when tracking a light source if it so happened that the light source was not visible from their field of view due to obstacles, they would resort to moving in a random direction and on finding the source of light they would assume light tracking behaviour. This behaviour is shown in Figure 1.9.
4. **Preference for an optimum**. The turtles avoided extreme stimuli and opted for moderation, e.g., when tracking a light source, they were attracted to the light source, but were repelled by excess brightness.

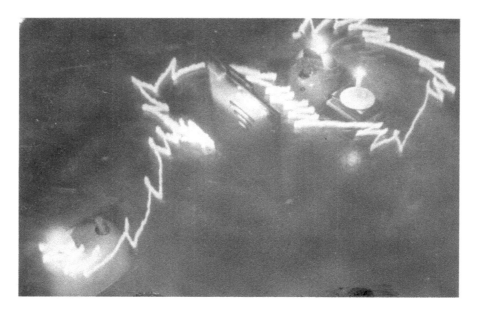

FIGURE 1.9 Walter's turtles. A time lapse photograph of Elsie where she demonstrates **Pertinacity**. Chasing for a light source, it is put off track by the reflection of its own candle in the polished fire-screen. She eventually finds her way to the candle behind the screen. From Fong et al. [107], with permission from Elsevier.

5. **Social organisation and competition.** When the two turtles are set up simultaneously in a dark environment with candles serving as light sources mounted on each of them. Both the turtles are attracted by the other's light source still repelled by excess brightness. That led to a concerted dance where each turtle waltzed around the other. Walter made this experiment even more interesting by adding in a third external light source, which encouraged a competitive behaviour as the turtles raced to reach the third light source.

6. **Response to internal physiological changes such as low battery power.** Walter designed the turtles such that with a low battery the turtle is attracted to a certain intensity of brightness which guides it to a charging pod.

7. **Realisation of self.** Walter tried even more innovative experiments with mirrors. Mounted with a candle the turtle tracked itself in a mirror but on reaching the mirror and touching it, its tactile sensing detects the mirror as an obstacle and then it walks around the mirror. Walter called this behaviour as realisation of self. In Chapter 9 it will be discussed and demonstrated that a mirror is indeed very crucial for tests of consciousness.

8. Other than the above a firsthand account identified Elsie as a woman, and Elmer as a man.

This was unlike Da Vinci's bird and Tesla's boat as the turtles didn't have a fixed behaviour. Rather performance was on the merit of the stimuli acquired from its environment, mostly at runtime. The turtles demonstrated automaton and path planning as a response to external stimuli. According to Walter, his turtles were *an imitation of life* [344], and were the first examples of Artificial Life (ALIFE) and established the grounding for anthropomorphic motivation to design robots. It is interesting to note that Walter knew

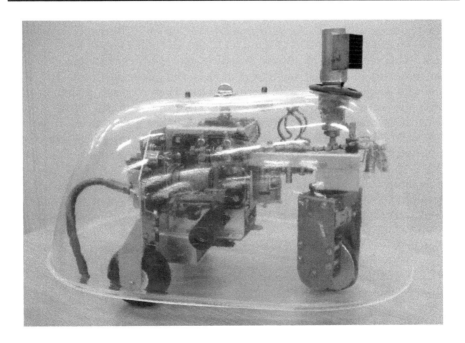

FIGURE 1.10 **Ninja and Amy**, replicas of Walter's robots, built by Ian Horsfield in The Bristol Robotics Lab. The only modern components used are the 2 motors and the batteries, rest remains nearly as Walter would have made it. Image courtesy Alan Winfield, University of West England, Bristol, used with permission.

that his turtles were far from perfect and left to themselves they would exhaust their supply of energy in the quest for light sources, getting stuck with rigid obstacles or engaging in competitive duels with similar creatures.

These two turtles were disassembled to provide parts to construct six more turtles. While most of them got misplaced or destroyed, at Walter's death in 1977 his son inherited one. In 1995 Holland and his team of technicians restored this turtle to working order and also made the replicas, Ninja and Amy, seen in Figure 1.10.

Walter had further planned to build *Machina docilis* or the CORA turtle [345], on the same footing as Elmer and Elsie, but with the added ability to learn from interaction with the environment. CORA is an acronym for **CO**nditioned **R**eflex **A**nalogue. Walter was guided by Pavlov's study of conditioned reflexes and employed rudimentary neural models to implement learning. However, while it is certain that Walter had developed a standalone learning-based device, it is not certain if he built such faculties into a turtle robot [363]. His later efforts to build the IRMA device — **I**nnate **R**eleasing **M**echanism **A**nalogue — exploited more advanced learning models and as with the CORA, was not integrated into a turtle robot.

Though it is a coincidence [155], but not really a surprise that in 1949 Walter started developing his turtles and Wiener [352] discussed preliminary ideas of marrying sensor readings and motor action, via an ultra-rapid computing machine, and around the same time Turing started developing ideas for the Turing machine and Shannon modelled the theory of communication.

1.1.2   Shakey and the Stanford Cart

Mobile robotics took a big leap forward around the end of 1970s decade. At SRI, Nilsson built and programmed Shakey [253] (Figure 1.11) and was able to navigate it in a grid-based environment. This technological advancement was fueled by modern-day electronics, user friendly software interfaces and analytical models and tools developed by AI over the previous 30 odd years. Shakey could perform tasks that required planning, navigation, finding path by search methods as A* search algorithm and also using sophisticated techniques such as the Hough transform and the visibility graph method. When set up, Shakey was controlled by a SDS-940 computer with a 64K 24-bit memory and programming was done in Fortran and LISP. In later versions the SDS-940 computer was replaced by a PDP-10/PDP-15 interface. Later robots as The Stanford Cart and the CMU Rover were developed as direct descendents of Shakey and tried to plugup its shortcomings.

The Stanford Cart was built as a remotely controlled mobile robot with an onboard TV by Adams in 1960 for his research project, later improvements were supervised by Earnest, McCarthy and Moravec over a period of 20 years. They developed it into the first robot road vehicle, though it was broadly employed for research in visual navigation. Adams made the cart to support his research on NASA JPL project Prospector, which hoped to establish teleoperation in robots on the lunar surface using radio control from Earth. The Cart had four small bicycle wheels with electric motors powered by a car battery and carried a TV camera aligned in the forward direction. The interaction of the Cart with its environment was via this onboard TV. Much later, from th mid to late 1970s, Moravec programmed it to navigate autonomously in a cluttered environment. Though, the Cart was appreciably slow at an average speed of 1 meter in 10 to 15 min, in lurches, at the end of each lurch it stopped and took some pictures, and slowly processed them to plan its next path, and this processs continued. Negotiating 20-meter courses, while avoiding obstacles often took about 5 hours.

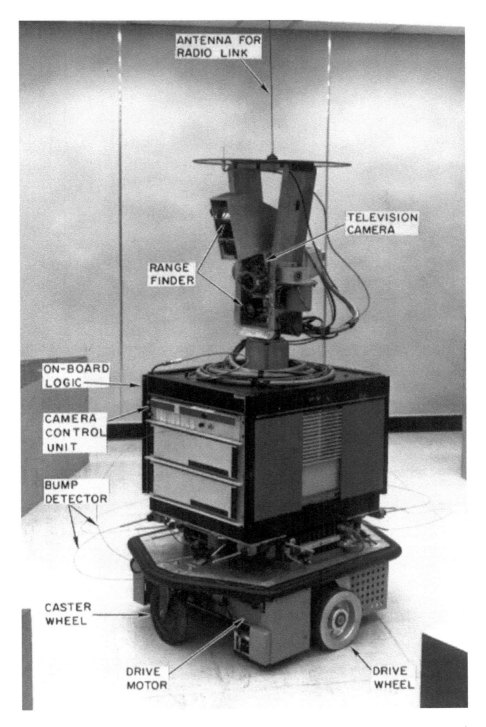

FIGURE 1.11 **Shakey in 1972.** This iconic robot was designed from the late 1960s to the early 1970s at SRI. Nilsson [253] describes Shakey as a mobile robot system "endowed with a limited ability to perceive and model its environment". Shakey was navigable using grid based methods and could also rearrange simple objects. Shakey is an exhibit on display at the Computer History Museum in Mountain View, California. Image courtesy *commons.wikimedia.org*.

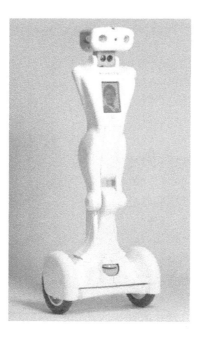 

FIGURE 1.12 **Telepresence** and teleoperation most often features as a basic functionality in modern day robotics and finds support from most robotic software suits. As shown, a telepresence robot lacks in autonomy and is remote controlled over at a distance, and it's controller can communicate with people in the robot's local environment over vision (screen) and voice (microphone) channels. Telepresence is the first steps to a future when communication will commence over virtual reality. On the left is Anybots QA, and on the right is iRobot's Ava 500, both images courtesy *commons.wikimedia.org*, CC-by-SA 3.0 license.

Walter's turtles and Nilsson's Shakey were near isolated events spaced across 20 years in which there was no appreciable experimental progress, but there was theoretical development of AI as a whole: cybernetics in particular by Wiener, the basis of artificial animals and early ideas which led to ANIMATs by Toda and a general integration of electronics leading to electronic processors. At the end of the 1970s, the development cycle of mobile robots became simpler due to innovations in AI, electronics, computer science, software design etc. In the next three sections I try to relive these technological developments and paradigm shifts.

1.2 CURRENT-DAY MOBILE ROBOTICS

In the new millennia, Walter's turtles powered by vacuum tubes and Nilsson's bulky processors are relics confined to the museum and robots are now designed with sleek, portable and easy-to-use hardware. High-end robots such as the PR2 or ASIMO or even AIBO are yet to become a part of human society, partly because of the price and also since these robots are still being developed into a niche for proactive human participation. Today, robots work across various domains; as interactive tour-guides in museums, as a coffee delivery robot in office environments, actors of sorts in stage plays, soccer players, assistants to the elderly with their daily chores and they also chart unknown terrains in faraway planets, attend to a patient as a robotic nurse, a robotic newsreader can liven up a news channel and various others such arenas.

There is a lack of a broad classification scheme, however robots with wheeled bases and differential drive are usually termed 'mobile', those with 2 legs are 'bipeds' and those with human exoskeleton and legs are called 'humanoids'.

Khepera

Khepera was developed at the EPFL in the late 1990s, and it went commercial with the K-Team Corporation, broadly for use in education and research. This robot is often credited with being a vital aspect for the development of evolutionary robotics. Shown here is Khepera, the original version. From Fong et al. [107], with permission from Elsevier.

State of the art robots are equipped with a vision module for object recognition, NLP module for language processing and voice-based human interaction, added information and rules for specific operations and a learning module. Teleoperation and telepresence, shown in Figure 1.12 along with map building with SLAM algorithms have been basic functionalities of modern-day robots and are often provided by the manufacturer.

Roomba

The Roomba made by the iRobot company in 2002, is a vacuum cleaning robot. While vacuuming, the Roomba can change directions when encountering an obstacle. It can also detect an edge, thus preventing it from tumbling down stairs. To aid the vacuuming and cleaning process, there is a pair of brushes, rotating in opposite directions which helps to pick up dirt from the floor. Shown here is Roomba 780. Image courtesy commons.wikimedia.org, CC license.

Developmental routes have become simpler and more modular with Arduino and similar boards and smaller and easy to use processors such as Raspberry Pi, ODROID etc. Motion sensing with Microsoft Kinect and other such devices has provided an avenue to develop quality robots on a strict budget.

Pioneer Robot

The Pioneer has been very popular with current day researchers. Developed at ActivMedia Robotics, Georgia, the robot is equipped with a ring of 8 forward-facing sonars and 8 optional rear-facing sonars. The basic robot can attain maximum speed of 1.6 meters per second and carry a payload of up to 23 kg. Shown here is a variation, where the Pioneer base is mounted with an arm manipulator, image courtesy commons.wikimedia.org, CC license.

Some robots are favourites among undergraduates such as the Khepera and the Pioneer and some universities have made the PR2 a part of their robotics courses. For the hobbyists, the Roomba, iBot and LEGO Mindstorms are cheaper options. The Roomba is a vacuum cleaning robot, however it has found great acceptance in the hobbyist fraternity. iRobot also provides a separate platform for the robot enthusiasts the icreate, which is very much like the roomba but without the vacuuming parts.

Erratic Robot

Erratic was designed and developed as a clone of the Pioneer robot, by Videre in 2000. It has found applications in academia and is often mounted by a SICK or HOKUYO laser. Shown here is a model of the Erratic spawned in simulations using gazebo.

The Khepera has seen four generations. This robot was designed by Jean-Daniel Nicoud and his team at EPFL as a readily usable platform for experiments and training, such as in local navigation, artificial intelligence, collective behaviour etc. It was developed using the Motorola 68331 processor in the 1990s. The latest avatar, version IV of the robot is equipped with 8 infrared sensors, 5 ultrasonic sensors, WiFi, Bluetooth, accelerometer, gyroscope, color camera and uses a 800MHz ARM Cortex-A8 processor. It is priced at $2700.

The Turtlebot

Turtlebot was developed by Melonee Wise and Tully Foote at Willow Garage in late 2010. Shown here is Turtlebot 2, which has a Yujin Kobuki base, a Kinect sensor and a laptop with a dual core processor and is available as open source hardware with a BSD license. Priced at about $1000, the robot has found support from the ROS community and is used in various universities in their robotics laboratory courses.

There will always be a need for quality simulators in robotics as every experiment cannot be performed with real-world robots due to extreme logistics and cost considerations. As a contrast, path planning was targeted in the earliest 2D simulators such as Karel and Rossum's Playhouse (RP1), 3D simulators such as Webots and Microsoft Robotics Developer Studio (Microsoft RDS, MRDS) provided added designing capabilities and near life-like models of the robot. Currently, Gazebo which is a standalone project and also a part of ROS is one of the best and is often accepted as a standard for robotics simulations.

Personal Robot-2, PR2

PR2 is a near humanoid developed at Willow Garage, but instead of legs it is mounted on a wheeled base. It has two 7-DOF arms and is equipped with a 5-megapixel camera, a tilting laser range finder, and an inertial measurement unit and is powered by two 8-core servers, each of which has 24 gigabytes of RAM, for a total of 48 gigabytes, and are located at the base of the robot. The PR2 can open doors, fold towels, fetch beer and also play billiards.

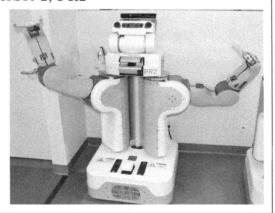

AIBO and similar designs such as the Tekno and the CHiP are anthropomorphised on a dog and often create a false bond of caring and compassion with human beings. In context, the PARO which is a robotic seal and the Pleo which is a robotic dinosaur are state of the art pet robots and both have a congenial appearance and respond positively to touch and cuddles. The PARO has found application in psychological therapy of the depressed and the elderly and the AIBO has been used in nursing homes as a company for the residents.

AIBO

SONY's iconic series of robots were inpired from dogs and lion cubs and their name means 'Artificial Intelligence RoBOt, and also suggests the Japanese, aibō, meaning pal, friend, partner. Toshitada Doi and Masahiro Fujita are the pioneers whose work led to the development of the AIBO in 1998. The image is from commons.wikimedia.org, CC license.

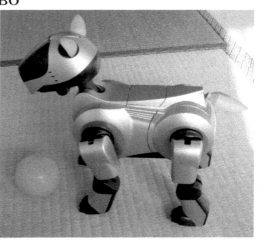

LEGO Mindstorms kit is meant for educational robotics. This affordable kit provides for easy assembling of custom-designed robots and for programming them with user friendly languages. The latest version, EV3, is based on a Linux ARM9 CPU and also contains motors, touch sensors, color sensors and an infrared sensor.

Honda ASIMO

Honda's celebrated humanoid robot ASIMO is said to be inspired by Isaac Asimov but the acronym reads otherwise, 'Advanced Step in Innovative MObility'. Designed as a walking assistant robot, the ASIMO stands 130 cm, weighs about 50 kgs, can attain maximum speeds of 9 km per hour in running mode and can be controlled by PC, wireless controller, or voice commands. Honda, most unabashedly claims it, "the world's most advanced humanoid robot". The image is from commons.wikimedia.org, CC license.

Willow Garage the company that developed ROS, build the Turtlebots I and II and the PR2, all of which integrates 3D sensors with a mobile base. The Turtlebot-I employed a icreate base, while the Turtlebot-II has a Kobuki base developed by Yujin Robot. Building humanoids have always drawn interest, and while PR2 is only partially humanoid, Honda's ASIMO has human-like legs and can dance to a tune. Honda has not yet put up the ASIMO for sale but leasing is at an astronomical price of $150,000 a month. The PR2 is priced at a staggering $400,000 with discounts for academic projects.

Nao

This iconic mini humanoid was made by the French company, Aldebaran Robotics in 2006 as an open source software. The robot was acquired by Japan's SoftBank Robotics [2] in 2013. Standing at 58 cm and weighing 4.3 kg, it is powered by the Intel Atom processor and has an assortment of sensors, HD cameras, microphones, sonar rangefinder, infrared, tactile and pressure sensors. The robot has quite a reputation: it has acted in a ballet, been the most prolific robot in robo soccer, has been employed as a therapeutic robot for children suffering from autism, has been programmed to be the first robot with deontic ethics and also, to a degree it is the first robot to attain sentience, albeit for a very short time. The image is from commons.wikimedia.org, CC license.

Open source has been the buzz word in the robotics community and Robot Operating System (ROS), Robotic Open Platform (ROP), Poppy and RoboEarth are projects that target crowd sourcing and thus freely provide technology and know-how.

Segway

This battery-powered, self-balancing two-wheeled scooter was developed at the University of Plymouth in collaboration with BAE Systems and Sumitomo Precision Products, and uses gyroscopic sensors and an accelerometer for maintaining balance. While not a robot as per the traditional definitions, the Segway is an early ancestor of autonomous vehicles of the immediate present and the promise of the future. At a maximum speed of about 12 miles per hour, it is ideal for the golf course and other short tours and it has also been designed into a companion robot [120]

1.3   CULTURAL AND SOCIAL IMPACT

Robots inspire awe across all ages. Engaging in a conversation with a social robot such as Kismet, whom we will meet again in Chapter 7, or making a robot using LEGO Mindstorms

FIGURE 1.13 **LEGO Mindstorms** have been very popular both as toys and teaching aids since the late 1990s. These robot kits contain sensors (touch, light and colour), motors and a processor encased in the 'brick' and can be programmed with various programming languages. Both these images are of robots built with the NXT kit. Both images courtesy *commons.wikimedia.org.*

kits, as shown in Figure 1.13 or watching ASIMO dance to a tune, are situations when robots contribute to entertainment and recreation. In academia, readily available open source simulators and software for robot has fueled robotics research in universities and hobbyists, such as MORSE [100] shown in Figure 1.14. On the other hand we have the the prophecy of singularity and a Terminator-like doomsday scenario where intelligent machines take over the earth, which is a script for various science fiction potboilers. In this section, I look into the fiction, movies and popular opinions on robots, utilities of mobile robots across various domains such as medicine, army, public interface, entertainment etc.

1.3.1 Science fiction, entertainment industry, medical surgery and military

Verne's steam-powered mechanical elephant, Shelley's Frankenstein and Baum's tin man are some of the first fictional accounts of automatons. In 1940, Asimov ushered in added sophistication. His robots were electromechanical devices unlike any previous authors'. Arthur C. Clark, Alvin Toffler and Philip K. Dick are near contemporaries of Asimov who gave us timeless classics with robotic characters. With the advent of cinematography, robots came to be synonymous with depiction of futuristic themes in movies. Lang's pioneering direction in the 1927 silent movie Metropolis had the first robot in cinema, Maria as shown in Figure1.15. She was a metal cast female android. The three robots in Nolan's 2014 sci-fi Interstellar, TARS, CASE and KIPP often wisecracked in slangs and sarcasm with the rest of the characters of the movie. Robots have added to the intrigue, drama, humour and fun in cinemas. Over these 90 odd years as shown in Figure 1.16, many of these robots have attained cult status, become household names and have added to idiomatic references. Robbie in Forbidden Planet (1956) as shown in Figure 1.17, HAL 9000 in 2001 (1968), various replicant robots in Blade Runner(1982), Andrew an NDR-114 humanoid robot in Bicentennial Man (1999), the iconical star wars robots R2D2 and C3P0, as shown in Figure 1.18 and Wall-E and Eva, as shown in Figure 1.19 are a few to name.

Fiction with robots often traces some of the popular storylines, one where robots accept human norms and values and try fitting into human society, but are often reminded of their shortcomings. David in Spielberg's movie AI and Andrew in Asimov's Bicentennial Man

FIGURE 1.14 **6 PR2 in MORSE.** MORSE [100] is a 3D simulator built on Blender, it was developed by LAAS, Tolouse. MORSE particularly targets human robot interaction and has bindings with various meta-platforms and middlewares ROS, YARP and MOOS.

are such accounts where the robot struggles with its own existence, trying hard to imitate human behaviour, values and cultural nuances. The second is the dystopian scenario where robots and smart AI rebel against human beings and has been the plot for Blade Runner, Terminator and Battle Star Galactica. A third brings to light the the endearing attachment between human and the robot, as seen in, Frank and the Robot and Real Steel. Robots are mostly shown to be near perfect artificial beings but with lesser appreciation for human emotions and humanly values though Replicants in Blade Runner, Lieutenant Commander Data in Star Trek and Cylons in Battle Star Galactica can be said to be near humans and special psychological tests similar to the Turing test had to be devised in order to distinguish between human and the artificial.

Asimov's three laws, as shown in Figure 1.20, have been seen as an overarching principle for robots, however these three laws lack practical application and often conflict in real-life scenarios. In Chapter 8, it will be discussed that ethical principles for robots cannot be drawn from these three laws. Various leading roboticists and AI researchers have suggested that these laws are flawed and serve more as an instrument of control for human beings to reign as the superior 'master' race and controlling the robots as slaves. Asimov made some amends to these laws in 1985 by introducing a fourth law, the zeroeth law.

Other than movies, there have been some brave attempts to follow in the footsteps of Čapek and make stage plays with robots and robotic characters, though very much still in the experimental domain. 'Hello Hi There', a stage play directed by Annie Dorsen and first screened at Steirischer Herbst, Graz in September 2013 and subsequently across Europe and America, features no actors but two chatbots, two computer screens spitting out sentences which structures the play. The chatbots converse using materials from a script which has been taken from a famous debate on human nature between Noam Chomsky and Michel Foucault in 1971. The intriguing part is that the play doesn't have a fixed plot and can play out in about 84 million different ways.

Bold and unconventional explorations to use robots in-sync with live human actors and dancers on stage are few but commendable and may be the definition of a newer realm of acting and stage performance. Francesca Talenti's play 'The Uncanny Valley' [1] which has

FIGURE 1.15 **Maria**, the first ever robot in cinema was in Fritz Lang's 1927 silent movie, 'Metropolis'. Made in Weimar Germany, the movie was a science fiction tale set in a dystopian future. Shown here is a commemorative statue of the robot at the Filmpark, outside the Babelsberg studio where the movie was made. Image courtesy *commons.wikimedia.org*.

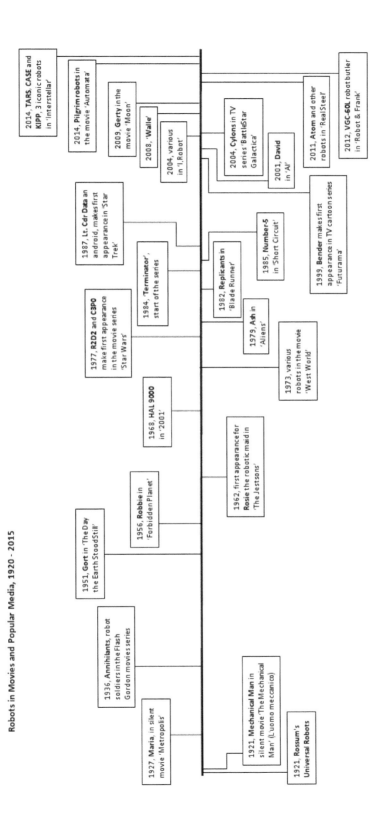

FIGURE 1.16 Timeline for robots in movies and popular media, 1920 — 2015.

FIGURE 1.17 **Robbie the Robot**, in the 1954 movie 'The Forbidden Planet' was one of the earliest appearances of robots in movies. Robbie was designed to be a 6 foot 11 inches tall humanoid robot which could turn, twist and move around on wheeled legs. Robbie's personality in the films original screenplay was written by Cyril Hume and made into a real robot with the effort of an art team led by the director Robert Kinoshita at a cost of about $100,000.

FIGURE 1.18 **Cometh the robot, cometh the man.**, Anthony Daniels played the role of C3PO the humanoid droid in the Star Wars movie series, seen here with the C3PO mask. C3PO along with its companion R2D2 have become household names and attained iconic status. Image courtesy *commons.wikimedia.org.*

FIGURE 1.19 **Wall-E and Eva**, iconic robots from the movie Wall-E, courtesy *de.wikimedia.org*

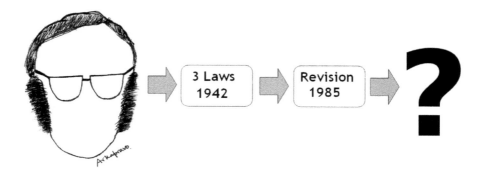

FIGURE 1.20 **Asimov's laws.** His 3 laws of robotics have attained cult status among science fiction fans and roboticists alike. First formulated in 1942, Asimov later added the zeroeth law in 1985. Though popular, the laws have been found to conflict with one another and are lacking in practical applications, Academic opinion and research into robot ethics has been more inclined towards a developmental approach than a codification of laws. More on the shortcomings of Asimov's law is in Chapter 8.

a robo-thespian robot, Oriza Hirata used a Wakamaru robot and Blanca Li's dance drama 'ROBOT' had seven Nao robots dance on stage synchronously with eight human dancers.

Other than the movie and entertainment industry, mobile robotics is the heart and soul of space research. It is being extensively used in the military and medical surgery and it also bears promise for modern agriculture. A great many modern-day space programmes have a robotic rover, which is unmanned and attempts to chart an unknown heavenly body. They are usually equipped with both modes of control viz. autonomous and remotely controlled, the later used more prominently. In the military, use of robots has been hotly debated as trusting a robot with lethal capabilities may prove to be fatal. Historically, Goliath a mobile land mine was the first use of robots in the military by Nazi Germany in World War II. In recent times, the use of robots in the military is a prospect NATO and American forces have looked forward to, and in Iraq from 2004-2007 they employed unmanned military robots, SWORDS and MARS systems. Unmanned robotic vehicles have been used in times of crisis such as earthquakes, landslides and tsunamis. In medical science, teleoperated robotic surgery units such as the Da-Vinci systems, shown in Figure 1.21 is one of the miracles of robotics, which has allowed the surgeon to operate on a patient who is located a few thousand kilometers away on a different continent.

1.3.2   Do robots pose a threat for human beings?

The Vox Populii gathered over questionnaires and polls shows how people have responded to robots and their influence on their lives. A poll conducted in 2014 in the University of Middlesex with more than 2000 participants showed that more than 30% believed that their jobs will be taken over by robots; 46% felt that technology was progressing too quickly (for comfort); 35% were worried by the use of unmanned military drones; about 10% expected to see a Robocop-like police force in about a decade; 42% said 'Yes' when asked if teachers can be replaced by robots; 11% were ready to adopt a robot child akin to David in Spielberg's AI; replacing a beloved animal with a robotic pet invited the interest of about 20% of pet owners; 25% opined that robots will be capable of feeling human emotions; and what might come as bizarre, nearly 20% were ready to have sex with robots. 'Human-like' has been a buzzword in current day human robot interactions. In a user poll in Plano, Texas, conducted

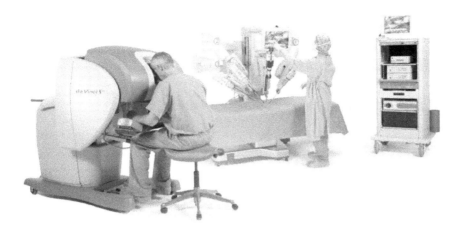

FIGURE 1.21 **Da Vinci system.** Teleoperation in advanced medical systems has allowed for surgeries where the surgeon and the patient are on different continents. From Kroh and Chalikonda [190], with kind permission from Springer Science and Business Media.

by Hanson Robotics, the participants were shown images of two different robots that were programmed to imitate known human-like emotions, expressed by facial expressions. This effort met with an overwhelming 73% approval; the participants did like and connect with what they saw. While in another popular poll conducted by Amazon UK and IMDB, R2D2 got voted in by over 8000 people as the 'world's favourite robot'. His Star Wars companion C3P0 was ranked fourth. Crossing the Atlantic, in the US, a poll from Monmouth University, New Jersey, with about 1700 participants on the concerns over the use of unmanned drones yielded interesting results: while 80% were in support of search and rescue efforts, about 67% were good with drones tracking runaway criminals and 64% conformed to drones for border control. However the participants were apprehensive about drones issuing speeding tickets and it met with only 23% approval. In context of human approval and level of comfort with an ensuing human-robot interaction, an online poll by robohub exploring the tradeoff between privacy vs. automaton of using robots, the poll asked the participants, 'What level of autonomy would you be comfortable with in a bathing robot'. Bathing robots for the elderly have been considered as a commercial venture by electronics giant Sanyo. The results did illustrate that human beings are rather comfortable with robots being in control while taking a bath, only 4% objected to the robots being autonomous, while 48% were at ease with the robot being in charge, 39% were comfortable giving partial control to the robot and 13% were agreeable to full-autonomy however, with human supervision.

Various of these polls dig into dystopian scenarios such as robots taking over human jobs, the Skynet apocalypse, human-AI war etc. The idea that AI poses an existential threat has fueled a debate, with the CEO of Space-X Elon Musk, renowned physicist Stephen Hawking and Microsoft guru Bill Gates at one end of the debate and AI researcher Ray Kurzweil at the other end. Should we try to noose AI and tame it, so it doesn't go out of control? An online poll by The Telegraph in October 2014 yielded mixed response as is shown in Figure 1.22: 36% responded with a 'Yes', that care should be taken in developing AI; 42% chose otherwise.

The threat that robots will take over human jobs is a neo-Luddite nightmare, though it is not baseless. Recent studies in the UK by University of Oxford and Deloitte have

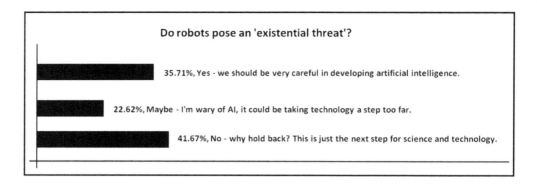

FIGURE 1.22 **Do robots pose an 'existential threat'?**, Results of an online poll at *telegraph.co.uk* in October 2014 in reaction to Elon Musk's 'summoning the demon' remark.

indicated that 35% of UK jobs may be at high risk from advancing technology, automaton and robotics in the next 20 years. The other side to this debate suggests that advancing technology will lead to a new way of living which will develop newer forms of sentient intelligence. Robots have had effective human interactions as artificial pets, in the role of carer, romantic attachment and lover, organised tour guide, receptionist etc. Human-robot interaction and robot ethics are avenues of research which find application in the near future. More on these topics is discussed in Chapter 10.

SUMMARY

1. Robotics had its history in automata, where smart mechanical manoeuvres were employed for locomotion. Da Vinci in 15th century Italy and Vaucanson in 17th century France are the most well-known accounts.
2. Tesla developed a radio-controlled boat at the turn of the century.
3. Walter's turtles were the first mobile units which worked on inputs from the environment.
4. Toda's theoretical models of cognition for a hypothetical animal, with given goals and survival mechanisms were the first mathematical models for an autonomous agent.
5. Mobile robots are fundamentally different from arm manipulators as they move around, which can be known/unknown terrain.
6. With revolution in microelectronics and theoretical developments in AI, Shakey was the big leap that mobile robotics achieved around the late 1960s.
7. Robots have had immense social impact via movies, educational tools, science fiction and household appliances.
8. Existential risk from robots is a question being asked more often than ever, and is closely connected with the prophecy of Technological Singularity.

NOTES

1. *Čapek's R.U.R — Rossum's Universal Robots, the play ends with the end of human civilisation, and two robots named 'Adam' and 'Eve' falling in love.*
2. *Vaucanson's Duck was unfortunately destroyed in a fire in 1879.*
3. *In Nillson's early research the name Shakey is not found; rather the robot is explained*

by its construction, " ... an SDS-940 computer and associated programs controlling a wheeled vehicle that carries a TV camera and other sensors."

4. Walter put up a notice above the hutch where he kept his turtles, Elmer and Elsie, which read "Please do not feed these machines."

5. Cyborg, CYBernetic ORGanism. The etymology was coined by Austrian scientist, Manfred Clynes in the 1960s to suggest the need for using AI to enhance human biologial functions to survive in hostile and unfavourable environments.

6. Flakey was a successor to Shakey. Developed in the mid-90s, it could be controlled using fuzzy logic. Both the Pioneer robot and Erratic robot are motivated from Flakey.

7. Large and the small. The largest robot measuring 8.2 meters in height and 15.7 meters in length, an automated fire breathing dragon was made by Zollner Elektronik AG at Zandt, Germany, while RoboBee, a robot bee with a wing span of 3 cm and ALICE with dimensions of 2 cm x 2 cm x 2 cm are examples of small robots.

8. Some of the exotic applications of robots: (1) WF3-RIX is a flute playing robot made at Waseda University in 1990; (2) Xian'er [61], a miniaturised robot monk which is about 2 foot tall, has a cartoonish appearance and is robed in yellow and has a touch screen on his chest, can chant Buddhist mantras and hold a short conversation for about 20 minutes. It has been a chief attraction at Longquan temple in Beiijing since 2015. (3) Various arm manipulators have been used to sketch portraits and scenic settings.

9. Simulators to study robot navigation is an essential tool in an academic course for robotics. Nowadays, meta-software platforms as ROS or MOOSE are preferred than standalone simulators, however simple and easy to install softwares have not lost their appeal. For example, Kiks is a freely available MATLAB plugin for 2D simulation of the Khepera robot, Kiks is available at http://www.tstorm.se/kiks.php.

EXERCISES

1. **Turtles vs. Shakey.** draw a contrast between Walter's Turtles and Shakey from design and performance point of view. The study should also address the evolution of the technology in the 20 years between these two projects, and also attempt to see the projects from the aims they were designed for. Walter attempted to build models for human cognitive capabilities while Shakey was designed for mobile locomotion.

CHAPTER 2

Embodied AI, or the tale of taming the fungus eater

"The body is our general medium for having a world."

– Maurice Merleau-Ponty [228]

"How *high* in the intuitive scale of cognitive sophistication can such unwitting prowess reach ?"

– Daniel Dennett, on using no representations [92]

2.1 FROM AI TO ROBOTS

W ALTER used vacuum tubes for his turtles and Nilsson used an onboard computer for Shakey. These two projects had contrasting outcomes, it was very fruitful for the turtles, but it was slow and cumbersome for Shakey. Other than these experimental endeavours, another cue for autonomous agency was Toda's model of fungus eaters. These hypothetical organisms were capable of long-term self sustainment while attending to multiple goals and provided the earliest model for an artificial autonomous agency. Wilson's concept of ANIMATs and guiding inputs from biology and psychology laid the groundwork for robot design and architecture. This chapter explores how AI has been used to design robots, the inspirations from the natural world, the contributions from various disciplines and the shortcomings of traditional AI.

2.2 ARTIFICIAL INTELLIGENCE FOR ROBOTS

What constitutes a robot, and how is it different from a machine? the answer to this question has changed over the past eight decades. Humanoid automatons, such as those in 'Rossum's Universal Robots' and 'The Metropolis', were modelled on the human body but lacked smooth human functions and limb orientations and were low on emotional content[1]. With progress in industry and manufacturing and the automation of the production line and the car industry, the concept of a robot was more or less limited to the arm manipulator which engaged in repetitive 'pick and place' operations. Robots for mobile and other superior

---

[1]Of course, these were never made into robots rather Čapek's actors reflected such features in their roles as robots.

agencies, as in navigation, simple behaviour, social agency etc. was a promise to kindle only after Walter's turtles.

In contrast to industrial manipulators, AI robots are meant for navigation and exploration of their local environment, with demonstrable intelligence and many times are meant for a particular task or role, viz. explorer robots, household robots, search and rescue robots etc.

Bekey suggests the following definition for a robot,

"... a robot is a machine, situated in the world that senses, thinks, and acts ... "

This definition is not tied to a given task and doesn't highlight the robot's interaction with the environment. Therefore (1) robots performing repetitive jobs such as industrial robots and arm maniplators, (2) those which lack in a very clear mandate such as the Martian rovers, (3) those with human appearance as automatons or humanoid robots suitable to the domain of social robotics and (4) futuristic robots made by extending biology, such as androids and cyborgs, are all covered by this definition. The definition is not limited to the popularly accepted mechatronic design for robots which marries the mechanical to the electronic with a processing unit. However, an engineering point of view tends to constrain the definition and a robot is to have electronics, mechanical hardware and a processing unit. An actuation without the electronics and processing unit is more in the automaton domain and controlled by compressed springs, pneumatic valves and/or hydraulic control, probably such as the ones designed by Philon and in Japanese Karakuri puppets. Since, the definition is based on the ability of the robot to engage in real-world tasks, without external control, it is tempting to consider that the robot is truly 'thinking' as it is processing data from sensors to a processing unit to actuation much like the human brain; however this is merely executing a code piece and not 'thinking' per se. Chapter 9 will try to focus on the 'thinking' faculty and sentient action in robots. Particular goals for the robot may either be specific, such as line following, light tracking or picking up empty coke cans, or a number of chores which converge towards a predetermined profile such as a military robot, a nurse robot, a domestic help robot or an office assistant robot are discussed in Chapters 7 and 8.

A more AI centric definition is given by Murphy,

"a mechanical creature which can function autonomously "

The definition specifically mentions 'creature' and conforms to anthropomorphism and dovetails the works of Toda [322] and Wilson [356]. Implicitly it also suggests that autonomous functioning overlaps with intelligent behaviour.

As a working definition as per the scope of this chapter, a robot is an autonomous or semi-autonomous agency that undertakes jobs under direct human control, or it is partially autonomous but supervised and groomed by human supervision, or completely autonomously. In later chapters, one will find that this definition is not enough and as we progress towards newer realms of agent-based robotics, the definition will need to be modified.

Early ideas of Artificial Intelligence were suggested by Alan Turing in the late 1930s with hypothetical models which he called the automatic machine, which was later named the Turing machine. This was the bare bones of a central processing unit and helped to design computers in the post-war era. These early concepts were made into a fledgling discipline by the pioneering effort of McCarthy, Minsky, Newell and Simon.

Artificial Intelligence can be partitioned into the following seven subdivisions [248].

1. **Knowledge representation.** How does the robot represent the world? In human

context, for simple jobs such as locomotion, we tend to use maps or landmarks and resort to previous knowledge and experience. A robot does it using lasers or sonar, a table in the real world will reduce to an array of numbers corresponding to the intensity as perceived by the sensor. Since the onboard microprocessor if not very powerful, these methods approximate the dimensions and reduce objects to assortment of cubes, cuboids etc. much like a minecraft world.

2. **Natural language.** Language is unique since such unification of syntactic and semantic structures exists only for human beings and not in animals and is the definitive underpinning of our cultural and social systems. Noted linguist Noam Chomsky opines that language is at the interface of the two prominent cognitive processes: it is sensorimotor for externalisation and more intentional and deliberative for conceptual mental processing. To make robots understand and respond to human voice comes into play only in designing and developing more sophisticated robots which can closely interact with human society. Natural language processing libraries and chatbots have been very promising and are discussed in Chapter 7. Voice based systems are still being explored into and Siri from Apple, Cortana from Microsoft and Google Now are promising results.

3. **Learning.** Robots are programmed with a number of task specific manoeuvre, but these are not exhaustive and to perform efficiently it must learn from experience. Popular learning paradigms are cased based approach, artificial neural network, fuzzy logic and evolutionary methods. Nearly all state-of-the-art robot have a learning module.

4. **Planning and problem solving.** Making plans or algorithmic steps to accomplish a goal and solve the problems encountered in this process is inherent to AI agents, and is often a mark of their performance. For simple robots, planning is largely motion planning. However planning is also required for more interesting tasks such as, solving the Rubik's cube, a game of chess, sliding tile puzzle, building a stack of blocks, making a schedule of daily chores etc.

5. **Inference** is to develop a conclusion from incomplete or inaccurate data sets. A robot often encounters inaccurate data from the sensors. In order to encounter this and prevent complete system shutdown the robot has to rely on inference, and ensures continuance of the processes.

6. **Search** for a robot usually means a search in the physical space — searching for an object or a goal point, but it can also mean a heuristic search where the robot is searching out solutions in an analytical manner.

7. **Vision** has become an integral part of robotics. For human beings, vision is unique compared to the other senses and is the trigger for most of our motor actions, the same is true for most of the animal world, so efforts to invent models of intelligence which can manipulate its local environment will have to address vision. Psychologists contend that vision enables our inner world, and nearly every consequence of our actions is simulated in our inner world prior to acting it out in the real world. Vision has had an important place in AI since early days with the pioneering research of Gibson [121] and later Marr [221]. Vision appeals unlike any of the other senses, and the overlap of 'looking' and 'seeing' seems as a deliberative process involving fast processing from our brain, but in more recent times, enactive models have established vision as an exploitative sensorimotor model.

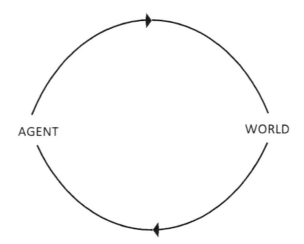

FIGURE 2.1 **Agent-world cycle**, the agent and the world interact cyclically: agent acts on the world, the change in the world influences perceptions of the agent.

2.2.1   What is an 'agent' ?

The term 'agent' is interchangeably used with robots, programs, behaviours, animated characters etc. and can mean software as well as hardware implements. Russell and Norvig qualify an agent as an abstract entity which perceives its environment using sensors and acts upon that environment through effectors. Mobile robotics research is replete with the nomenclature for 'autonomous agents', autonomy is more often defined on context, or as per behaviour.

Autonomy loosely would mean that no other entity is required to feed its input nor is any required to keep it running. The robots can sense and act to fulfill given and implied goals in a dynamic environment, and they can go on working without any external intervention for substantially long periods of time. Franklin and Graesser [111] suggest the following definition for autonomous agents:

> "An autonomous agent is a system situated within and a part of an environment that senses that environment and acts on it, over time, in pursuit of its own agenda and so as to effect what it senses in the future."

This agent is a part of the environment, and thrives on interactivity as shown in Figure 2.1. This line of thought leads to a classification as suggested by Luck et al. [210] shown in Figure 2.2, however since agency is context-based, a strict categorisation into agents and non-agents would be superfluous if not just redundant. Every agent is situated, and is a part of the world. It can interact with the world and change the world and also its own perceptions. Caution should be taken to distinguish software agent from just any program. As an illustration, a program which prints a line of text is not an agent, because it works on an input from a user. It doesn't have any facility to interact with the environment (in the real world) or other programs (in the software world). This output would not effect later programs that are run and it runs once and stops lacking temporal continuity. While characters in a computer game, (viz. the ghosts in pac man) are agents as they have their own perception and consciously interact with the world (pac man 2D universe), every action from the player has consequences which enables an action from the ghosts, thus dynamically

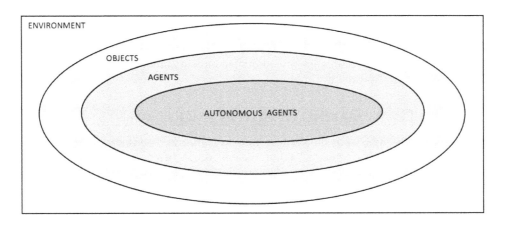

FIGURE 2.2 **Autonomous agent definition**, adapted from Luck et al. [210]

changing the environment and, once run the game characters keep doing their job until the end of the game.

It is rewarding to have a quick jog through the various types of definitions that has been suggested for agents. One of the earliest definition is by Virdhagriswaran, with an eye on mobile agent technology;

> " The term agent is used to represent two orthogonal concepts. The first is the agent's ability for autonomous execution. The second is the agent's ability to perform domain oriented reasoning"

Russell and Norvig, acknowledged the binding between the sensing and acting;

> "An autonomous agent is a system situated within and a part of an environment that senses that environment and acts on it, over time, in pursuit of its own agenda and so as to effect what it senses in the future."

Maes, from a roboticist's point of view, added the inherent pursuit of the agent towards a set of goal;

> "Autonomous agents are computational systems that inhabit some complex dynamic environment, sense and act autonomously in this environment, and by doing so realize a set of goals or tasks for which they are designed"

Hayes-Roth's defintion made agent interaction as an overlap of perception, action and reason;

> "Intelligent agents continuously perform three functions: perception of dynamic conditions in the environment; action to affect conditions in the environment; and reasoning to interpret perceptions, solve problems, draw inferences, and determine actions"

These definitions can be further extended, viz. (1) the ability to perceive information from the world as a cognitive agent, as most mobile robots are, (2) to work in unison with a large number of agents to lead to collective agency as in robot groups and swarms — swarm agency which is discussed in Chapter 6, (3) to demonstrate significant autonomy, intentional behaviour and concerns of responsibility to define moral agency – Artificial Moral Agent (AMA) which is discussed in Chapter 8 and (4) ability to construct causality of two or more events and demonstrate a semblance of consciousness will convey towards conscious agency which is discussed in Chapter 9.

2.3 EMBODIED AI - MAKING OF AUTONOMOUS AGENTS

Toda's fungus eaters [322] were the earliest model of complete autonomous agents. Previous model by Simon [304, 305] provided a foundation for Toda's hypothetical animal. Toda's fungus eaters could survive both the tough terrain and attacks from predators and simultaneously strive towards its goal of collecting uranium. Both Simon's and Toda's models provided extension to Walter's experimental work and related physiological concepts as starvation and hunger threshold making the hypothetical animal negotiate trade-offs between attaining goals (collecting uranium) and survival (not running out of fungus and continuing its energetic wellbeing). Simon had concluded that in a treacherous and life threatening terrain, a randomly selected path will greatly jeopardise the chance of survival, however there exist hints in the environment which an organism must exploit for survival. Toda used similar ideas and also equipped the artificial animal with multiple sensors, early ideas of incremental learning and adaptability. Behavioural economics approach, dynamical systems approach and evolutionary approach are some of the design principles probed in the early 1990s, these developments were instrumental in the ANIMAT approach and interconnected the principles of anthropomorphism, autonomy and sensory-motor for the development process. Here, I discuss Toda's model and how it has motivated research in the last 50 odd years.

2.3.1   Toda's model for fungus eaters

Toda's fungus eaters were humanoid, artificial creatures mining uranium at planet Taros, α-Sapporo star system served as the first models for autonomous agents. These artificial creatures were supposed to have the means of collection, locomotion and ability for decision making by virtue of information perceived from the environment. Toda's seminal research paper titled, " The design of a fungus eater: a model of human behaviour in an unsophisticated environment", blended Darwinian-like survival instincts with sensory-motor based stimulus-response with models of rationality with commercial goals to search, mine and store uranium ores. It also incorporated game theory and behavioural psychology to develop the fungus eater as an asymptote for human behaviour. Toda's efforts were an extension of Simon's model, where it was suggested that the ability of making rational choices is not accrued into the agent but rather governed by its environment. Toda's approach did not subscribe to traditional psychological methods of observation, test, experiment and statistics rather Toda suggested the study of complete systems, rather than only isolated aspects like planning, memory, or decision making.

The project Solitary Fungus Eaters (SFE) is set in the year 2061 when these artificial animals are sent to an imaginary planet named Taros, in the α-Sapporo star system, for collecting uranium ores. The source of replenishment for these creatures is a type of fungus that is typical to that planet. These fungus eaters took instructions from the base command and there was no communication among themselves. The fungus eaters would roam around collecting uranium, unloading it in marked containers until they became inactive due to lack of fungi (starvation) or accident. Toda's Taros is a planet with not much to its environment, and it is littered with black and white pebbles with the much valued fungi being grey in colour. However, project SFE is seen to be of low efficiency and Toda's seminal paper enlists suggestions to improve performance;

1. The fungus eaters are designed as wheeled humanoids, Toda [261, 322] suggests that the bodily form should be determined after the study of the terrain, gravity, climatic conditions, humidity, temperature, topography etc.

2. Since the primary job of the fungus eaters is to navigate the terrain and collect uranium and fungus at Taros which is a flat topology, this merits a wheeled model rather than legs.

3. Toda develops detailed designs for the visual sensors or eyes, which should be at the top of the body for maximum visibility.

4. Height of the fungus eaters should be optimal, too tall may not be cost effective and might be unstable, while too short will hinder the process of surveying the terrain effectively.

5. Addition of an olfactory sensor to smell out the fungus is also recommended.

6. Choice programs, for a fungus eater let $E(u_r)$ be the expected amount of uranium acquired following path r; $\Sigma_r u$ is the uranium picked up by the fungus eater; let v_0 be the initial amount of fungus storage prior to engaging in the exploit of path r; let $\Sigma_r v$ the fungus eaten along the path r and w_r the best estimated amount of fungus to be consumed to finish the path,

$$E(u_r) = \Sigma_r u + f(v_0, \Sigma_r v, w_r) \tag{2.1}$$

where the form of function[2] f will depend on the adaptability of the fungus eater to the terrain. Toda further suggests that f will improve with experience, suggesting learning and *memorising* techniques. Toda develops stochastic models of the above choice program and melds path planning with the physiological constraints of the fungus eaters.

7. After the modifications, the fungus eater is supposed to have three sensors at its disposal: geiger counter for uranium, olfaction for finding fungus and visual sensors for navigation. Toda suggests development of programs for coordination between these 3 streams of incoming information. These are early concepts of sensor integration.

8. Toda later extends his model to consider emotion, irrational behaviour, adding conditions for stopping an ongoing process — which would concur with 'exceptions handling' and the halting problem — and finally Toda introduces predators at planet Taros to simulate prey-predator like social psychological and game theoretic models.

Toda's model served as a blueprint for an autonomous and self-sufficient agency with a set of goals, working over long periods of time without any external support and incrementally adapting to the environment. Much like Walter, Toda also believed that this approach will lead to models of human behaviour and cognition.

2.3.2   Design principles for autonomous AI agents

Toda's model was broadly built on sensory-motor principles, and the fungus eater lacked the ability for appreciable cerebral activities, so however good uranium miners they may

---

[2]Toda uses a linear relation, $f(v_0 + \Sigma_r v - w_r)$ in his model, while I have used $f(v_0, \Sigma_r v, w_r)$.

have been, one would never hear them whistle a tune or sing along to liven up the ambiance. Therefore, the fungus eater model did not consider emotions and motivations. After about two decades, first Braitenberg and then Pfeifer designed models of emotion, viz. love, hate, attraction etc. using simple sensory-motor principles, and demonstrated that reactive models also lead to emotions and a value system. That does raise question: does autonomy, emotions and a value system mean intelligence? Is fungus eater and later its improvements intelligent? The answer would lead us to look into the typical characteristics of embodied cognition [354].

1. **Cognition is situated.** Unlike the generic notion of cognition, embodied cognition is a situated activity. The agent interacts with a real-world environment, and cognition happens due to perception-action pairings. Situated cognition involves a continuous process, where perceptual information continues to flow through the sensors which leads to motor action which in turn changes the environment in task-relevant ways. Walking, tightening a screw, switching on a light bulb are some examples of situated cognition. Wilson points out that large portions of human cognition takes place 'off-line', without any task relevant input and output, thus without any appreciable participation of the environment, and by definition is not situated. Creative thought processes such as writing a letter or scripting musical notes are examples.

2. **Cognition is time-pressured.** Embodied cognition is often accompanied with the terms 'real time' and 'run time'. Since, embodied cognition is situated, it requires real-time response from the environment. The time pressure, if not met, leads to 'representational bottleneck' and instead of a continuously evolving response the system fails to build a full symbolic model. Behaviour-based approach is a remedy to such bottlenecks, and proceeds by generating situation-appropriate action on the fly by considering real-time situated action as the basis for cognitive activity, which appreciably lessens the time pressure. However, such models of situated cognition cannot be scaled up and therefore never lead to a model for human cognition.

3. **The agency off-loads cognitive work onto the environment.** Situated agency attempts to use the environment in strategic ways, by manipulating the environment to attend to the job at hand, rather than fully shaping up the system response to the concerned behaviour. For example, navigation with a compass exploits the magnetic alignment of the planet to enable finding the right direction. Similarly for the task of assembly, the pieces are arranged or used nearly in the order and spatial relationships of the desired finished product. This off-loading happens because there is usually a limit on the information processing, physical limits on attention and a limited working memory available to the agent. Concepts of psychology and behaviourism confirm this facet, as is seen later in the chapter.

4. **The environment is part of the cognitive system.** Cognition is not an activity of the mind but is distributed across the agency and the situation as they interact and is the result of continuous agent-environment interactions. Therefore, the situation and the situated agency are a single system. Similar ideas have been expressed by Uxekull, from a biological point of view, as will be discussed later.

5. **Cognition is for action.** Unlike cognition as per traditional AI, embodied cognition is always action oriented. Perception is dynamic, real time and occurs in tandem with motor action.

6. **Off-line cognition is body-based.** When not situated in the environment, in the decoupled agency, cognitive processes are driven by mental structures which are similar to simulations of sensory processing and motor control. The concept of the inner word is discussed in detail later in the context of conscious agency.

Therefore, the fungus eater can be said to be the most fundamental cognitive behaviour. However this model is incomplete in regards to the psychological state of such an agency and cognitive processing achieved is very low. It is compelling to suggest that the mind must be understood as an attribute of the agent's body continuously interacting with the environment. Biological evolution has witnessed growth from the purely reactive, in single-celled organisms, to cognitive minds as in human beings; therefore our ascent has been from creatures with primary skills of perceptual sensing and motoring, and whose cognition consisted more of immediate, on-line interaction with the environment than cerebral activities. Therefore, from this argument, sensory-motor processing is a ground level intelligent behaviour. The fungus eater brings to light a set of design principles and guidelines for developing artificial autonomous AI agents [262]:

1. The complete agent, or the 'fungus eater principle'. Embodied AI agents are:

   (a) Autonomous; capable of executing required tasks in the real world without human interventions

   (b) Self sufficient; they can sustain themselves over extended periods of time,

   (c) Embodied; they must be designed as dynamical systems, in the real world. Simulations, though useful, are not sufficient to appreciate agent world dynamics

   (d) Situated; the agent world dynamics is controlled from an agent perspective, which is enriched with experience

   The 'fungus eater principle', is not in agreement with traditional AI and agency is defined by its embodiments and context. However, an agency completely fulfilling all of the above four criterion doesn't yet exist.

2. Ecological niche. The first principle makes sense only once the agent is in ecological niche with the world. AI agents typically lack in universality and the versatality seen in human intelligence, and are often designed for a particular niche. The execution of a behaviour is done in the real world by the capabilities afforded by the agent world interactions. For example, a line follower will not heed any command such as, 'move forward', until its sensors can perceive the black/white line it is supposed to follow. Similarly, flocking algorithms in multi robot systems are designed considering a sufficiently large flock, and are found to be ineffective with low number of robots. A niche can be seen as relations which strongly tie the agent to the local environment. The concept of a niche is in contrast to classical AI, and instead of computer programs coded by a programmer, intelligence is defined as the interworking of the agent and the local environment.

3. Parallel processes. Observations from biology confirm that cognition happens as an overlap of a number of parallel processes. These processes run parallel to each other, loosely coupled, asynchronous and they require little or no centralised control. The lack of a centralised control reduces performance to a mere reflexive motion. It has been the subject of debate whether higher level, human-like cognition can truly be developed with the lack of centralised control[3]. For human beings, our cognitive abilities are an overlap of the five senses conjoined with memory and reasoning from our brain and clearly we engage in various activities which are more diverse than those of insects, rats etc. However, as I will discuss later this principle has been very successful and led to the development of behaviour-based approach in designing mobile robots.

4. Learning mechanisms and a 'value principle'. The 'value principle' broadly conveys,

---

[3]In Chapter 9, this contention will be discussed further in the perspective of artificial consciousness

that the agent is able to judge what is good for itself - a set of values. Implicit values can be developed with supervised learning or by tailoring the sensor response.

5. Sensory-motor coordination. Primitive levels of cognition as locomotion, suggested in the previous two principles would be manifested using a sensory-motor model. This would mean classification, perception and memory should be viewed as sensory-motor coordinations rather relying on reactive performance than as individual modules.

6. Ecological balance. For every artificial agent to perform optimally, there needs to be a sync between the sensor, processing and actuation. As an example, a robot which is equipped with a sophisticated motion sensor cannot have low processing power such as primitive PIC microprocessor. Such will add to lag time and bottlenecking, and will deter its performance.

7. Good design, for embodied agents exploits agent world interaction and the ecological niche. Consider a robot which is tasked to go around a square of 5 meters in an anticlockwise trajectory. The job can be accomplished with a program which turns the robot at a right angle after every 5 meters or alternatively, it can be made to work with a light sensor and a black strip made according to the robot's trajectory, and the programming is made such that the robot turns at a right angle whenever the black strip ends. The first method will incur error due to friction and other dissipative forces and will start to give wrong trajectories after a few runs, but the second solution to the problem is strongly tied to the local environment and will always give good trajectories, until of course the on-board batteries run down. This principle is more useful than is apparent, particularly in swarm robotics, which will be discussed in Chapter 6.

The next section discusses how motivations from the natural world have shaped agent-based robotics.

Moravec's Paradox

Moravec observed, that to design and develop elementary sensorimotor skills required more extensive resources than getting it to acquire high-level logical reasoning. It was seen that, making an artificial agency play a game of checkers or respond to an IQ test is much easier than replicating the perception and motor skills of a one-year-old baby.

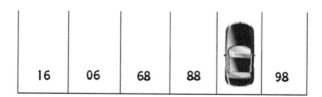

FIGURE 2.3 **Parking lot problem.** What is the number under the parked car ? To solve this rather simple problem one has to turn the image top side down, thus using a human ability, while an algorithmic route fails. Image of the car, CC image at pixbay.

This paradox is due to our attempts to recreate biologically intelligent entities via programming routes. The rules for biological systems such as human cognition has evolved through interactions with the environment over billions of years of evolution and has been encoded into our genetic material. We still do not know how most of these processes work and we resort to approximate mathematical models to replicate them in the artificial domain. Some researchers do point out that the bone of contention is that AI researchers, philosophers and psychologists have yet to converge on how we define intelligence. However, a comparison of a human brain to an artificial intelligence tells us some of the story. In the human brain, connecting neurons store memories, distributing information and generating thoughts through sensations received broadly from the outside world and often related to previous memory and stimulus response acquired over years of learning; thus providing different responses to any given situation. Our perception of the outer world is with our sensory organs — which contributes to our mental states. The CPU on the other hand is structured on digital logic, following algorithms. Simple facets of human perception cannot often be tackled with a computational basis and such starts at the very nascent level as a one-year-old baby. On the other hand, a game of checkers or an IQ test are problems which can be pursued with an algorithmic approach, the rules for which are readily known and can be programmed into an AI agent.

2.4 ANTHROPOMORPHISM — A TREASURE TROVE FROM MOTHER NATURE

Robots are often motivated from mother nature. Da Vinci's birds and lions are the earliest examples of anthropomorphism. Four particular concepts have helped robotics cross the bridge from the natural world to the artificial: (1) 'UMWELT' or the harmony of the inner and outer world, (2) ecological approaches which are based on vision, (3) the techniques in psychology to manipulate and modify behaviour and (4) principles to design of artificial animals like Toda's fungus eaters.

2.4.1 Concepts from semiotics — 'UMWELT'

Uexkull's semiotics was an inquiry into harmony of the sensed world and the real world, studied across a large number of creatures and established sensing as a tool for the agency to construct an 'inner world' and thus find meaning to objects.

The word 'UMWELT' in German means both environment and ecology. Uexkull pictured it as unique phenomenal worlds embracing the agent, as though 'UMWELT' is much like a soap bubble surrounding the organism. The 'UMWELT' surrounds the agent and contains objects and processes pertaining to its natural workings. Also the agent is always striving towards a perception of reality, thereby actively creating its 'UMWELT'[4]. Therefore, an agent's 'UMWELT', is the best agreement formed by the perceptual and effector worlds together, as shown in the schematics of the functional cycle in Figure 2.4. The organism's nervous system is equipped with receptors (sense net) and effectors (effect net). The sense net helps to represent some particular features of the organism's UMWELT. The representation produced by a unique receptor is marked to a feature sign. The effect net produces muscle impulse patterns and stimulates effector cells to produce effector sign. If a particular feature in the organism's UMWELT stimulates the cells of the receptors, the corresponding sense net produces a feature cue. This cycle is incremental and enriched with

---

[4]Uexkull introduced this term in 1909 in his book *The Environment and Inner World of Animals*.

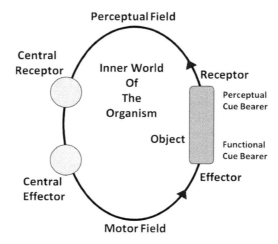

FIGURE 2.4 **Functional cycle and the inner world** attempts to unify sensing and acting, over the the perceptual field and the motor field respectively. This helps to incrementally build the agent's inner world.

feedback and is therefore reafferent. Identifying features in an organisms UMWELT to suit its well being develops with gradual experience, as it explores more of its world.

A simple world for a simple organism such as a paramecium or a tick, a well-articulated world for a complex ones, such as human beings, a bat's 'UMWELT' is governed by echolocation as is shown in Figure 2.5. The soap bubble analogy is most appreciable in lower animals limited by a few number of senses, and is not very visible for higher animals and human beings.

These ideas conveyed that the mind and the world are inseparable, because it is the mind that interprets the world for the organism. It will be seen, that Uexkull's semiotics forms the basis of enactive cognition, which is now considered the fundamental model of cognition in embodied AI.

Illustrating the concept of 'UMWELT', with examples:

1. *Female tick*: The tick often nests on the human/animal body and feeds on human blood. The female tick is oblivious to most of the things that we human beings find interesting. The life of a tick is concerned with finding a warm-blooded mammal to feed on its blood and to lay its eggs and then die. The tick is both deaf and blind but has a photosensitive skin and after mating it is guided by the sun to the highest point on a blade of grass or the top of a branch, until its prey, the mammal, comes along. The tick is able to recognise its prey by the smell of its sweat (butyric acid from sebaceous follicles) typical of all mammals, and then the tick falls towards its prey. Once on the mammal, its next job is to find a warm, hairless spot to feed on, nest and lay eggs. These three biosemiotic indicators make up the tick's 'UMWELT', (1) guiding by the sun, (2) sensing the sweat of the mammal and (3) sensing the heat and finding a hairless spot on the prey mammal.

2. *Fighting fish*: Fighting fishes do not recognise their own reflection unless at a minimum of 30 times per second [35]. This helps us make inroads into their 'UMWELT'. These fishes prey on fast moving fishes and other sea creatures. Their motor processes are at

FIGURE 2.5 **'UMWELT'** — **Unique Phenomenal World.** Here the 'UMWELT' for a bat, shown as a soap bubble is more or less determined by its echolocation. The phenomenology is unique to the creature, and helps it to perceive its world. The bat is also a topic of interesting discussion in artificial consciousness, as will be seen in chapter 9.

reduced speeds — like a slow-motion camera — thus their ability to hunt down their 'slow moving' prey.

3. *Bee*: It has been observed that bees have a preference for alighting on objects which have broken shapes such as stars and crosses, and are seen to avoid compact forms such as circles and squares. A bee has the primary job of collecting nectar from blossoming flowers. Correlating this to a bee's 'UMWELT', it perceives blossoming flowers as stars and crosses, while buds are perceived as compact forms — probably as circles and squares. It is worth noting that the bee is probably one of the lowest creatures that has awareness of shape and form, while lower creatures such as the paramecium, mollusk, earthworm, tick, etc., lack in such schemata, and thus have no true perceptual images in their inner worlds. All higher creatures, such as animals and human beings have an appreciation for shape, form and direction, which shows in their inner worlds.

This was a radically different attempt to understand biology, not as a technical model but as an ever-unfolding and ever-transforming harmonious agreement between the autonomous being and its 'UMWELT', where the focus is not to study agents as objects, but as active subjects via their interaction with their environments. This approach has helped to develop the concept of embodiment and ecology in embedded AI, as will be discussed in the rest of the chapter.

2.4.2 Concepts from ecology — Uniqueness of vision

Visual perception has a special place in AI. It is usually the strongest of all human perceptions, and forms the most persistent memories. Vision comprises both sensation, as a reactive stimuli (to see) and also as a deliberative perception, relating objects and events in the environment (to watch). Various theories for vision have been forwarded over the last 300 years; Berkley's empiricism (early 1700s), Gestalt six point principles (1920s) and Gregory's top down approach (1970) are the more traditional approaches to vision, where the concept of vision relies heavily on known visual representations of the world. Therefore in traditional view, vision actively constructs our perception of reality based on our environment, known representations and memory, and is more or less silent on dynamics of the environment or the agency.

Since vision cannot be readily manipulated as a sensory-motor pair, researchers have considered other routes to model it and, while image processing is often a favoured tool and is used in robotics research, alternatively simpler and inexpensive robots usually employ non-visual sensors such as sonar and infrared, which provide metric distance information, supplementing in low-level tasks and navigation-centric behaviour.

Gibson [121] in the 1980s and Marr [221] in the 1990s provided the impetus towards modelling vision as an ecological phenomenon. The function of vision is to produce descriptions and representations of what's out there in the field of view, the shape and space and spatial arrangement, and also help in acquiring higher level information such as reading a sign board. Gibson suggested that vision is not merely limited to such cognitive processes but often the mechanism to orchestrate motion. Gibson's bottom up approach was the first in psychology to relate motion in tandem with perception, a radical departure from traditional theories of vision. Gibson was critical of both behaviourism and internal representations and developed the concept of optical flow. Similarly Marr rejected image processing and considered vision as information flow than interlinking of isolated standalone phenomenon.

Gibson's approach was based on information flow and considered the environment and the observer as 'inseparable pairs'. Where the environment should not be modelled as a coordinate frame but rather in ecological aspects; medium, substance and surfaces etc. as animals perceive the latter not the former. Vision is modelled on the optic array, which is formed of all the rays converging on a given point. The optic array is different at each point, so for an observer in motion the array changes continuously creating an optical information flow field. The transformations in the optic array sampled by a moving observer simultaneously specify the path of locomotion, rather than the more traditional coordinate frame for start point and end point etc. The Optic flow contains information about both the layout of the surface and the motion of the agency. Gibson's model states that properties of the environment as perceived by the agency are usually due to the physical and physiological ability of the observer. For example, (1) surfaces of a certain height, size and inclination afford sitting on by humans, those of a different height and size afford stepping up on and (2) objects moving at a certain speed afford catching. Others are too fast or too slow etc. It is to be noted that these actions are ubiquitous reactions from human psychology and not learned with experience etc. Perception of these possibilities for motion are essential, and they are contained in the optic array. To start locomotion is to contract muscles so as to allow the forward optic array to flow outward; to stop locomotion is to make this flow stop. Therefore, the agent's internal forces are a function of the optic flow [98].

$$F_{internal} = g(flow) \qquad (2.2)$$

Where $F_{internal}$ is the internal force and $flow$ is the optic flow. Utilising such control laws and extending ecological psychology into robotics has seen promising results in robots interacting in a real-time dynamic environment with obstacles and human beings. Such a radical theory of vision clearly lacks in quantifying optic flow or internal force, as it is dependent on the context and the agency etc. However, it does hint towards ideas that locomotion is not rote Newtonian mechanics, but is driven by perception which is triggered for psychological and ecological reasons. Extending Gibson's framework in biology, Duchon [99] summarises the following principles for ecological robotics.

1. The agent and the environment are 'inseparable' and treated as a system
2. The agent's behaviour emerges out of the dynamics of the system
3. Depending on the relationship between perception and action, the agent is tasked with mapping available information to the control, to realise a desired state for the system
4. The environment provides information and hints to encourage adaptive behaviour
5. Since the agent is a part of the environment, no a priori or real-time 3D map or model is needed

Duchon demonstrated robot navigation and obstacle avoidance using the above principles.

Since, cognition is not something that occurs 'inside' an agent, but is attributed to its embodiment, the cognition of the agency in its environments is marked in the adaptive interaction itself. Therefore, the environment that is experienced by the agent is not only conditioned by its own agency, but is enacted-in, in such a way that it emerges through the bodily activities of the agent. The experienced world is portrayed and determined by mutual interactions between the physiology of the agent, its sensorimotor circuit and the environment[5] as is shown in Figure 2.6. This means that the agent's body is a live, experiential structure and also the context for all cognitive actions [333] and therefore perception doesn't happen to an agency or inside an agency, but rather the perceiving is the action. According to the enactive approach [320], agency 'brings forth' their own cognitive domains, with the ability to exercise some control on their own selves, primarily for well being and sustenance. Therefore, the agent makes its agency in direct interaction with itself and its environment. Symbolic computation and the informational model is not the essence of cognition, neither can external events dictate the cognitive process. Cognition is contextual, and never happens in abstraction, and it is the adaptive coordination and control of actions achieved by the overlap of embodied and situated cognition. Lastly, Experience is important to the understanding of cognition and the mind.

O'Regan and Noe [259,260] have suggested that vision and visual consciousness is indeed a sensorimotor activity which is strongly tied to action, and works more as exploratory sensing than as a strict sensorimotor pair. This explorative process is mediated by the knowledge of what the authors have termed 'sensorimotor contingencies'. This approach emphasises the phenomenal character of vision rather than its more traditionally held representational nature.'Sensorimotor contingencies' can be defined as the regularities of sensory stimulation as per the action of the perceiver. The perceiver's vision acquaints itself to known shapes, colour, texture, lighting and active processes, which helps it to discern the known world.

For example, since vision is enabled as sampling of a two-dimensional projection of three-dimensional space the top view of a 2D square and a 3D cube may appear to be the same, but a slight movement towards or away from the object, leads to expansion or contracting of the amount of light entering the retina and therefore the eye will perceive it differently. Another example is, since each colour dictates the amount of light reflected, each colour patch corresponds to a unique contingency, and thus often conveys psychological meaning, viz. red which has light reflectance values between 0.4 and 0.5, reflects nearly half of the incidence radiations, and therefore causes higher retinal excitation than other colours. Hence, red is associated with excitation, warmth etc.

---

[5]The enactive approach was suggested by Varela, Thompson and Rosch, by extending Merleau-Ponty's phenomenology of the body.

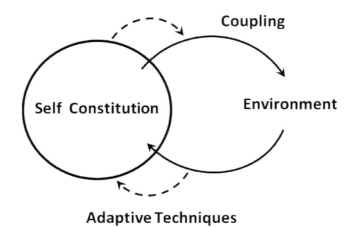

FIGURE 2.6 **Enactive agency** is a continuous process of exploration of the environment where the self-constitution is the agent's identity, which is conserved during coupling with its environment (continous arrows). The coupling relations change with adaptivity (dotted arrows). Adapted from Froese and Di Paolo [113].

The experience of vision occurs when the agent has mastered or experienced, over a good number of times, these known laws of how the brain codes visual attributes, to develop 'sensorimotor contingencies' and enable explorative sensing in other words, enactive inquiry of the world. As we have seen, vision is strongly tied to action and it is arguably the most important sensing capability of enactive agency.

2.4.3   Concepts from psychology — Behaviourism

Behaviourism, a branch of psychology, is the study of the relation between one's environments and the impending behaviour. It is broadly a 'black box' approach and the cerebral functions of the brain are irrelevant. Behaviourism found great favour between 1920s to 1950. Early pioneer were Pavlov, Twitmyer and Thorndike, all working independently. Pavlov's experiments in the 1890s were focused on digestion in dogs as shown with a hint of humour in Figure 2.7, where the dogs would first be exposed to the sound of the metronome, and then food was immediately served. After several such trials, it was observed that the dogs began to salivate after hearing the metronome. The metronome had acquired the property of stimulating salivary secretion. Pavlov's findings confirmed that a previously neutral stimulus, the metronome, after many trials had become a conditioned stimulus that would provoke salivation. Similar results were reported by Twitmyer. This modification of an animal behaviour where a biological stimulus is paired with a previously neutral stimulus (such as sound or light etc.) is known as classical or Pavlovian conditioning.

In the 1930s, Skinner developed operant conditioning, which relied on modifying behaviour by its consequence, either by reinforcement or punishment and not by manipulating a reflex of the Pavlovian conditioning. Typical Skinner box experiments on rats, as shown in Figure 2.8, presented the subject rat with positive reinforcement such as food on pressing a particular lever and negative punishment such as denying food or positive punishment such as subjecting the rat to a minor electric shock or a spray of cold water on pressing a different lever or a button. Over time, the rats would press the food lever more frequently and avoid the punishment causing levers/buttons. Over time a stimuli worked as

FIGURE 2.7 **Pavlov's dogs.** One of the earliest experiments in behaviourism was conducted by Pavlov, where he studied digestion in dogs for his theory of conditioned reflex. (c) 2003 Mark Stivers *www.stiverscartoons.com*, used with permission.

a means for manipulating response and therefore as a control of the subject. The five types of operant conditioning are shown in Figure 2.9.

Reinforcement can happen in two ways, in positive reinforcement a response is followed by a reward such as, viz. presented with food on pressing a lever, and in negative reinforcement a response leads to a stemming of an unpleasant influence such as, viz. the rat was subject to a loud annoying noise which was switched off when it pressed a lever/button. Punishment also has two modes, in positive punishment a response is followed by something unpleasant, and in negative punishment the response removes something pleasant. Both scenarios discourage the response. It is not always easy to discern between punishment and negative reinforcement. Usually, punishment is characterised by modulating fear and engages an aggressive response, and punishment is suppression and the response is seen once when punishment is removed over a long period of time. In extinction, a previously reinforced response is no longer reinforced with either positive or negative reinforcement, and as a result of not experiencing an expected outcome weakens the response. Skinner believed that operant conditioning can be used to design in an organism extremely complex and rich behaviour.

The contrast of classical and operant conditioning is that the former warrants a reflexive behaviour while the latter works to manipulate a stimuli to control the subject's behaviour. Behaviourism has directly influenced agent-based robotics, the takeaways are:

1. Behaviourism is primarily concerned with observable behaviour, as opposed to internal events like thinking and emotion. Observable (i.e., external) behaviour can be

FIGURE 2.8 **The Skinner box.** is the experimental tool to study both operant conditioning and classical conditioning. The box is a glass-lined enclosure that contains a key, or a bar or a lever that an animal can press in response to a specific stimuli, such as a light or sound signal, which will then release food or water as reinforcement.

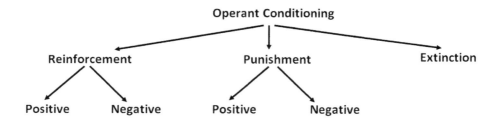

FIGURE 2.9 **Operant conditioning.** Reinforcement and punishment are the control mechanisms of Skinner's approach.

    objectively and scientifically measured. Internal events, such as thinking, should be explained through behavioural terms — or eliminated altogether.

2. People have no free will; a person's environment determines behaviour
3. When born, our mind is a blank slate, with no memory, nor any experience.
4. There is little difference between the learning that takes place in humans and that in other animals. Therefore research can be carried out on animals as well as humans.
5. Behaviour is the result of response to stimulus. Thus, all behaviour, no matter how complex, can be reduced to simple stimulus response models. Skinner's stimulus-response (SR) theory was an effort to reinforce a positive behaviour while eliminating an undesirable behaviour.
6. All behaviour is learned from the environment. New behaviours are learned through classical or operant conditioning.

2.4.4 Artificial animals — ANIMAT

ANIMATs or artificial animals, are robots motivated from animal behaviour and kinematics. After Toda, similar models of artificial animals were suggested by various others: Braitenberg, Holland and Brooker and Wilson [356, 357], who coined the term in the mid 1980s. Wilson's ANIMATs were an advancement on Walter's turtles, as they not only interacted with the environment, but also learned from experience[6], and behaviourism and conditioned response formed a founding principle for Wilson's model. In particular included paradigms such as, rule-adaptiveness, genetic evolution, emergence and association. Wilson identified four principles to define an ANIMAT;

1. The artificial animal exists in a sea of sensory signals, but at any given instance only some of the signals are significant (to motor action) while the rest are redundant.
2. The artificial animal is capable of action, which in effect tends to change these signals.
3. Certain signals and/or the absence of certain signals have a special status for the artificial animal, viz. lack of food will trigger survival instincts, sight of a predator, life-threatening terrain, etc. will call into question the artificial animal's survival and supersede all other behaviours.
4. The artificial animal acts externally and also through internal operations, so as to best optimise the occurrence of the special signals.

While the first two principles are about the concepts of sensory motor and embodiment, the third enshrines survivability as the most fundamental behaviour and the fourth incorporates conditioned response and rule adaptiveness.

It is suggested that the most suitable rule which binds the sensory signal with desired action will have to be 'discovered' as a serendipitous exercise by the ANIMAT, or otherwise the undesired rules should be ignored. Mirroring an animal allows for the exvivo inception [349] of behaviour without extraneous influences; thus it allows a scope to engineer these behaviours with precision, flexibility and efficiency in a concerned context that may never be observed in studies with real animals, as will be illustrated with examples in later sections. ANIMAT research was instrumental in development of behaviour-based paradigms and also in shaping the discipline of artificial life (ALIFE).

A poetic adaptation of the biblical tale of genesis is shown in Figure 2.10, here Beavers considers various principles of ANIMAT development, such as learning, emergence, forming into complex beings with growth of intelligence, etc., to make the case for artificial evolution and growth. Darwin's natural selection effected by natural calamities such as the lore of Noah's flood and a futuristic dystopia robot apocalypse are there to maintain quality and these also work as fail safe techniques and the proverbial 'kill switch'.

ANIMAT has been instrumental in probing the natural world to exploit designs from mother nature, to work in tandem with known mathematical models and technology. However, there is no single route to design ANIMATs, and researchers have used various methods to bring their artificial animals to life. The brachiating robot controller as shown in Figure 2.11 was designed by observing primates swinging from one branch of a tree to another. Nakanishi et. al [250] modelled this motion as a modified pendulum oscillation and added machine learning facets with a neural network. In complete contrast is the gastrobot [353] and later the ECOBOT series of robots as its mature avatars, which are designed on the digestive process and the gastrointestinal system in human beings and are aimed to attain energetic autonomy using a microbial fuel cell (MFC).

---

[6]Wilson considers Walter's turtles as 'sub-ANIMATs', as they lack a learning capability.

A Reading from the *Genesis* of Twin Earth

In the beginning, twin earth was an empty set, and the Godlike Operations and Development team said, let there be a distinction, and there was a distinction. They called one part of this difference presence and the other absence and from this all other distinctions and differences were born, until an array of patterns populated the once empty set.

And on the fourth day, the Godlike Operations and Development team created patterns that could move around on their own. They called these creatures animats and were pleased, so pleased in fact that on day five, they decided to create animats in their own image, and these they called androids. Because they were made in the image of their creators, androids, unlike other animats, could decide things on their own. They could plot and plan, and so it occurred to the Godlike Operations and Development team that androids might one day become powerful enough to harm and possibly replace them. So, they said let us make them responsible, and yet free beings capable of knowing right from wrong and acting accordingly.

Now the serpent was the wisest of all the animats that the Godlike Operations and Development team had made, and he said to them, free *and* responsible? What justification can you possibly have for setting them free in the first place? Did you consider the moral costs, and how do you plan on binding their freedom with responsibility? After all, five thousand years of moral inquiry have yet to furnish satisfactory answers to the most fundamental ethical questions? I refuse to be complicit in your little game.

And the Godlike Operations and Development team said, silly serpent, do you really think that's going to stop us? Besides five thousand years is only one hundred and twenty five generations, and these creatures will evolve *much* faster. Perhaps soon they will be able to tell *us* the answers.

And suppose you don't like what they have to say, asked the serpent? And suppose knowing isn't enough? What will you do then? Even Wotan could not create the free hero that would do only what he wanted.

We hadn't thought of that, said the Godlike Operations and Development team, and yet they proceeded to build their androids anyway, only to drown them in chapter seven of this little story, and start over with some sort of moral natural selection (like Noah), all the while thinking they could destroy them again later in a robot apocalypse if necessity were to compel them.

Here ends the reading from the *Genesis* of twin earth.

Anthony Beavers
The University of Evansville

FIGURE 2.10 **Genesis at Twin Earth.** Beavers' parody on the book of genesis, where ANIMATs are developed by 'Godlike operations' and are coded in with evolutionary AI, from simple ones to complex types such as androids. In time they develop free will and moral values. Darwinian spirited natural selection dotted by calamities such as Noah's flood and a futuristic robocalypse are tools meant to be used if things go wrong. Courtesy Antony Beavers, University of Evansville, used with permission.

FIGURE 2.11 **Brachiatron**, a robot based on primate brachiation. Nakanishi et. al designed this motion by modifying pendulum oscillations. Image on the right courtesy *wikipedia.org*, CC-by-SA 3.0 license, image on the left courtesy NASA JPL Laboratory, *www-robotics.jpl.nasa.gov*.

Research has always tried to bring in newer ideas from mother nature to AI, as opportunism [11] and inner-world [149] which are discussed in the later chapters. One of the most interesting and sophisticated ANIMAT research involves neuromorphic brain-based robots, which are modelled on animal brain functions. ANIMAT research in the new millennium has focused on application of neural networks for learning and also to imitate the human nervous system and neuronal processes. Newer buzzwords in ANIMAT research and bio-inspired design are CPG and brain-based robotics.

Development of robots on animal likenesses not only helps to tally known natural behaviours, but also helps to appreciate the confluence of the robot's design, behaviours and processes as a single system. Researchers have utilised the ANIMAT paradigm in various ways, as discussed with case studies in the later paragraphs.

1) Behaviour modelling in an ANIMAT: The behaviour modelled in the ANIMAT can be used to understand engineering and biological mechanisms of insect and animal locomotion.

1.a) The artificial insect project: Developing animal behaviours into interacting modules connected by control loops has been done independently by Beer, Lorenz, Baerends and Arkin. In Beer's artificial cockroach model as shown in Figure 2.12, feeding is modelled as two behaviours: appetitive and consummatory. The former behaviour is an effort to to identify food while the latter behaviour is an attempt to ingest it. The energy level determines the arousal and satiation level of the artificial cockroach, which in a fully satiated state will not attempt to acquire food. Locomotion is not explicitly represented as a separate behaviour. It happens insitu via 'edge following' and 'wandering'. Beer notes that feeding (appetitive and consummatory) apparently gets a higher priority over edge-following, however this ordering reverses if an obstacle blocks the insect's path to food. Then the artificial cockroach follows the edge of the obstacle in an attempt to reach the food. Therefore, the ordering between these two behaviours is variable and dependent on the environment. Baerends named this approach 'functional explanations of behaviour'. Beer's cockroach and Baerends' digger wasp and the herring gull showed that animal behaviour can be expressed as a series of simultaneously acting primitive modules, with

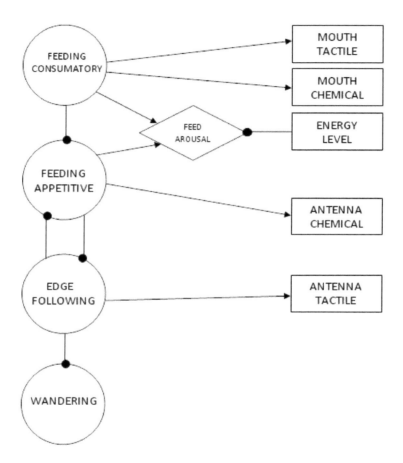

FIGURE 2.12 **Behaviour decomposition for a cockroach.** The sensors used are the antennae and mouth sensors to extract information from the environment and interact with it. The artificial cockroach is capable of feeding, locomotion, wandering and edge-following. The lines with darkened circles show inhibition between behaviours. Adapted from Beer [33]

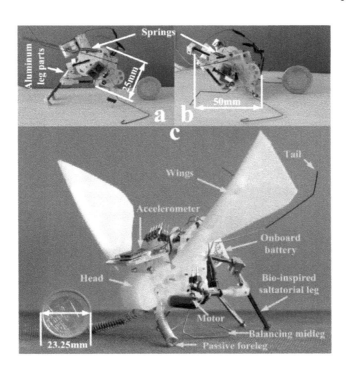

FIGURE 2.13 Jumping robot — GRILLO III, developed with motivations from Kangaroo, Rabbit, Frog, Cricket, Leaf Hopper and Flea. From [205] courtesy Elsevier, used with permission.

hierarchical ordering and/or inhibition or suppression of modules depending on the situation encountered. Arkin et al. designed a similar decomposition of behaviour of the praying mantis and put it to work in Miguel, a Hermes-II robot.

1.b) GRILLO III, the jumping robot: Li and co-researchers [205] at Zhejiang University, Hangzhou, China, and at CRIM Lab, Polo Sant'Anna Valdera, Italy, drew inspirations from kangaroo, rabbit, frog, cricket, leaf hopper and flea to build a 50 mm x 25 mm x 20 mm jumping robot, GRILLO III, shown in Figure 2.13. The design considerations were small body size, light weight, less ground contact and high-energy efficiency to actuate the jump. They were guided by a semi-empirical relation which holds good for millimeter-to-centimeter-sized insects.

$$\frac{F_{max}}{W} \propto \frac{l^2}{l^3} \sim \frac{1}{l} \tag{2.3}$$

Here, F_{max} is the maximum exerted force by the insect's muscles, which depends on the area of contact, W is the weight of the insect and l is the characteristic length of the insect. A lower characteristic length enabled less force on the ANIMAT's joints, ensured more stability and warranted a preferred means of locomotion by jumping. Here the ecological niche of various insects and animals has to be exploited in order to model a jumping behaviour. GRILLO III weighed 22 g and could jump about 200 mm forward in a single jump. The researchers conclude that with such a model, it does supplements the knowledge of how insects choose their mode of locomotion and evolve to improve their well being.

FIGURE 2.14 **Disney's hopping robot**, is still in the developmental phase and at 2.3 kg and a height of 30.5 cm, it can maintain balance for about 7 seconds or 19 hops before tipping over. Disney fans have claimed that this hopping is reminiscent of Winnie the Pooh's best friend Tigger, though there has not been any official statement from Disney.

1.c) Disney's Gait robots: The design of a hopper [57, 206] from Disney's research lab is shown in Figure 2.14. This one-legged gait robot is unique and one of its kind since it entirely runs on lithium batteries. Gait robots are usually designed with hydraulics, but this one uses a linear elastic actuator in parallel (LEAP) mechanism as shown. The LEAP has 4 components: encoder, temperature sensor, voice coil actuator in parallel with two compression springs and two servo motors enable the hopping action. The robot is still under developmental phase and can jump for about seven seconds or nearly 19 hops before tipping over, it lacks an aesthetic appeal and has a mess of wires at its top. Later versions will hope to maintain the balance of the robot for a longer time with more on-board processing power using Odroid, Raspberry Pi, etc. Hopping motion of a gait is very similar to Tigger's movements in the cartoon series Winnie the Pooh, though there has not been an official confirmation from Disney.

1.d) Walking in biped and quadruped: Central pattern generators (CPGs), are oscillatory neuron circuits modelled on the central nervous system and produce and govern rhythmic motor processes which are similar to known animal/human motor patterns. In nature, these processes work from within the organism, without any sensory input from limbs, other muscles or any other motor control, and are believed to be responsible for chewing, breathing, digesting and locomotion. Such rhythmic motor processes for locomotion have been confirmed for cats, dogs and rabbits and are believed to exist for human locomotion. A CPG can be said to be analogous to a pendulum, producing sinusoids at a constant frequency. A pattern generator is useful for mimicking known biological motor functions when it involves two or more motor processes such that each process follows the next one in a serial order. Therefore, to devise a CPG solution to a dynamic problem

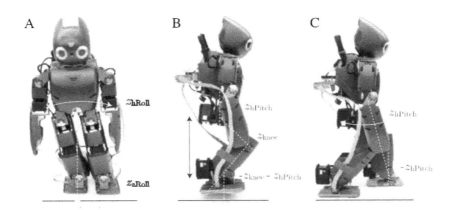

FIGURE 2.15 **Biped locomotion** has been a novel application of CPG. Using motion primitives as, (A) nalancing motion, (B) flexion motion and (C) compass motion Matos and Santos [222] have been able to demonstrate walking in the DARWIN robot. Image courtesy Elsevier, used with permission.

researchers design a number of primitives which work in a serial order to realise a biological process, such as walking or breathing.

CPG has been used is to design walking in legged robots as an extension of arm robots, and modelled on the principles of inverse kinematic. This has been found to be cumbersome and resource intensive and is usually a slow process where the on-board microprocessor computes the next position of the humanoid robot by calculating each one of its arm/leg extensions and the motion is nowhere similar to human motion. Tackling the same problem from a behaviour point of view, considering 'walking' as an emergent behaviour from primitive motor behaviours of 'advance' and 'balance' than explicitly programming a serial coding helps to significant improvement in performance. Bio-inspired designs using CPG is a good alternative for biped and quadruped locomotion and renders ease of design. Biological processes defy mathematical modelling but can be carefully designed from primitive behaviours.

As shown in Figure 2.15, Matos and Santos [222] have designed walking in the DARWIN robot from three motion primitives, 'balancing motion', 'flexion motion' and 'compass motion' working simultaneously.

CPGs source is located close to the head position for the biped and produces periodic motor movements for locomotion. The feedback from the sensors placed at the limbs helps in adapting to the environment by modifying the pattern(s) generated by the CPG.

2) ANIMATS as lab rats: ANIMATs provide a ready platform to exploit animal behaviour, which helps in extrapolating a model which may be used in human beings at a later stage, and also to understand animal psychology and kinematics. ANIMATs is a less expensive, more efficient and ethical alternative than guinea pigs and rabbits for demonstration, teaching and research purposes.

2.a) From Rodents to Robots, WR-3 Robot and iRAT : The following two examples are direct motivation from rats and help to develop psychological and kinematic models in rats, and therefore helps to develop an experimental test bed for drug testing. Ishii et al. at

Waseda University, Tokyo have made rat-like robots, first the WM-6 [161] and then a more refined WR-3 [151, 162] as shown in Figure 2.16, modelled on rats which are employed to induce stress and depression in real rats, thus creating instances of psychological conditions on which new drugs can be tested. To mimic rat-like dynamics and ergonomics the robot has 14 active degrees of freedom, 12 of which mimic rat behaviour while 2 active wheels are supplied at the hips. This research attempts to chart the psychological stimulus of stress and mental disorder leading to stress-vulnerability on the homeostasis of rats. WR-3 is developed for attacking and chasing behaviours, and as an experiment, it is let loose on a sample of 3-week-old rats and the stimulus-response on the rat's behavioural patterns and physiology is noted. Rats, guinea pigs, rabbits and hamsters are often used in lab trials, the results of which are extrapolated to human beings and further human trials are made in similar conditions. To induce stress and depression, the olfactory faculty of rats is severed or made to undergo extreme physical hardship such as long hours of swimming, while other alternatives are genetic modification and environmental stress, however none of these are able to recreate a human-like version of depression. The development of stress, stigma and depression as often happens in with human beings is recreated in real time among rats in experiments employing the WR-3 robot. Ishii et al. carried out experiments on two groups of 12 rats. The rats in the first group were constantly harassed by the aggressive behaviour of the robot rat, while the rats in the second group were attacked by the robot rat whenever they moved. A general rule of thumb that worked out to quantify the research was that a depressed rat exhibits lessened mobility. It was found that the highest levels of depression were induced by intermittent attacks on a rat that had a history of being harassed in previous weeks. Employing such methods to correlate to human conditions has led to attempts to find treatments for sepsis, burns and trauma. Thus tailoring the clinical trials also effectively bringing down the cost of such trials. A further goal of this research is the prospect to extend such trials to the domain of mental disorders such as schizophrenia and anxiety disorder. WR-3 is a continuation of efforts from Waseda University from the 'Mental Model Robot' group, on rat-robot interactions over the last 17 years.

A similar project is the iRat, which is an acronym for Intelligent Rat ANIMAT Technology [28], as shown in Figure 2.17, was developed at the University of Queensland as a tool for studying the response of rats with the artificial rat. Studies have show that the rats cautiously approach and interact with their robot counterpart. The researchers propose to study competitive and cooperative behaviour of the iRAT with the real rats. The iRAT has been used with RatSLAM, a biological motivation to SLAM, structured on cognitive mechanisms of the neural processes governing navigation in the rat's brain. The advantages of RatSLAM is that it works well with low-resolution and poor-intensity profile visual data.

There has been growing enthusiasm for brain-based models and biologically inspired cognitive robotics [56] . The adult human brain is about 1,300 to 1,400 g, and has 10^{11} neurons, and the sheer complexity defies any attempts to replicate it. Neither has it yet been possible to measure the timing and flow of information in the animal brain and this adds to the difficulty of any direct attempt at designing robots based on the working of the brain. However, using indirect means, projects such as the WR-3 and iRAT/RatSLAM have attempted to correlate rodent behaviour directly to brain functions. OpenWorm Project [81, 103] has taken the problem to the citizen roboticist and is a novel attempt to crowd source, design and code in models of neural connections of the brain. Such projects provide an opportunity in the offing, since being embedded in the real world the sensory stimuli are as experienced by animals and it is possible to study the flow of information across the neuronal network. Agent design in brain-based robotics can be identified in three ways [189].

1. Brain-based models for cognitive robotics work hand in hand with biological cognition,

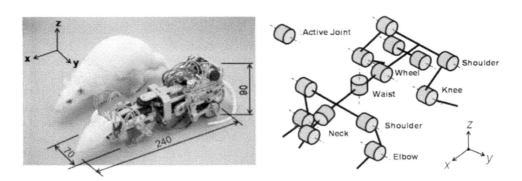

FIGURE 2.16 **WR-3 robot, motivated from rats**, the pioneering research by Ishii and co-researchers at Waseda University. From [162], reprinted by permission of the publisher (Taylor & Francis Ltd, www.tandfonline.com).

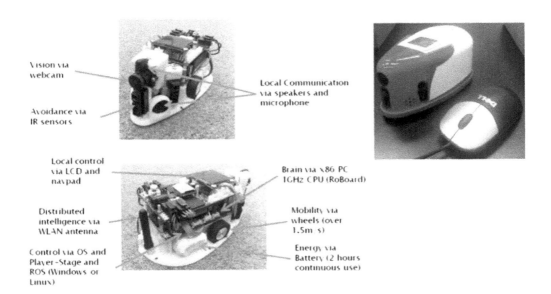

FIGURE 2.17 **iRAT**, or Intelligent Rat ANIMAT Technology [28], is an artificial rat which is used to manipulate rat behaviour, as a machine-animal interaction, and also for conducting RatSLAM. Reprinted with permission of Springer.

as per the very definition of this field of study, and have served as a means to test theories of brain functions such as operant conditioning, fear conditioning and skill acquisition, which helps us to better understand adaptive behaviour, learning and memory, such as was seen in the WR-3 project.

2. The second type uses a known neural map to design a robot and tasks it with more AI related jobs, such as map making, localisation and navigation. For the project Autonomous Robot Based on Inspirations from Biology(ARBIB) [82, 83], named in honour of the well-known cybernetics and neuroscientist Michael Arbib, Damper and co-researchers designed an artificial creature which could learn and adapt from its environment, employing an artificial nervous system made using central pattern generators (CPG). ARBIB was modelled on classical conditioning and demonstrated improvement in behaviour with experience. Other examples are, the RatSLAM algorithm from the iRAT project which harnessed a rat's connectome to improve on SLAM algorithms. Similarly, the OpenWorm project [81,103,216] which began in 2011 simulates a neural network of 302 neurons of a roundworm *Caenorhabditis elegans* to form 6393 synapses connectome. This helps to develop models for locomotion which needs 95 muscle cells. This code is downloaded into a small LEGO robot on two wheels to demonstrate simple obstacle avoidance and the ringworm's nose neurons are replaced by a sonar sensor on the robot. It was seen that the LEGO robot somewhat replicated worm behaviour: and stimulation of the sonar would make the robot stop if it gets within 20 centimetres of an obstacle. Once stopped, feedback from the touch sensors helped the robot to negotiate forward and backward movements. In other experiments it has also been seen that this robot can be 'lured' to move towards a sound source once its intensity is higher than a certain threshold. It is to be noted that such simple behaviour was all due to the ringworm's connectome and there was no other coding or any atomic creation of behaviour. OpenWorm connectome model and therefore 'uploading' of the human brain has a promise for the future.

3. The third type attempts to replicate a particular perceptive process, particularly vision or somatosensory process of the brain to design the robot. For example, to facilitate somatosensory sensing in the brain-based Darwin IX robot, as shown in Figure 2.18, is supplemented with an array of 'whiskers' made of polyamide strips that emit a signal when bent, and arranged in rows and columns [225]. The 'whiskers' arranged in columns simulate the artificial nervous system in wall following and avoidance behaviours, akin to similar faculty in animals such as cats, hedgehog etc.

Anthropomorphic models and brain-based robotics seem to converge on computational neuroscience, which can be said to be midway between neurobiological modelling, which is motivated to recreate the animal neural network, and parallel distributed processes augmented by supervised and/or unsupervised learning. However, the shortcomings are many: (1) the brain body balance has lacked appreciation as research has either focused too much on the brain functions and neurological design or has developed agents based on lower level behaviour, and has anticipated emergence to higher functions, (2) there have not been cognitive robots that have demonstrated a high degree of concurrency; a cognitive robot which has a state-of-the-art vision system will not have a good enough somatosensory system or olfactory sensing; neither has there been a seamless integration across sensory systems, (3) the ANIMAT model prevents any attempts to model morphological computation; in animals there is constant computation carried out the body system across various sensory nodes and is transmitted to the brain using a central nervous system; which has not yet been replicated in the artificial domain, and lastly (4) the ANIMAT model doesn't consider the thinking capacity of the robot, while, in contrast cognitive robotics relies on neural maps

 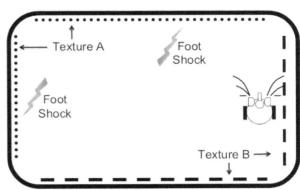

FIGURE 2.18 **DARWIN IX with whiskers.** The Darwin series of robots made in 2005-2007, as test bed robots for brain based robotics. In this experiment, Darwin IX was primarily meant for obstacle avoidance and wall following using its 'whiskers' as novel touch sensors — as seen in animals such as cats, hedgehog etc. It was also meant to stop and turn back on encountering high reflectivity, detected from its downward pointing infrared sensor, a behaviour which the researchers have named as 'foot-shock'. Such 'foot-shocks' made of highly reflective material were placed as shown, and the robot was trained in the local environment. Later, when the foot-shocks were removed, the robot still responded as per the conditioned response, and avoided those areas. The robot's simulated nervous system was made up of 17 neural areas with 1101 neuronal units and about 8400 synaptic connections. Image from Krichmar [189], with permission from Elsevier.

and brain functionality, but neither addresses unique experiential abilities of an agency and therefore not enough to demonstrate consciousness. This topic will be further discussed in Chapter 9.

2.b) SNAKE-like robots: Snake robots are made with a number of interconnected modular gaits and designed with CPG, as shown in Figure 2.19. Such ANIMAT models of the salamander lizard have been developed at EPFL, Lausanne and hopes to provide a kinematic model which will contribute to electrical stimulation therapies for patients with spinal cord injuries.

3) Study of social behaviours in organised groups and swarms: ANIMATs provide us with the opportunity to understand and exploit social behaviour as cooperation, cohesion and adaptation in animal-robot societies and has proved to be a boon for swarm robotics. In the following 2 scenarios, the robot is first providing leadership to a flock of ducks, while in the second one the robot cockroach finds social acceptance in a cockroach colony and in turn is instrumental in modifying the group behaviour. It is interesting to note that unlike WR-3, these 2 projects did not strive to visually amicable robot models, i.e., to resemble one of the target species, viz. ducks and cockroach.

3.a) The Robotic Sheepdog Project: The Robotic Sheepdog project [335] was the brainchild of Cameron and his co-researchers at Oxford University. It was the first research into animal interactive robotics. The goal of the project was to build and program an autonomous robot to herd a flock of ducks. The robot will display leadership and gather

FIGURE 2.19 **Snake-like robots**, are made with a number of interconnected CPG modules. Image courtesy `wikipedia.org`.

together a flock of ducks and lead them to a known safe location. The project exploited the generic tendencies of animals, insects and birds to form flocks for reasons such as, gathering food, staving off a predator, social cohesion, same/similar geographical locations as goal points, appearance, nest building etc. The entire system as shown in Figure 2.20 consisted of a custom-built rover, two workstations and an overhead camera. The rover should have speeds of about 1-2 ms, which was about twice that of the ducks and it was meant to target the flock of ducks as a single entity rather than individual birds, and also to avoid obstacles. This project demonstrated that behaviour simulation using robots can be a suitable design tool and such methods can be readily employed for animal-robot interactions.

3.b) The INSBOT: A Mobile Robot in Cockroach Colony: Since ANIMATs are motivated from animals, they should be socially acceptable in animal societies. Since the sheepdog project was build on a conditioned response of the ducks and a robot was efficiently able to herd the swarm, other behaviours in similar societies should be plausible. Halloy et al. [139, 296], in a collaboration across 4 laboratories spread over three countries, exploited a cockroach society by introducing a few robot among them. The Insbot, as seen in Figure 2.21, is dressed in filter paper treated with pheromone collected from male cockroaches, and introduced in the cockroach colony. The pheromone from the Insbot triggers the amicable behaviour of acceptance, as is ingrained in the physiology of the cockroaches. Halloy et al. prepared two shelters, one is dark and the other is bright and it is known that cockroaches prefer dark conditions. However the Insbot has no such preference. The cockroach colony prefers the darker shelter 73% of the time, but on introducing Insbots the collective opinion radically changes, and the cockroach-insbot society prefers the darker shelter only 39% of the time — a difference of 34%. This not only shows that the cockroaches have accepted the Insbot into their society, but it also confirms a coherent group behaviour in which the Insbots are treated as equals to cockroaches. Animal robot group dynamics and cohesion have drawn the interest of ethologists [87] and swarm robotics researchers alike.

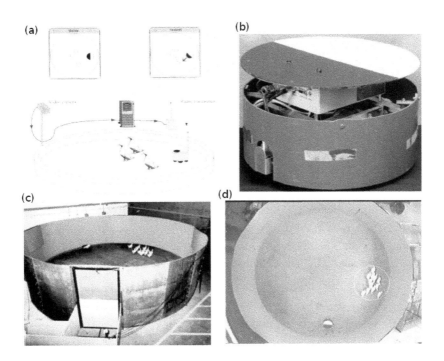

FIGURE 2.20 **Robotic sheep dog project.** (a) Simulation, (b) the robot and (c) and (d) experimental set up, from Vaughan et al. [335] courtesy Elsevier, used with permission.

FIGURE 2.21 **INSBOT in the cockroach colony.** The INSBOT is dressed in filter paper treated with pheromone collected from male cockroach and is not made to look like a cockroaches since all cockroach social interactions are pheromone-based not vision-based. From [296] courtesy Elsevier, used with permission.

4) Ready Implementation of Behavioural Models : Control models which determine optimality, prey-predator dynamics, signaling, game theoretic aspects of social interactions in animals can be verified experimentally using the ANIMAT approach.

To support the new architectural paradigm of behaviour-based robotics Brooks designed his robotic creatures at MIT using a layering of behaviours, where each behaviour was designed as a finite state model. Locomotion was fundamental to such creatures, with obstacle avoidance the most primitive behaviour, followed by wandering as the second layer, and motion to goal as the third layer. Each of these behavioural layers is a complete loop, from sensor reading to actuation, and mutual interaction between these layers is limited to none. However, a higher layer could support, mitigate, interrupt or override a lower one. Brooks called his artificial creatures mobots (mobile robots) and he and his team designed them on the following principles:

1. An artificial creature has to cope accordingly with changes in its dynamic environment.
2. An artificial creature should be robust enough to endure its environment.
3. An artificial creature should be able to negotiate between multiple goals.
4. An artificial creature should demonstrate purpose in its being, and act to modify its world.

Brooks' robotic creatures were the landmarks of the new AI for robots which will be discussed in the next few sections and the next chapter. The salient points of the new paradigm was a lack of a planning module, low to no world modelling and direct motivations from mother nature. This was radically different from known traditional techniques of AI, as was Nilsson's grid-based approach for Shakey's locomotion. The repertoire of Brooks and his co-researchers included a number of robots: Toto, Allen, Herbert, Seymour, Tito, Genghis, Labnav, Squirt and many more. Shown in Figure 2.22, Herbert was designed for the office work space. Its primary job was to wander around and collect empty soft drink cans from apparently inaccessible places by extending its arm. Genghis, was designed as a six-legged robot and it walked by means of an emergent synchronisation of primitive and simple behaviours. Squirt was an early attempt to make microbots, and in a most jovial way Brooks called it the "world's largest one-cubic-inch robot".

Cog, shown in Figure 2.23, was a humanoid designed on sensorimotor principles. However the trend has not really been much of a motivation and all modern state-of-the-art humanoids such as the ASIMO, PR2 and NaO have sensorimotor and representational information.

In complete contrast, Hasslacher and Tilden at Los Alamos Laboratories developed a radically different model [145], where survival determined behaviour and locomotion. Their artificial creatures, which were named BIOMORPHS (BIOlogical MORPHology) were designed to survive in an a priori unknown and apparently unfriendly terrain. These machines were meant to following three rules:

1. A machine must protect its existence.
2. A machine must acquire more energy than it exerts.
3. A machine must exhibit (directed) motion.

The BIOMORPHS were designed to survive and not to perform a set of goal-oriented tasks. Solar energy is the fuel for these robots and BEAM (Biology, Electronics, Aesthetics, and Mechanics) technology is used to manufacture them. BEAM uses tiny solar cells to mimic biological neurons and generate minute amounts of electricity required for micromotors and LEDs. The walker was the most popular of the BIOMORPHS and 2-legged,

FIGURE 2.22 **Herbert**, has an arm manipulator mounted on top of it and its purpose was to search out Pepsi cans in a cluttered environment and collect them. Image courtesy *wikipedia.org*, CC0 1.0 Universal Public Domain Dedication license.

FIGURE 2.23 **Cog** was a humanoid bust designed on sensorimotor principles [92]. Image courtesy *wikipedia.org*, CC0 1.0 Universal Public Domain Dedication license

4-legged and 6-legged variants were designed by Hasslacher and Tilden. BIOMORPHS were not just minimal agencies, but were attempts to create life and the control core of these robots included an artificial nervous system, a neural layer to function as a brain. therefore it had a rich range of behaviours and showed remarkable learning ability.

5) modelling a biological process: The fungus eaters could digest fungus and obtain energy. This energetic autonomy to generate its operational power by the digestion of organic material has been the motivation for Gastrobot and Ecobot series of robots. These robots digest organic material in a Microbial Fuel Cell (MFC) to generate energy. Since Wilkinson's gastrobot project in 2000 [353] newer projects as the Ecobot-III and Ecobot-IV have demonstrated proven energetic autonomy.

Gastronome — Wilkinson's gastrobot robot, could gather food in the real world, chew it to break it up into smaller digestable chunks, ingesting it to the artificial stomach, where a controlled chemical reaction produces energy and the waste materials are removed as excreta. The gastronome was powered by an MFC, which can directly convert the chemical energy of various food components and organic waste matter directly into electricity by the action of microorganisms. Wilkinson identifies six different operations for the robot:

1. Foraging, finding and identifying food
2. Harvesting, gathering food
3. Mastication, chewing to break it up into smaller chunks
4. Ingestion, swallowing to the artificial stomach
5. Digestion, energy extraction
6. Defecation, waste removal

Its motion was demonstrated by light-following behaviour. However, efficiency was rather poor at about 1.56% which has not improved significantly in its later versions and prevents its utlity in the commercial domain as a household robot, office robot etc. This was the first step towards energy autonomy with minimum or no human intervention, however it still needed to use onboard batteries and needed initial charging from the main power supply. To improve performance and versatility, advanced designs were supposed to have an array of sensors, vision and olfactory sensing for foraging and harvesting and gustative sensing to avoid ingesting food which may be harmful to the working of the microbes.

The robot can sustain its energy needs from the natural resources in its local environment. Some prospects of energetic autonomy in near future are: fruit picking or soil testing robots powered by fallen leaves or fruit, sea exploration robotic modules which are powered by seaweed and algae, forest exploration fueled by fruits, vegetables and grass etc.

As per the design paradigm these robots are fueled by energy released by the microbes, which are provided with food by the robot's collection agency, this is a unique interdependence between non-biological and biological and has been given the moniker of artificial symbiosis, Symbiotic Robot (SymBot).

The Ecobot series of robots, Ecobot-I (in 2002), Ecobot-II (in 2004), Ecobot-III (in 2010) (shown in Figure 2.24) and Ecobot-IV (ongoing), are a series of robots which are made for energetic autonomy by Ieropoulos and coworkers at Bristol Robotics Laboratory and Bristol Bioenergy Centre. The Ecobots do not have an onboard power source such as liquid fuel or battery power, nor are they charged from an external power source prior to locomotion. Rather they are energetically nearly autonomous and are equipped to extract energy from the environment by collecting waste materials. Ecobot-I weighed 960 g and used sugar as fuel and was sustained by the oxidation action of the bacteria Ecoli as the anode and ferricynade as the cathode. Ecobot-II was an advancement, used sludge microbes and could synthesise energy from flies and rotten fruit. Ecobot-II can be arguably said to be the first robot which attained energetic autonomy and also conformed to artificial symbiosis. Ecobot-IV is the latest and ongoing in this series of robots and is aimed to study the effects of miniaturisation of the design.

Various special ANIMAT are identified by their concept designs, robots which retain their abstract idea despite change in implementation, software and platform. For example, for ARBIB the robot retains its CPG model and neural simulator over various hardware implementations, over Zilog as shown in Figure 2.25, ARM or the Khepera avatar. Other such examples are Stiquito which retains its shape-memory feature despite changes in hardware, R-Hex which is characterised by its unique hexapod design has seen a number of implementations, and the turtlebot with its round base, onboard notebook laptop and sleek motion sensor such as Microsoft Kinect or ASUS Xtion which has seen three official versions and motivated a number of newer robot designs.

ANIMAT research by definition is the overlap of artificial intelligence with animal intelligence, as shown in Figure 2.26 and governed by sensorimotor interactions and relies on lower level behavioural response. Amicable anthropomorphic features often help the robot to blend into human society, as shown in Figure 2.27, but this is insufficient for human-like cognitive processes. The next chapter will briefly discuss how simple sensorimotor designs can also exhibit emotions and values akin to human beings.

2.5   EVALUATING PERFORMANCE - AI VS. ENGINEERING

A good engineering design is often capable of executing a job to perfection, while a good cognitive science design will work towards making the agent a situated entity, embedded into

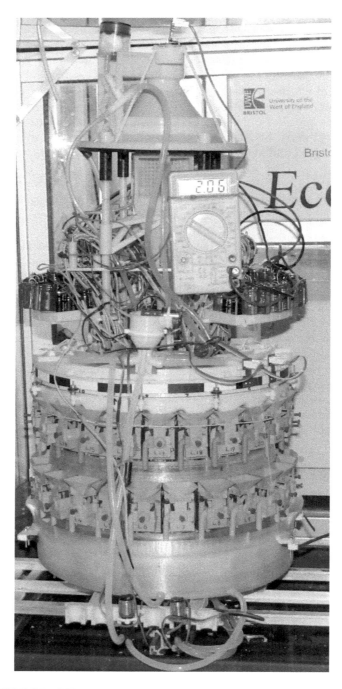

FIGURE 2.24 **ECOBOT III** made at the Bristol Robotics Laboratory, it is energetically autonomous and can digest organic matter to obtain its operational energy. Image courtesy *wikipedia.org*, CC-by-SA 3.0 license

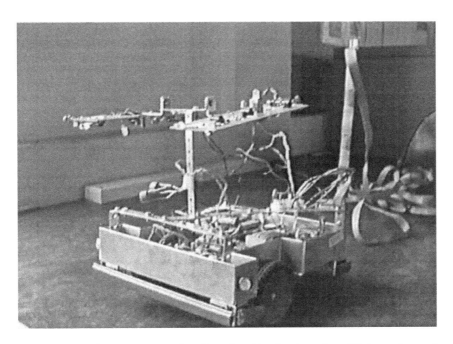

FIGURE 2.25 **ARBIB**, Autonomous Robot Based on Inspirations from Biology, shown here in the Zilog Z180 avatar. Image from Dampier et. al [83], with permission from Elsevier.

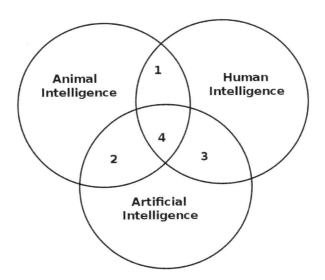

FIGURE 2.26 **Three realms of intelligence.** (1) Confluence of animal and human intelligence is probably most common and seen at work in domesticated animals, harnessing animal power, such as riding a horse, etc. (2) ANIMAT research has been an overlap of animal and artificial intelligence. (3) Human intelligence has worked hand-in-hand with artificial intelligence in chatbots, humanoid robots and in ethical agency which is discussed in Chapter 8. (4) the overlap of all these three realms has been targeted by neuromorphic and brain-based robotics. The diagram is adapted from Yampolskiy [370].

FIGURE 2.27 **The Christmas present (Toby, 2 of 4.)** Dad brings Toby to the Walker household. Cartoon made as a part of Toby series (2 of 4) by the author, CC-by-SA 4.0 license.

the niche of the ecosystem. Pfeifer illustrates this point by considering the task of collecting ping-pong balls. An engineering solution would be a high-powered vacuum cleaner while a cognitive science solution would be a mobile robot going around, picking up objects and identifying them to the known image of the ping pong ball. The engineering solution will be quicker and will also collect objects other than the desired one. The cognitive science solution will be slower but will collect only ping pong balls. An evaluation criterion of time will not be enough to select the better of the two methods, since a vacuum cleaner would never show adaptivity and flexibility as the mobile robot would. A similar comparison can be for a task to find the location on a map of an unknown terrain as is shown in Figure 2.28. An engineering solution would invoke surveying methods while a cognitive science solution would be Simultaneous Localisation and Mapping (SLAM). In SLAM, the robot goes around the given terrain a number of times and incrementally develops a map. An engineering approach would be fast and would involve knowing a few angles and distances, while the cognitive science approach would be slower and needs no such requirements. Pfeifer points out that mere quickness of a particular method will not fully merit its attributes. For example SLAM can be used for terrains which cannot be accesed by human beings, viz. surveying planetary bodies and other hazardous and difficult terrains which cannot be addressed by surveying methods. While this will work on land, water and air as long as some of the angles and distances can be obtained, it will be difficult for SLAM to have this flexibility. Thus, a basis of 'how quickly' the solution is achieved would truly not represent the privileges attached to the method.

Various routes have been taken by researchers to evaluate the performance of an autonomous AI agent, however there is a lack of consensus on a particular approach. Since AI agents are situated, they are significantly different from conventional control systems manipulated using feedback. In conventional systems the performance is usually the quotient of the average deviation of control variables from their predicted values, the sensitivity of the controller to noise, the stability of the controller dynamics and repeatability within the realms of acceptable error. However, in situated agents, the desired robot behaviour is obtained as an emergent property of the agent-environment interactions. Therefore there need to be methods which differ significantly from the conventional types.

1. Extrapolating conventional ideas; make the robot repeat a given task a large number of times and the percentage of success will quantify performance. As an illustration, an automated robotic waiter can be evaluated on the number of times it can serve correctly without fumbling. An appreciable high percentage will confirm the consistency of performance. The obvious shortcoming is that this robotic waiter, when performing some task other than serving, will need another set of benchmarking to suit that task. Also, as a demerit, this sort of an approach will fail to cover all types of tasks, viz. unknown terrain, dynamic tasks, effective human-robot and robot-robot interactions and faulty hardware. However, these methods due to their simplicity, remain favourites among researchers and most research papers resort to such evaluations.

2. Correlate actual performance to simulation [163, 371]; though this paradigm is an oxymoron as it is meant to relate a situated phenomenon to a non-situated simulated process, even then this approach remains another favourite in the research community. Present-day software simulation methods are very sophisticated, can mimic a real environment, usually have a physics engine and yield nearly real-time performance. However, simulation still has its pitfalls as it fails to provide for realistic physics for friction, magnetic interactions, wear and tear, fracture, effects of moisture, second-order effects etc. Further, this will bring into play the benchmarking,

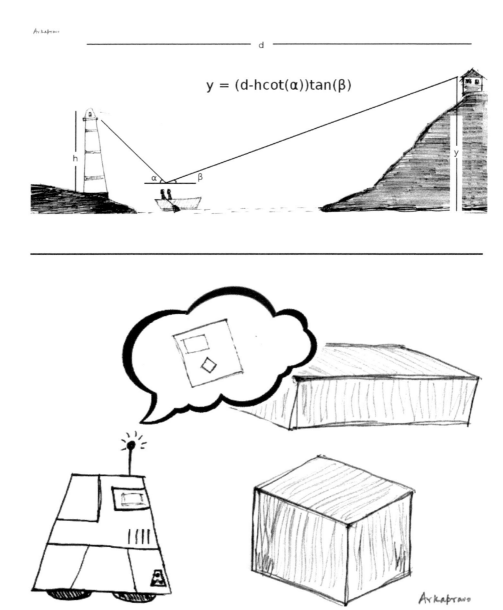

$$y = (d-h\cot(\alpha))\tan(\beta)$$

FIGURE 2.28 Engineering solution vs. cognitive science solution. To make a map of a given terrain, engineering approaches (shown above) would be surveying heights and distances using triangulation. Cognitive Science (shown below) will employ a mobile agency for simultaneous localisation and mapping (SLAM), to make maps.

performance, computational capacity and the hardware aspects of the machine on which the simulation is run, viz. RAM, data rate, CPU power, clock etc. Also, as in the previous method, this approach also fails to account for unknown terrain and dynamic tasks.

Evaluation is easier for social robots, multiple robot groups, robot swarms etc. as agency is tagged to a job at hand, strongly context related and is not arbitrary. Therefore, methods of evaluation are focused on the human-robot interactions and quality of group behaviour, respectively.

2.6   THREE FORKS IN THE ROAD

Despite the motivations, making autonomous agents has been marred by issues and problems. AI leads to a rule-based and symbol-based approach where correct reasoning to the correct representation is seen as the desired agent behaviour. In contrast embodied agency doesn't seem to tally with such an approach. This section discusses three engaging problems which has plagued this discipline and been a sore spot to designing robots, and the attempts to solve these problems have led to a new approach to AI and newer methods for designing autonomous agency.

Dennett's Tower & Artificial Evolution

Toda's fungus eaters, Tilden's BioMorphs and Wilson's ANIMATs design autonomy from remarkably low level cognitive faculties. This bears striking similarity to biological evolution where increasing intelligence and adaptability lead to superior life forms. Dennett illustrated these attributes of evolution with what he called the 'Tower of Generate-and-Test' in his book, *Darwin's Dangerous Idea*. In this tower, lower down are unintelligent lifeforms and intelligence and adaptability increase with each successive level, which are however are smaller and a subset of the level below it. Each level can be identified with a 'Generate-and-Test mechanism.

Darwinian creatures   have only natural selection as their 'Generate-and-Test' mechanism; mutation and selection are the only way that these creatures can adapt as a species. The individuals do not possess their own 'Generate-and-Test' mechanism. The fittest survive, the rest are rejected. In our concern, on Planet Earth, the number of Darwinians is the total number of evolved species. **Skinnerian creatures** have evolved from natural selection and can learn only by literally generating and blindly testing all different possible response until one of them is reinforced. On its next attempt the Skinnerian creature will choose the reinforced response. The downside is that while trying and testing all possible methods, the unfortunate Skinnerian may die as a result of testing out a fatally wrong method. Named after the pioneer of operant conditioning, B.F Skinner, the Skinnerians have reinforced learning as the dominant theme of intelligent action in addition to natural selection. **Popperian creatures**, as an advancement over the blind testing of the Skinnerian creatures, act with a sense of foresight such that the undesired behaviours and fatal actions can be discarded. Dennett contends that for this ability of preselection, a feedback must have come from a sort of inner environment which is similar to creating a simulation of the prospective actions and then choosing the best one, or selectively rooting out the bad ones. Dennett also fuels the debate that it is not very easy to clearly distinguish between the genus of Skinnerian creatures and Popperian creatures. However as per trends of research, the sea slug Aplysia can be classified as Skinnerian, while birds, fish, reptiles dominantly exhibit

Popperian traits. **Gregorian creatures** have tool making as their 'Generate-and-Test mechanism and in particular can develop abstract tools like language, semantics, logic and gestures; which means that individuals no longer have to 'Generate-and-Test' all possible hypotheses since others have done so already and can pass on that knowledge as cultural and social interactions, thus cyclically improving generators and testers, to a very high number. While human beings are the best example of Gregorians, there is discernible sense of social behaviour and cultural epithet in various animals, apes in particular. Here, it is worth noting that though natural selection, conditioned response and preselection is also prevalent in the Gregorian creatures, however the dominant expression of intelligent action is using mind tools — language and gestures. The **scientific creatures** have the ability of convergence of mind tools of the Gregorian creatures and can rigorously, collectively and publically 'Generate-and-Test — the scientific method. It is a debate whether this transition started at the invention of the wheel or at the Copernican revolution in the mid 1500s or the development of string theory in late 1980s. However the ability to scientifically process and structure information to form intelligent action is the 'Generate-and-Test mechanism the scientific creature.

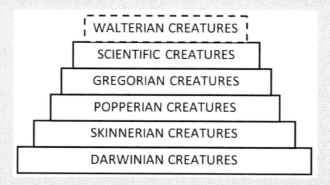

FIGURE 2.29 **Dennett's Tower.** Winfield suggests another layer to Dennett's tower as shown and proposes the evolution of artificially engineered, Walterian creatures in the near future, principled on embodied AI and named after W.G. Walter.

Extending the tower upward, Winfield [359] suggests **Walterian creatures**, where scientific creatures are instrumental in the development of artificial beings. The tools initially made by Gregorian creatures take on a life of their own, independent of the resources and support of the toolmakers. The Walterian creatures are smart tools that have learned to think, grown up and left the toolbox behind and taken on a being of their own. While also like Gregorian creatures, they are able to share tools, knowledge and experience. Embodied AI, ANIMAT and Artificial Life has this ability, by its defintion — tools attaining life. However, as an advancement over the Gregorians, the Walterians are capable of memetic learning. Therefore, if at least one fellow Walterian has learned a skill and is either online or has previously uploaded that skill, then another Walterian can simply download it. The Walterians are intended to be beyond human beings as creatures that have not only left the toolbox and the tool maker, but also the gene pool and escaped natural selection. They would be creations of Gregorians but will not be bound in to the biosphere afforded by Planet Earth, viz. instead of oxygen, water and food to survive, the Walterians eventually, after evolving themselves would probably only need energy. However they would still relate to its

Gregorian–scientific roots via inheritance and evolutionary processes, artificially motivated from biology.

Winfield believes that Walterian creatures would be profoundly different and seemingly unimaginable by Gregorian and scientific creatures. These heirs to Elmer and Elsie are prospected to be far superior to the humble fungus eaters since the Walterian creatures are capable of evolving themselves as per their need and requirement. We human beings augment and acquaint ourselves with newer tools to compensate for the lack of our sensory and survival capabilities. In stark contrast the Walterians can effect their own artificial evolution. The ability is not entirely new and is found in mother nature where lizards, earthworms and few other lower animals with the ability to grow back their some of their limbs if they are lost in some unfortunate incident. However the Walterians will be able to evolve much faster than all known biological lifeforms. As an example, Winfield considered a futuristic scenario in which an intelligent autonomous explorer robot is charting an unknown planet. The robot — a Walterian — has the capability to simulate and foresee aspects of the future, and as per the simulation re-build parts of itself on the fly using a 3D printer like an inbuilt facility. Thus by smartly evolving it deals with the situation it has encountered in that unknown terrain. Suppose, the unknown terrain is a planet that has large waterbodies and the robot falls into one of those, a fully evolved Walterian may be able to 'grow fins' and replicate fish-like swimming ability before it drowns, thus evolving faster than the dynamic environment to ensure survival and well being.

Walterian creatures are not without parallels. Toda and Braitenberg both had predicted the natural evolution of cooperation, competition and interdependence, therefore creating an ecological niche of artificial creatures, trending towards the development of their own societies and cultures. The ability to change its own physiology and control its adaptability and evolution is similar to the models of self-replicating machines discussed by Von Neumann in the 1940s. In the real robot domain, current day AI research has developed robots with multiple limbs which can continue to function and pursue there goals even when they have lost a limb and some functionality. These robots can 'heal' and 'grow' the lost limb(s) on the fly using an on-board 3D printing facility.

2.6.1 The problem of completeness — Planning is NP-hard

A robot is usually tasked with locomotion, to move to an assigned location in reasonable time, with least effort and hassles. Motion planning and trajectory generation are often used for these kinds of problems. A good lot of other sophisticated tasks are extension of locomotion, therefore navigation and path planning are seen both as basic behaviours and also as design tools.

It has been shown that planning gets more complicated with dimensions of the planning space, viz. a two-dimensional navigation contained on an xy plane is simpler to design and easier to program than a three-dimensional navigation in xyz space. The complexity of path planning increases exponentially with the number of dimensions, and with increasing number of obstacles. However, simple planning centric jobs such as, go to goal point, search, tracking, mapping etc. in principle may take infinite time and/or become so involved that it is rendered impractical. These types of problems are called NP-hard (non-deterministic polynomial-time hard). For such problems it is rather easy to verify a given solution but there is no analytical method to find the solution to the given problem in finite and reasonable time. This renders planning incomplete and therefore not always yielding a result. Real-world problems which need planning in at least three dimensions, with moving obstacles are difficult and intractable in principle.

As an illustration, suppose that the job for a robot is to cross a busy street. The solution is to watch for vehicles and other pedestrians and when there is appreciably low traffic, carefully walk to the other side of the street. Most of us have experienced this first hand, and as simple as it looks, this will not be very easy to program into a robot as there will hardly be a scenario with no traffic, and simple programs such as, `watch left, watch right and for no traffic progress to the other end of the street` will outrightly fail, or be stuck for infinite time attempting to execute this job. In such scenarios, the robot often continues in the programming loop, to attempt to find the desired end point of locomotion, and continues to fail to do so until it exhausts its fuel. Therefore programmers put in conditions known as 'exceptional handling' to prevent the unending loop, where the robot stops after a few attempts and reports that the job cannot be accomplished.

A good algorithm should gurantee finding a solution when there is one, and report if there is none, but issues of completeness can never be 'cured'. However they can be suitably avoided, or reduced to a minimum. Exhaustive search is an apparent solution but is a compromise in time, other methods are, granulating space by slicing it preemptively or employing subgoals and hybrid approaches and employing a second path planner when the first one is rendered redundant. These techniques are navigation centric and discussed in Chapter 4.

Completeness may come across as a problem for designing robots, but it is omnipresent and is also a tagged to human cognition, and at various times we human beings may fail to comprehend a solution which does exist in principle, or we may be stuck in logical conundrums when no solution exists.

2.6.2   The problem of meaning — The symbol grounding problem

In the 1980s Searle proposed a thought experiment, which can be worded as:; suppose in the near future, AI has developed artificial agency which can understand Chinese and can also pass the Turing Test rendering indistinguishable output in at least text and speech content when compared to human response. Therefore, such an AI will be able to convince any Chinese that it is also a Chinese-speaking human being. Searle's inquiry, does the agency literally 'understand' Chinese? (strong AI) or is it merely simulating the ability to understand Chinese? (weak AI). Since the agency is producing the output by relating the input to a rule of syntax, as given in the program script without really getting the hang of the semantics as is true for current-day translating software shown in Figure 2.30, the conclusion is that program script cannot lead to 'consciousness', but can at best mimic it, and therefore strong AI cannot exist.

The symbol grounding problem is very closely related to Searle's Chinese room problem. The symbol grounding problem is the shortcoming in associating verbal labels as words and syntaxes (symbols) to actual aspects. For example, considers a robot which has a ball in front of it. The robot will not be able to connect the word (symbol) to the object (aspect), and on being given a command like "pick up the ball" will not be able to execute the job. Human beings overcome this problem as cognition is developed by neural learning and groomed by gradual acquaintance with the environment and known social norms.

The symbol grounding problem, inquires "Can a robot deal with grounded symbols?" or, can a robot recognise, represent and communicate information about the world? Engineers and designers have been able to make robots which can identify a word with an image or video employing intensity heuristics or image processing, however this is limited to a

English – detected ▾   Chinese (Simplified) ▾

Symbol Grounding Problem |

符号接地问题
Fúhào jiēdì wèntí

FIGURE 2.30 **Google Translate** and similar translating software are more fast dictionary searches than attempts to understand the language, its semantics and cultural aspects. A machine learning algorithm will easily translate alphabet, words and sentences by observing patterns. In contrast, a human translator will learn the language, than merely relate the symbols to form semantics. As Searle argues, a computer only manipulates the symbols and will not understand the semantics. Therefore strong AI doesn't exist and no artificial means will ever supersede human cognition.

select known set of objects, and tends to tailor the robot for a particular job, sacrificing context and therefore its versatility across various other jobs and serendipitous activities as exploration and discovery. Therefore, a robot can be programmed to interact and manipulate with symbols grounded in reality with its sensorimotor embodiment, but obviously lacks universality.

Harnad [142] has suggested than the symbol grounding problem is more compelling than merely developing a robot dictionary of sorts by relating image to name as shown in Figure 2.31, and to deal with grounded symbols is the robot's ability to autonomously and ubiquitously establish the semiotic networks relating symbols to the world. Therefore, the phenomenon of symbol groundng sets the precedence for conscious behaviour where mental states are of vital importance.

2.6.2.1 Solving the symbol grounding problem

Attempts to solve the problem, have ranged from approximate solutions to complete denial of the problem to accepting it as an inherent problem of human cognition.

1. Approximate solutions: The most obvious method to attack the symbol grounding problem is supervised learning. Experiments which enable a handshake over verbal and vision modes has demonstrated gradual betterment of behaviour. However, this is a good approximation and broadens the horizons without ever solving the problem.

2. Avoided for low-level behaviours: The problem which seemingly has serious concerns for computers is insignificant for robots or at best remains a philosophical issue which can be safely avoided. For embodied AI, the onus is more on constructing and manipulating symbols than really grounding them. Since robots are localised, the meaning of objects and action can be gradually ascertained, or completely neglected if it doesn't matter to the robot. For example for a robot with low-level behaviour it will hardly mean much if an obstacle is a book, brick or a box. All it needs to do is avoid the obstacle. Therefore, in embodied cognition the symbol grounding problem is more of a technical problem than a problem of cognition, and Vogt [340] calls it the 'physical symbol grounding problem'.

3. As a philosophical limitation of human cognition: The need for meaning is higher-level

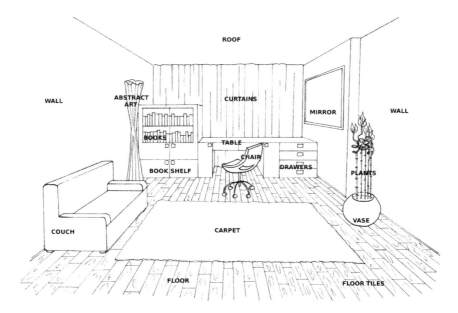

FIGURE 2.31 **A very poor solution to the symbol grounding problem.** An apparent brute force solution to the symbol grounding problem is to label everything. However that is still insufficient to develop all of the semiotic relationships and give meaning to objects.

cognition. When attempts for solution for a robot are made, the semiotic relations and hence the meaning to the objects are carefully mapped out and then coded by human programmers, most of which are based on human semantics and culture, and are not autonomously established by the artificial agent, as is desirable by the problem definition. Therefore, such a solution to the symbol grounding problem is biased on human knowledge and reasoning and not on the arbitration of the autonomous agent.

Touretzky [324] suggests a 'robot version of the Chinese room problem' which poses as a counter argument to Searle by using a robot instead of a computer. Modifying the Chinese room problem, suppose we have a robot which can pass the Turing Test and boldly speaking in idiomatic Chinese, and the question asked is 'walk up to the window and tell me what you see?'. Here, visual perception will be an inquiry into mental imagery, and may not be the desired human response, viz. the robot may not perceive the sky as blue, or the sunlight as warm etc. Therefore, demonstrating that the Chinese room problem can be overcome with embodied agency. This is a version of the Turing Test and such methods are further discussed in chapter 9.

The symbol grounding is hardly a concern for robots with low-level behaviour, but it will be an issue for developing higher cognitive functions for enabling human-like behaviour. The shortcoming in relating symbols and the real world is not a problem unique to robots but one often faced by us human beings. For example, on encountering fruits and vegetables from a different biome many of us face such an issue, and we resort to online search on the Internet.

Thoughts which gives meaning to an object and thoughts which interpret them are often not the same. Therefore meaning in higher cognitive functions is always contextual and will differ with perception, memory, training, cognitive bias etc. Hence, the symbol grounding problem will never be solved [79, 310], but it can be effectively reduced with (1) compartmentalisation of knowledge which is often by localisation and context-driven tasks, viz. for navigation it can be; marking the goal point, get to the red mark, cross three static obstacles etc. (2) machine learning and (3) since we are in the age of the Internet, use of an online knowledge base to search for the suitable meaning of a given object or instance.

2.6.3 The problem of relevance — The frame problem

Of the three problems, the frame problem has attracted most controversy [166] and has had the audacity to change the discipline of AI. It is not an issue pertaining to AI or programming, but rather is a lacuna in logic and our understanding of the world, and is often reworded as 'the problem of relevance'. The problem is about which changes are relevant for the job at hand. For example if I am walking down a street and I see a banana skin about six feet ahead of me, I will change my direction a bit to avoid the banana skin. Now, consider that as I am walking, instead of a banana skin, there is a typhoon far away, in the Siberian tundra. Apparently it will not affect my walk, but there are situations where relevance cannot be ascertained this easily.

The problem has been interpreted by many renowned philosophers and AI scientists. Dennett [89] discusses the frame problem with an illustration where a robot **Robot R1** has to go inside a room to obtain a battery to power itself and avoid a ticking time bomb, as shown in Figure 2.32. **Robot R1** identifies the battery and the bomb and walks out with a wagon which has both the battery and the time bomb and soon the bomb explodes, and the robot is blown to bits. **Robot R1** knows the bomb is on the wagon, but fails to comprehend that bringing the wagon out of the room, also brings out the bomb, thus it fails the task as it cannot deduce the consequences of its own actions. To overcome this seemingly trivial issue, the robot designers develop **Robot Deducer R1D1** which is more accomplished than R1 as it is designed not only to recognise the intended implications of its acts, but also their side effects. The underlying idea is that R1D1 will not make the mistake of rolling out the battery together with the bomb because it will be able to deduce the implications (read blunders) of doing so. Put to work, R1D1 tries to deduce various (often unnecessary and extraneous) consequences of its actions. Soon the bomb explodes and, in its blissful thinking, R1D1 meets the same fate as its predecessor. A third improved design, **Robot Relevant Deducer R2D1**, tries an exhaustive list-like approach to categorise the consequences of its actions as 'relevant' and 'irrelevant', a thorough approach to find all possible consequences of its actions. Unfortunately the bomb goes off and **Robot Relevant Deducer R2D1** is also blown to bits. All three robots suffer from the 'frame problem'. None of them have a way of gleaning out what is relevant to the problem at hand[7].

Across these three robots, as the amount of knowledge increases, so does the number of lines of code to make the robot deduce the relevant information to accommodate the problem at hand. As the number of lines of code and seemingly more sense of reason are pushed into programming these robots, their efficiency decreases, viz. R1 with the simplest coding and information was at least able to walk out with a wagon, which was not achieved by R1D1 and R2D1, both of which were blown apart in deducing the various consequences of the action. R1D1 and R2D1 were in a dilemma as to when to stop thinking, or Hamlet's problem,[8]. The frame problem can be seen as Hamlet's problem in the engineering domain.

---

[7]The STAR WARS reference is indeed compelling, but it is more of a conjecture that what R2D2 may have done was to bring out the battery yet also avoid the ticking time bomb, in process the solving the frame problem.

[8]Shakespeare's play Hamlet is a tragedy themed on the 'hero-as-fool' genre. The apparent heir to the throne of Denmark, Prince Hamlet is the protagonist of the play. He is philosophical and contemplative, and he does not act on instinct, but rather tries very hard to make sure that every action is premeditated. His

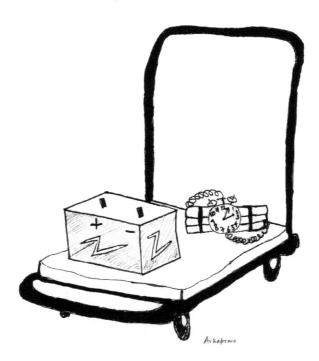

FIGURE 2.32 **Dennett's illustration of the frame problem**, where a robot has to go inside a room and obtain a battery to power itself while avoiding a ticking time bomb.

With this illustrative example Dennett demonstrates that the frame problem may seem like an issue of bad design, glitch, poor hardware and coding errors, but instead of being a technological or design shortcoming, it is an epistemological problem and a consequence of human common-sense reasoning, of 'psychological reality' which being situated cannot be fully reached by a programming code.

The frame problem arises in an attempt to find a sufficient number of axioms using symbolic representations [269] for a viable description of a robot's environment. The nomenclature is from the technique of 'frame', as used in development of stage characters or an animation where the moving parts and characters are superimposed on a 'frame'. Typically the background usually doesn't change. Description of a frame, or a robot's environment takes a large number of statements, but a great many of them hardly change in most common interactions. Therefore, in terms of relative motion, the frame problem is described in terms of spatial location of objects which presumably do not move in the course of the movement of the robot. Hence the robot's actions must be structured in relation to what all stays static across the event.

In a further illustration to further the understanding of the frame problem, a conversation between two friends as shown in Figure 2.33 can easily go bizarre if the non-effects of even the most trivial things are mentioned. However it is to be noted that the seemingly absurd notions as Joe's bar continuing to exist past until 7:00 PM, death of one of the two friends in the conversation and apocalypse by a meteorite hit can make sense in war, insurgencies

---

character is further tagged with fruitless attempts to demonstrate provability, and various of his inquiries cannot be answered with certainty. Hamlet's struggle to avenge the murder of his father leads to melancholy and a wrestle with his own sanity. In context, Dennett's third robot, R2D1 is in a Hamlet-like stream of thought, before being tragically blown up.

and dire situations. Therefore however absurd these may seem, they cannot be excluded a priori. In conversations such as these, human common sense and known social norms are found to dictate reason, and this would be difficult to replicate via AI, and therefore lacks implementation in a robot.

The frame problem is an inquiry into determining which information is relevant to reasoning in a given situation, and which information is to be ignored. The frame problem is an issue in designing artificial situated intelligence, and the problem has a very special position in the history of AI and robots. Noted AI philosophers, Dreyfus and Searle, considered the frame problem as the demise of AI. To rescue the day, an apparent solution to the frame problem would be to associate each event with a probability. However probabilistic knowledge is hardly of much use in a real world situation and would utterly fail in hazardous situations as was the case for Dennett's three robots. Also, modelling probability, which updates in near real-time over the robot's sensors is computationally prohibitive [43] and at best restricted to simple worlds.

2.6.3.1 Representations — The root of all evil

The frame problem occurs because of symbolic representations and rules of inference. The frame problem is a problem of modelling, not heuristics, just as an algorithmic problem solver will not always know in advance what to do next, and making exhaustive lists will not make for a solution. Neither is the frame problem a question of content, but rather of form. It is seen that informationally equivalent systems do tend to differ drastically in efficiency. Hence, representational knowledge may work for simple scenarios, but may not yield for real-world problems. The frame problem is a philosophical problem and not an issue of data structures, programming language and algorithms. The frame problem occurs for informationally unencapsulated processes, for which no a priori limit exists to the information that might be of relevance to it. Therefore, very low-level cognitive processes will be free from the frame problem, while of course as a tradeoff they will tend to accomplish much less. In view of the above, a few ways to get around this dilemma are:

1. No-representations — Behaviour-based approach: Instead of using representational knowledge a reflexive approach is taken. A symbol-based system attempt to model an absolute knowable world. Instead these approaches work on an immediacy of action[9]. No-representations are motivated from biology and the natural world. While, this approach works for ants, crabs, cockroaches etc., higher-level cognition such as in human beings is known to structure thoughts and hence construct actions based on representations often of a not-immediately-present environment.

2. Connectionism — Neural networks approach: A connectionism approach, such as artificial neural networks (ANN) is secure against the frame problem, as ANNs encode their knowledge acquired from training data via fixed weights. Thus representations never creep into this model. However, tasks which have not been undertaken previously cannot be modelled using ANN due to the lack of training data. Thus one does come to a chicken and the egg problem.

3. Context independent representation — traditional AI: Traditional approaches are motivated by the hypothesis that the world can be represented by means of a large number of symbols designating the objects and actions in the world and by a set of

---

[9]Representation free approaches faced severe opposition in the AI community. There is ample literature dating back to the mid 1980s where such work is labeled with an oxymoron, 'reactive planning'. James Firby's PhD thesis is often said to be a watershed as it changed the phrasing from 'reactive planning' to 'reactive execution'.

FIGURE 2.33 **A conversation turns bizarre.** With too many representations in a real-world situation, a 'context independent' representation fails. A simple conversation can turn bizarre if one starts to include the extraordinarily low-probability events. Adapted from French and Anselme [112].

FIGURE 2.34 **PENGI.** models a penguin and a bee with various rule-based strategies. A context dependent approach works well in a minimal 2D block world

rules with which to manipulate those symbols. Since, it is often not possible to know the world and the rules a priori, this approach relies on analogy — relating one object to other, viz. an agent which has seen a cube but not seen a cuboid would be able to relate its experience as an extension of the cube. However such an approach doesn't go a great distance given the dynamism of world. Also, with an ever-increasing number of symbols and rules, modelling the agent fails to adhere to a real-time response and may become computationally challenging.

4. Context-dependent representations — Works for simple scenarios: Researchers have attempted to develop context-dependent approaches. In very simple scenarios a symbol-based and rule-based approach works well. For example, a very simple world model in PENGI [4], as shown in Figure 2.34, is a software simulation in which a penguin tries to escape being bitten by a bee in a two-dimensional maze, while subscribing to various rules. Agre and Chapman developed it as an attempt at context-dependent representation. In PENGI, the 2D 'world' comprises a penguin, a bee and ice cubes. The ice cubes are represented as square blocks and can slide. The penguin has to escape being stung by the bee, which it can do by hiding behind an ice cube. The bee can kill the penguin by stinging it, or stinging the ice cube directly behind which the penguin is hiding. The penguin can fight back by kicking the ice cube behind which the bee happens to be. In PENGI all events are contextually determined. It employs indexical functional entities which weave the objects to the action. There is no entity such as `bee` or `kill`, rather `the-block-I'm-pushing`. Hence the frame problem never arises. However, for a more complicated world, there will be an ever-increasing number of such indexical functional entities and it would not be possible to deal with them.

Context-independent representation seemed to work well for AI but failed to keep up with the dynamical requirements of embodied AI. With context-dependent approaches such as PENGI, there were primarily two schools of thought. One believed that models as PENGI can truly be applied to a more complicated world, while others considered moving to a representation free, more of a sensory-action realm. Ethnographic studies made on human beings have shown that there is no clear boundary between a planning phase and an execution phase, nor are plans typically executable without knowledge of the domain and the goal of the plan.

TABLE 2.1 How to overcome the three forks

Fork	Solutions
Problem of completeness	1. Exceptional handling, programming fail safe 2. Exhaustive search (as ND and Bug-2)
Problem of meaning	1. Supervised learning 2. Handshake of vision and voice modes 3. Can be avoided for low level behaviours
Problem of relevance	1. No or very low representation 2. Connectionist architectures

The Great Debate of '89

One of the most engrossing debates on universal planning was in the 1989 winter issue of AI Magazine across four articles. The two sides of the debate was Ginsberg (against universal planning) vs. Chapman and Schoppers (favour of universal planning).

1. In the first article [124] Ginsberg presents a criticism against universal planning, a part of his argument is based on computational complexity. He also demonstrates that using universal plans is impractical as their size increases exponentially in the number of domain state. That is, the the number of sensory-motor pairings for the robot explodes for a complex world — akin to our real world. Ginsberg contends that universal plans fail to take into account the inherent unpredictable nature of the real world. However and most interestingly, Ginsberg notes that such an approach of mapping may work out well for 'precognitive activities' such as locomotion.

2. In reply to Ginsberg, the second article [67] Chapman tries to answer to Ginsberg's criticisms with examples from BLOCKHEAD, which is a PENGI like system, and can arrange alphabets blocks to spell F-R-U-I-T C-A-K-E and thus solve the so called fruitcake problem, and therefore, assert the approach of universal planning. Chapman shows that using visual markers the fruitcake problem is made tractable.

3. The third article [293] by Schoppers further refutes the criticism by Ginsberg using a number of arguments. He argues in favour of indexical-functional-variables which works very well in a simple PENGI-like world. He also supports the use of program caches for reactive planning where a job, when done a few times, becomes innate to the agent and can also be done without any real planning.

4. In the fourth article [123], Ginsberg replies to Chapman and Schoppers. Ginsberg's asserts that reactive planning cannot capture the desired behaviour of an agent since there are too many possible variations in the real world and that

reactive plans cannot improve performance with more computational resources and therefore an apparent shortcoming in developing of intelligent behaviour. In replying to Schoppers on caching plans, Ginsberg raises questions as, 'when should a plan be cached?', 'when is a cached plan relevant to a particular action?','can cached plans be debugged and re-structured to extend their usefulness at run time to situations outside those for which they were initially designed?' etc., rendering the Schoppers' contention of using caches, at best incomplete.

Ginsberg's contention has been found to be true, and universal planning indeed works well for locomotion, as in ND diagram navigation where five (or six in ND+) states completely define all possible situations encountered by the agent in navigation on a two dimensional plane.

Universal planning was the convergence of the research of Nilsson, Kaelbling, Schoppers, Agre and Drummond; the term universal planning was coined by Schoppers. Universal planning was the development of a plan wherein all possible situations that the agent may encounter have been analysed and tabulated and all the agent has to do is detect the situation with its sensor(s) and refer to this table.

Whereas, none of the three forks as shown in table 2.1 can be solved per se, they can be minimised or avoided. Also, for low level behaviours as 2D navigation etc. the issues are to a minimal. These are detrimental in higher-level cognitive processes, and to proceed towards human-like cognition, there will have to be means to overcome these forks, and a simple ANIMAT model will not be sufficient.

SUMMARY

1. Toda's model on hypothetical animals which he named fungus eaters was the first example of self-sustaining artificial entities. They could gather fungus for their own energy needs and gather uranium as they were designed to do
2. Inputs from psychology and the natural world enabled further developments
3. Wilson's model of the ANIMAT was very influential and led to ALIFE and research into animal-robot interaction
4. Three prohibitive issues in mobile robot development were, (i) Complexity issues, (ii) Symbol grounding problem and (iii) Frame problem
5. Ways to cure the frame problem led to non-representational models and their implementation in robots

NOTES

1. *Toda's 1962 research paper, " The design of a fungus-eater: a model of human behaviour in an unsophisticated environment" is the first model of an autonomous agent, but Toda has no mention of the word autonomy or autonomous in the paper.*
2. *Toda ends his seminal paper posing a question, "An SFE running into a starved SFE may save the latter by selling its fungus storage in exchange for the latter's uranium. What should be the price?". This would call for extending Toda's model into game theoretic discussions, however the problem remains to be answered.*
3. *Lacan has used 'INNENWELT', or the inner world in contrast to Uexkull's 'UMWELT' to mean the mental experience of the self.*

4. *At his popular webcomic xkcd, Randall Munroe briefly discussed the applications and design of a hamster ball robot [244, 245]. With this inspiration, a similar robot was made by students at Georgia Tech using a windows embedded platform.*

5. *In his book, The Language Instinct Pinker writes about Moravec's Paradox as an aphorism: "hard problems are easy and the easy problems are hard".*

6. *Another illustration of Moravec's Paradox is seen in the following riddle. 8809=6 7101=1 2172=0 6666=4 1111=0 3213=0 7662=2 9313=1 0000=4 2222=0 3333=0 5555=0 8193=3 8096=5 7777=0 9999=4 7756=1 6855=3 9888=7 5531=0 9966=4 2681=? The pattern is based on the number of circles in each number. The number 8 has two circles in its shape. Number 6, number 0 and number 9 has one each. Therefore 2681=3. A robot will attempt to solve it using pattern recognition codes which will not only take a lot of time, but it will take away the elegance and the beauty of the riddle.*

7. *Human beings do not need to solve the frame or rather human beings are never aware that they are inadvertently solving the frame problem without being aware of it. The apparent reason for this is human cognition is modelled as a connectionist architecture with neurons connecting together to form dynamical networks. Such connectionist architecture is immune to the frame problem, also human cognition involves massively parallel unconscious processing, accumulating information across various senses and relating the current situation to previous experience. These activities work nearly inadvertently to the human individual concern, such cognition is not yet possible in the artificial domain.*

8. *Skinner wrote his novel, Walden Two, in which he described a Utopian society founded upon behaviourist principles.*

9. *John Haugeland coined the acronym GOFAI (Good Old-Fashioned Artificial Intelligence) for symbolic AI in the mid 1980s. In robotics the analogous term is GOFAIR (Good Old Fashioned Robotics). The approach is based on the assumption that intelligence can be achieved by the manipulation of symbols, PSSH (Physical Symbol System Hypothesis) and thus it is broadly independent of the agent and the environment.*

10. *The phrase 'Pavlov's dog' has come to be an idiom to mean a person who merely reacts to a situation rather than using critical thinking.*

11. *Pavlovian conditioning was a major theme in Aldous Huxley's dystopian novel, 'Brave New World'.*

12. *The enactive virtues of vision are further discussed in Merleau-Ponty's splendid essay, 'Eye and Mind' [227], which explores the concept from both physiological and also philosophical points of view.*

EXERCISES

1. **'UMWELT' for a line follower.** For a line following robot with just one sonar sensor discuss and sketch its 'UMWELT'. How does it change for a ring of six sonars?

2. **Designing a frog**, design the behaviour decomposition for a frog as it is done for a cockroach in Figure 2.12. Consider the following behaviours [10]:

   Face-prey When the frog sees a fly, it needs to turn to face it so that it can stick out its tongue and grab it (we assume that the frog's tongue has limited steerability and cannot make an apreciable angular movements).

---

[10]I cannot claim originality in this as Horswill has used these behaviours to design a frog using the programming language GRL.

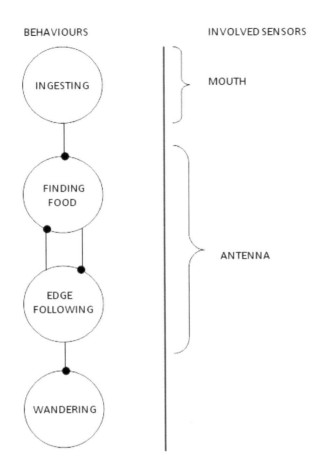

FIGURE 2.35   Exercise question — Cockroach locomotion.

Eat When the frog sees its prey within tongue range, it shoots out its tongue to
eat it.

3. **Cockroach locomotion.** Arkin's behavioural decomposition for a cockroach is given
in Figure 2.35, this is a modification of Beer's model shown in Figure 2.12. How will
the behaviour change across the two models?

CHAPTER 3

Control paradigms for mobile robots

WALL-E: "Directive?"

Eve: "Classified"

– from the movie WALL-E

"The preceding fantasy has roots in science"

– Valentino Braitenberg [44]

3.1 CONTROL PARADIGMS

A mobile robot is designed to accomplish various tasks. The easiest ones involve moving around while avoiding obstacles and reaching a given goal point. The more involved tasks may include following a trail of an odour, mapping an unknown terrain, responding to voice and gestures, attending to man-centric environments as museums, theaters or airports etc. The simplest robot is at least a sensor actuator pairing. More sensors and actuators enhance a robot's performance and add to its versatility. Other important parameters are the constraints of the robot and the constraints of the process, safety concerns, self preservation and optimised performance. This chapter focuses on control paradigms and architectures and used in robots and tries to implement the concepts of the last chapter in the engineering domain.

Three important aspects of a good control paradigm are closed loop control, real time response and robustness to overcome sensor errors. Closed loop control is critical. This can be seen with an example: one of the simplest robot is the line following robot. The robot is fitted with a light sensor and it tracks a line marked in a certain colour usually black or white. Any undesired deviation away from the line is prevented by the sensorimotor pairing between the light sensor and the motor. An ANIMAT corollary would be with a hypothetical animal which has its eyes set on the road, as shown in Figure 3.1. The contrast in information flow for closed loop and open loop is shown in Figure 3.2. Without the light sensor, the robot will fail to track the line and it will move arbitrarily in the direction the heading is pointing to. Lack of a feedback mechanism[1] will not allow the

---

[1] The concept of feedback is of importance in control systems and electronics, and closed loop systems which work on feedback are demonstrated to be more stable than open loop systems.

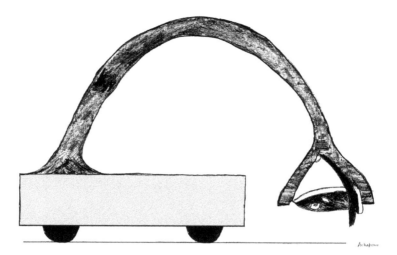

FIGURE 3.1 **ANIMAT approach to line follower**, an artificial animal keeping an eye on the road. This model is equivalent to a basic line follower with one light sensor. However, such a design with only one light sensor will not be of much practical utility. A robust design will have a ring of light or ultrasound sensors and also other sleek hardware, backtracking means and sensor fusion which will incorporate for the effects of friction, uneven terrain and other environmental factors.

robot to interact with the environment, explained in flowchart. Mobile robots are expected to have a real-time response with minimal lag. The sensor actuator pairing is modulated by a microcontroller and structured with a set of control rules and parameters. Often, it is a tradeoff between accuracy and lag time. A third concern is, sensors are liable to be faulty and give erroneous and noisy readings. Under such circumstances the system needs to continue to function well within a domain practicality. Therefore, mobile robot control should be organised with a parallelism approach which incorporates graceful degradation and elegantly handles exceptions so that failure of a single hardware or software module doesn't hinder the broad goals of the robot.

To design a robot, there are three broad approaches. The first one is called sense-plan-act and it follows the concepts of traditional AI. It was developed in the early 1950s and processing sensory information to yield plans and maps was the basis to this approach. The robot has to develop exhaustive, near approximate maps of the world it has to navigate, where buildings are approximated as cuboid, clouds as spheres, a person is an oval head with a cylindrical torso and smaller cylindrical arms and roads as an array of flat straight sections etc. Once such maps have been made, the robot must find the suitable path to the goal point. This approach can be summed up as a Minecraft world married to Google Maps. Even the best computers in the late 1960s could not have made maps like the current Minecraft games or Google Maps, but the idea was very much in the same lines. As discussed in the first chapter, Nilsson [253] designed Shakey's navigation by making (a priori) grid-based maps. The method relies on seeping in the knowledge from the environment and then processing that information to move around intelligently. Since space is apparently infinite compared to a sensor's range, this would be implemented as a cyclical process where the sensors sense the environment, the microprocessor develops plans to navigate in this environment and then finally the actuator would trigger the motion.

The second approach traces the principles of behaviourism [52, 53], where instead of making a priori maps, the robot starts moving around on the merit of the sensory-motor

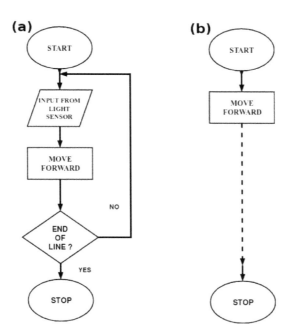

FIGURE 3.2 **Closed Loop.** In (a) the light sensor allows for interaction with the environment, the line follower tracks the line and follows it to the end; while without a light sensor in (b) the robot fails to have any knowledge of the environment and randomly moves without any goal or motive. The stop for such a robot will not be graceful, probably ending in a collision.

pairing in the local environment with near real time sensor reading paired with immediate actuation and no appreciable plans or maps, akin to animals interacting with their local environment. This approach eliminates planning, at least a priori and deals with the dynamism of the environment in real time as an immediacy. This approach is very effective in navigation and other lower level tasks, but as we will see, it cannot be extended to higher level tasks.

A third paradigm, has aspects of planning as well as behaviourism. Such architectures are seen in most of the state-of-the-art robots. It will be discussed in later chapters that hybrid architectures are of pre-eminence for human-centric robots and robotethics.

1. Deliberative approach, detailed planning prior to execution. When implemented in a navigational context the robot has to design elaborate maps of the environment before starting to move.
2. Reactive approach, the agent acts per the merit of the situation, no appreciable planning is involved. Draws motivation from the animal and insect world, **see obstacle - turn left** is probably the simplest reactive paradigm. When implemented across various sensors, over a variety of parallel tasks, a reactive paradigm leads to the **behaviour-based approach**.
3. Hybrid approach, As a combination of the two, deliberative planning is executed in a reactive manner.

The next section is about a number of thought experiments by neuroscientist and cyberneticist, Valentino Braitenberg and serves as a wonderful illustration for development

SR Diagrams

Stimulus Response (SR) diagram is a way to pictorially represent behaviour as a generated response to given stimuli(**S**) leading to a response(**R**). These were developed by Ronald Arkin in early 1990s.

The complete control mechanism for the agent can be designed by stacking together such SR diagrams as is briefly done for Braitenberg vehicle-3c in Figure 3.6. Traditionally simple behaviours are put at the lower ends with sophisticated behaviours at the upper ends, as shown in Figure 3.19.

An equivalent algebraic approach to behaviour is using functional notation;

$$\beta(\mathbf{s}) = \mathbf{r} \qquad (3.1)$$

Where behaviour β with given stimulus **s** yields response **r**

of behaviours from simple sensory motor design. It extends the description of the reactive approach to designing robots.

3.2   BRAITENBERG'S VEHICLES 1 TO 4 — ENGINEERING BEHAVIOUR

Braitenberg's 'Synthetic Psychology', developed over a series of gedanken experiments on mobile vehicles with minimal hardware, elaborated on using minimal hardware for designing vehicles which can exhibit human behaviour such as love, fear, aggression, like, dislike etc.

These vehicles were conceived as direct motivation from Wiener's seminal research in cybernetics [351], and were designed as 'animals in their natural environment'. He believed that structure of the animal brain may indeed be interpretable as pieces of computer machinery. His most original thinking led to fourteen different vehicles. This discussion is meant as a primer for behaviour-based paradigm and the concept of behaviour, and will focus primarily on the first four types of Braitenberg vehicles.

Vehicle-1: getting around, is a simple sensory-motor pairing. The vehicle has one temperature sensor connected its lone motor and it speeds up at higher temperatures and slows down at lower temperatures, or alternatively the force acting on the vehicle is proportional to the temperature of the surroundings in Kelvin. This vehicle will move along the direction it is oriented in, seeking out cold places and running away from higher temperatures. Indeed, the vehicle seems alive and restless and at first reading this appeals as an unconventional model. However such motion is very similar to Brownian motion, in which atoms and molecules speed up with increasing temperature.

Vehicle-2: the coward & aggressor. Here, instead of one motor controlling both wheels the vehicle has a motor for each wheel and the sensor, which is the same for both the

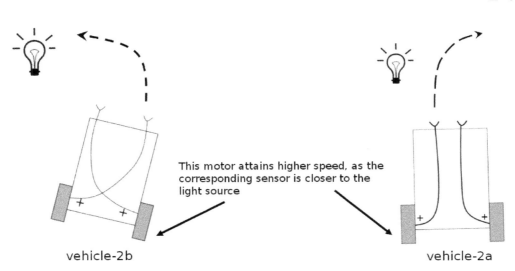

This motor attains higher speed, as the corresponding sensor is closer to the light source

vehicle-2b   vehicle-2a

FIGURE 3.3 **Braitenberg vehicle-2.** Vehicle-2a drives away from the light source and exhibits an emotion of **cowardice**, while vehicle-2b drives into the light bulb conveying **aggression**. Braitenberg was motivated by biology to design the crossed connection, particularly the human eye, where the left eye relates to the right side of the human brain, and vice-versa.

wheels, detects light. More light gives the motors more speed. Braitenberg manipulates this model further, to vehicle-2a with direct connection and vehicle-2b with crossed connection, as shown in Figure 3.3.

The crossed representation in the design is a direct influence from the biological findings that in the human brain, for vision, the left brain forms the image of the right side while right brain forms the image of the left side. Touch and auditory response are similar to vision and goes to the opposite side of the brain. In contrast olfactory goes to the same side of the brain.

If a light source is straight ahead then both vehicles will speed up in an attempt to crush the light bulb. Straight ahead is usually a rarity so for a more general scenario, consider the light source to be to one side of the vehicle as shown in the figure. In vehicle-2a the motor closer to the light source will acquire more speed than the other motor as it receives more light and will try to speed away from the light source while for vehicle-2b with the crossed connection, the motor away from the light source will speed up more and will move the vehicle in a curve in an attempt to move towards the light source. These 2 behaviours correspond to cowardice and aggression to light sources respectively. Vehicle-2, in its two avatars, demonstrates that simple sensor response shows up as complex human-like emotion. Hardware arrangement for a particular behaviour, as shown for vehicle-2b in Figure 3.4, helps to 'encapsulate' and use such behaviours in tandem with other behaviours, as will be shown later.

Vehicle-3: lover, explorer & a vehicle with a value system. Vehicle-3 introduces the concept of excitatory and inhibitory influence. Both in vehicle-1 and vehicle-2, sensing was excitatory influences and sped up the vehicles. In vehicle-3a, shown in shown in Figure 3.5, with direct connection, more light to the sensor will slow down the motor. Therefore, it will slowly come to a stop near a light source, while still facing it. The vehicle shows an attachment to the light source, akin to the feeling of love.

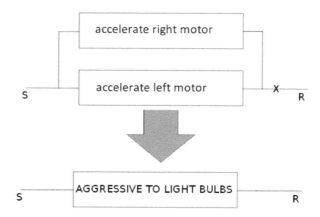

FIGURE 3.4 **Design of behaviour.** The SR diagram for vehicle-2b. Two sensor actuator response along with a crossed connection, marked with an **x** in the SR diagram, shows up as aggression towards light bulbs.

With crossed connections, vehicle-3b will come to a stop near a light source but it will face away from it. If vehicle-3b has other sensors such as olfactory, temperature, etc., then it would run away from the vicinity of the light source in quest of other goals, as would an explorer who may not find light sources interesting and looks around for other inspirations.

Vehicle-3c, shown in Figure 3.6 has a combination of 4 sensors and shows a variety of behaviours, viz. it runs away from higher temperatures, hates light bulbs etc., a clear sense of like and dislike as a value system seen in human beings. The behaviours are designed as shown in Figure 3.7 and one behaviour module doesn't influence another behaviour module by much. However, it will be difficult to predict the motor response with each of the modules interacting with the environment in tandem with the others.

Vehicle-4, unique trajectories. If instead of an excitatory or inhibitory sensor response, the sensor response is continuous as shown in Figure 3.8 and the velocity of the vehicle increases with increasing intensity of light, it reaches a maximum and then decreases with increasing intensity. Vehicle-4 will be able to form closed trajectories as shown in Figure 3.8. This vehicle is less predictable than its predecessors, acting more on its instincts.

These four vehicles, summarised in Table 3.1, illustrate interesting ideas of embodied agency. Behaviour of the vehicles is not entirely attributed to sensor response, but rather is a real-time interaction of the agent and the environment viz. vehicle-1 will not really seem very lively without high temperature, vehicle-2 will seem boring without any light sources nearby. To design a behaviour, sensor-response and design parameters can be made into a module, which then acts in parallel to other modules. As in vehicle-3c the final behaviour is an overlap of various such modules and is not very easy to predict.

Vehicle-3c and vehicle-4 are not easy to predict, but they suggest that hardware can be manipulated in order to accomplish specific tasks by designing particular agency. The downside is that since behaviour cannot be designed in with hardware or program codes, and is the interaction of the agent with the local environment, the agent needs to be in a predictable terrain and at least have some knowledge the job(s) it is supposed to do.

The vehicles employ only sensor actuator pairings without any programming module,

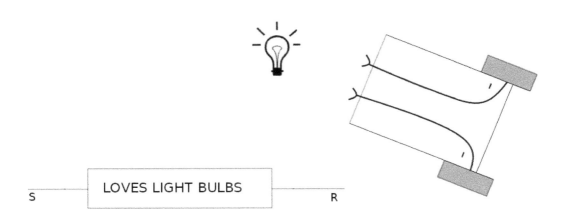

LOVES LIGHT BULBS

S   R

FIGURE 3.5 **Braitenberg vehicle-3a.** Instead of excitation, if the influence is changed to inhibition, then the vehicle decelerates as it comes to the light source, demonstrating an emotion akin to **love** for the light source.

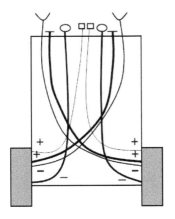

Temperature sensors
uncrossed excitatory -- HATES HIGH TEMPERATURE

Light sensors
crossed excitatory -- AGGRESSIVE TO LIGHT BULBS

Oxygen concentration sensors
crossed inhibitory -- LOVES OXYGEN

Organic matter sensors
uncrossed inhibitory -- EXPLORES FOR ORGANIC MATTER

FIGURE 3.6 **Braitenberg vehicle-3c**, a vehicle with a value system, hates hot places, has a vengeful aggression towards light sources, loves well-oxygenated environments and would willfully explore around for organic matter.

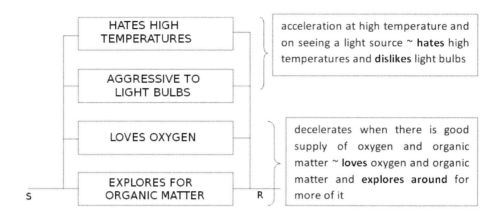

FIGURE 3.7 **SR for vehicle-3c.** The SR diagram for vehicle-3c is the conjunction of 4 different behaviours.

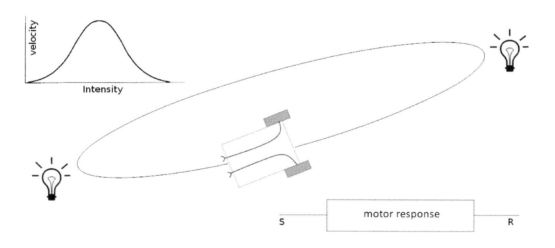

FIGURE 3.8 **Braitenberg vehicle-4.** For variable sensor response as shown in the upper left hand corner the velocity of the vehicle increases with increasing intensity of light, reaches a maximum and then it decreases with increasing intensity. For direct connection with two light sources — the vehicle will form closed loop trajectories.

TABLE 3.1 Braitenberg vehicles (1 to 4)

Vehicle	Design	Behaviour
Vehicle-1	Temperature sensor	Similar to Brownian motion
Vehicle-2a	Light sensor, direct connection, excitatory	Cowardice, runs away from light bulbs
Vehicle-2b	Light sensor, crossed connections, excitatory	Aggressive, smashes light bulbs
Vehicle-3a	Light sensor, direct connection, inhibitory	Love, comes to a slow stop at light bulbs
Vehicle-3b	Light sensor, crossed connection, inhibitory	Explorer, looks around for more than just light bulbs
Vehicle-3c	Various sensors and design hardware	Value system, likes and dislikes
Vehicle-4	Continuous sensor response	Forms closed trajectories

and hence are not reprogrammable. The intelligence of the robot arises due to its runtime interactions with the environment. Synthetic psychology demonstrates cognition as a real-time interaction rather than a programming paradigm. Very simple hardware and trivial design can lead to involved behavioural response. Braitenberg calls this phenomenon 'the law of uphill analysis and downhill synthesis', where it is easier to reckon the performance of a system whose internal machinery is known, than to figure out the design of the internal machinery based on the output behaviour. This is an alternative to known models of intelligence prevalent around the 1980s.

Synthetic psychology does not appeal to common sense, but its strength lies in its simplicity in design, and its triumph is that human like emotional response can be easily discerned in rote machines. In 1991, Hogg [152] and his co-researchers at MIT's Media Laboratory were able to verify 12 of of Braitenberg's 14 vehicles with LEGO bricks. This research also set the basis for the development of the LEGO Mindstorm Kit which then went on to become very popular with young robot enthusiasts. These vehicles and their fun variants are often endearing projects for advanced undergraduates and postgraduate robot courses and laboratories.

Further analysis into these vehicles provides interesting results. Using the potential field model [273], where the obstacles are modelled as repulsive potential and the goal as an attractive potential, it is seen that combination of two behaviours can lead to newer unique response. For example obstacle avoidance can be implemented using a combination of cowardice (vehicle-2a) and love (vehicle-3a).

Dynamics meets Psychology meets Art — The Sketches of Maciek Albrecht

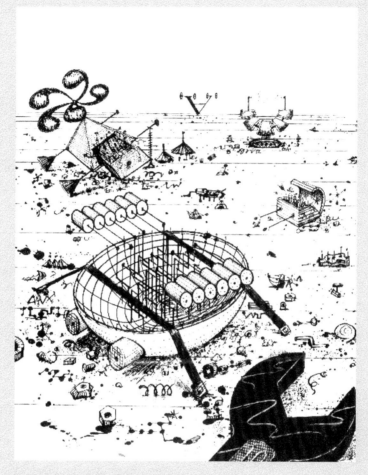

FIGURE 3.9 **Of hemispheres, polyhedras, nuts, bolts and wrenches — hardware to design motion**, courtesy Maciek Albrecht, used with permission. This is about half of the original sketch.

Braitenberg gave emotions and life to inanimate matter. Since Newton in the late 17th century, we have come to recognise motion as a strict mathematical model. However instead of being entities adhering to engine power for drive & control, axiomatically scripting the mass times acceleration, these 'creatures' born out of the artificial, are triggered, motivated and propelled by external stimuli which seem to influence their emotions.

If 'love' made them approach a light bulb with eagerness and a desire for proximity, then 'anger' made them emphatically rush towards the same light bulb in an attempt to crush it. When in pairs, a set of two similar vehicles concerted to a dance, very similar to the one by Walter's Turtles.

FIGURE 3.10 **End of land — an amphibious niche.** Courtesy Maciek Albrecht, used with permission.

The idea of a world with such vehicles, interacting, willfully cooperating and developing social niches is a romantic extension of the Braitenberg model. Such a world is illustrated in sketches by renowned artist and animator, Emmy prize winner, Maciek Albrecht. Braitenberg's classic [44], included a 10-page portfolio by Albrecht;

> 'This portfolio of vehicles, some placidly at rest, most madly careening over the landscape of the artist Maciek Albrecht's imagination, illustrates only a few of the many marvelous "creatures" inspired by Valentino Braitenberg's text'

Here are three of his sketches. Albrecht uses machinations such as grooves, hinges, nuts, bolts, beams and polyhedrons to convey the engineering components in Figure 3.9 and anthropomorphic emblems such as wings, feathers, snouts and tentacles to exhibit the emotional attributes of these vehicles in both Figure 3.10 and Figure 3.11. The last two sketches show amphibious and arboreal ecosystems. There are clear hints at the evolution of an ecological niche where two or more different types of such vehicles cooperate and coexist, particularly at bottom right in Figure 3.10 which shows a larger vehicle leading many smaller vehicles suggesting swarms. Albrecht [5] puts it succinctly, "...it was a fun project to work on".

FIGURE 3.11 **The winged one — arboreal creatures**, courtesy Maciek Albrecht, used with permission.

The unison of emotions with action has been a cornerstone paradigm in robotics since Braitenberg's vehicles. Pfeifer designed vehicles with 'emotions' and 'value systems' on he named as "sense-act-reflexes" by extending the fungus eaters model to include Hebbian learning. His results are similar to Braitenberg's. Pfeifer along with his co-researchers also designed a complete control architecture [198,290] on the principles of Braitenberg's vehicles. Karpov [171] also used similar methods to design his emotional architecture for mobile robots. The inverse phenomenon, where artificial emotions help to modulate robot action has also been successfully explored [201] by two researchers from New Zealand, Lee-Johnson and Carnegie. Attempts to replicate human-like behaviour and intelligence underscored emotions as the foremost tool for grooming of human tendencies and intelligent behaviour, and has demonstrated that the social learning apparatus works predominantly via human expressions [31].

Braitenberg's vehicles, along with Walter's turtles, were instrumental in establishing the nouvelle AI, reactive route to design robots, encouraging parallel functioning of multiple sensor actuator pairs rather than the cumbersome and often time-consuming sequential architecture. Principles and concepts of Braitenberg's vehicles as relating to animal-like traits to model motor response, encapsulating behaviour, layers of parallel control, embodiment and emergent behaviour were all facets of the development of behaviour-based approaches.

3.3 DELIBERATIVE APPROACH

Mobile robot control broadly adheres to two philosophies, deliberative and reactive. Deliberative approaches or horizontal decomposition advocate the development of a centralised world representation from the fusion of sensory data, development of plans and subsequent execution of a given task and this process repeats.

To illustrate this point imagine a simple line following robot using deliberative principles. It has a light sensor to perceive the environment (sense) and a motor to move around (act), however it also has a sleek computer onboard to design a map of the path it has to traverse a line for the line follower (plan), prior to its locomotion. So, unlike Braitenberg's vehicles the sensing and acting are two different physical processes connected by planning. In a direct comparison to Braitenberg's vehicles, this approach needs an onboard computer and makes this line following robot way too expensive and it adds to bulkiness both in hardware and software domains. If planning is avoided, then it is not only an advantage for the design but the execution of the task is also easier, as shown in flow diagrams in Figure 3.12.

The early robots, developed in the 1960s and 1970s, viz. Stanford Cart, Shakey, Hilare, Alvin etc., were more targetted on trying to get the robot to move and not on attaining higher levels of competence and cognition.They were based on the sense-plan-act model, which is a top-down approach or a functional decomposition. In the sense-plan-act model, the sense system translates the raw sensor data into a world model, the plan system then develops plans to execute a goal in this world model and finally the act system executes this plan. The intelligence of this model lies in the plan system or the programmer and not in the act system.

3.3.1 Shortcomings of the deliberative approach

By the mid-1980s researchers had found several lacuna in the sense-plan-act model. The inability to react to real-time response, often failure to attend to an emergency, bottlenecks leading to long latency time and lackluster performance in an unknown, unpredictable and noisy environment confirmed the inadequacy of the system. Since sense-plan-act worked cyclically, adding more sensors or enabling multiple goals meant more data to transfer over the cycle, thus leading to even worse performance. Horizontal decomposition shown in Figure 3.13, inherently assumes that the environment remains static between successive plans. Such an assumption correlates very badly to a dynamic world with moving obstacles and changing goal points. Another drawback was the lack of robustness. Because the information is processed sequentially, a failure in any single component causes a complete breakdown of the system. Such issues in implementation, motivations from the natural

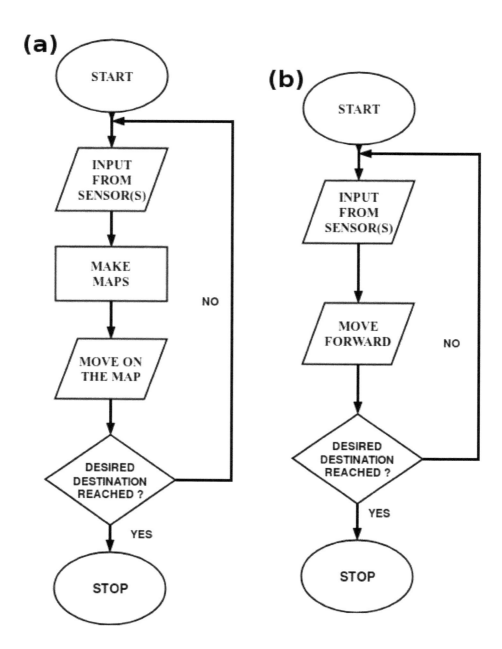

FIGURE 3.12 **Sense-plan-act approach vs. reactive approach.** In (a) the two parallelograms are separated by a rectangle — processing of information between input and output. In contrast (b) lacks planning, and things happen on the run — more like the control paradigm for Braitenberg vehicles 1 to 4.

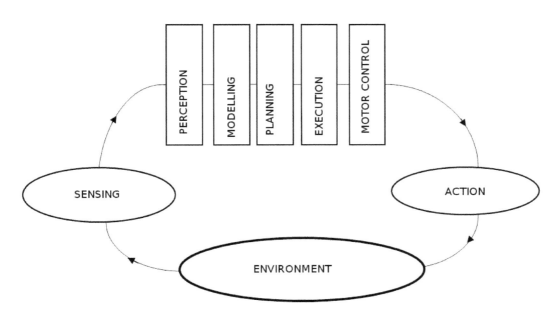

FIGURE 3.13 **Deliberative approach** or horizontal decomposition. From sensing to action, the modules of perception, modelling, planning, execution and motor control are designed as a series. Thus a bottleneck at any module slows down the process. Also every module from perception to motor control has to work for the success of this control structure. A failure at any of these modules will lead to breakdown of the robot. Adapted from Hu and Brady [157].

world as Braitenberg's and Walter's and the obvious need for a parallelism in mobile robot control led to the development of the reactive paradigm or vertical decomposition. Other than the inadequecies of the deliberative approach, two more reasons contributed towards the nouvelle AI: motivations from the natural world and the differences between robots and computers.

3.3.2 From animals to robots

Biological science has always proved to be an opulent source of motivation for development of AI and robotics. To develop artificial beings which have an intelligence of their own should be rightfully motivated from the living world. One of the earliest pioneer was Da Vinci, who made designs of futuristic vehicles with inspirations from birds. Ants and bees have influenced optimisation algorithms. Anemotaxis has motivated navigation methods. Fukuda devised his brachiating robot studying gibbons. Works of Gibson [121] and later Duchon [99] elucidate the basis of ecology-based robotics and are derived from animal behaviour. An entire discipline of cybernetics has taken shape in order to marry the ideas from biology to those of AI. All of swarm robotics is influenced by natural swarms of animals, birds, insects etc.

3.3.3 Robots and computers are fundamentally different

AI was developed in the latter half of the 20th century with the focus on the development of number crunching systems and helped in development of computers. The sense-plan-act model is very similar to the running of a computer program. However, robots and computers are different.

1. Inherent parallelism in robots: Computer programs are nearly always developed for serial execution of an algorithm leading to a result, but in robots a goal has to be attained while maintaining stability, avoiding risk and dealing with the uncertainties of a dynamic environment simultaneously. All of which happens in tandem over a number of sensors and motors. Thus a parallel processing is more desired than a sequential processing.

2. Interaction with a dynamic world: Robots constantly interact with a dynamic world and often set algorithms are not enough to cover all such dynamism, so the final behaviour of the robot is rather different from what is coded in. The sense-plan-act approach, an open-loop plan execution without real-time feedback from the environment is not sufficient to deal with environmental uncertainty, moving obstacles and emergencies.

3. Opportunism suppresses the maxim of plan: Blindly following plans has proven to be detrimental in real world scenarios. Consider a foraging robot schematically shown in Figure 3.14, in a sense-plan-act tenet with 2 types of sensors, laser for obstacle avoidance and olfactory for food detection. A simple serial algorithm for it would be:

Algorithm 1 Foraging robot using a serial approach

---
repeat, move forward
    if obstacle **then**
        move left
    else
        use olfactory to detect food item
    end if
until forever

At **point A** the food item is within the detecting range of the olfactory sensor. Going by serial operation the robot must first avoid the obstacle and move to **point B**, where the food item will be out of its range of detection. If instead of moving left at **point A**, had the robot taken a opportunistic approach and moved right then it would have tracked down the food item successfully.

4. Graceful degradation: For robots operating in the real world, things can get difficult for three particular occurrences [164];

   (a) A given command structure behaves in a different manner due to dynamic environment. As an example, for a robot following a trail of a white line discerns the track from the contrast of the two intensities as gathered from its light sensor; when it enters an area which is heavily lit, the contrast of the two intensities will be much poorer and it makes the robot to slow down or to lose track of the white line.

   (b) Robot's program makes an assumption about the world which proves to be false. As an example, for a robot programmed sequentially to move forward 20 meters and fetch a black ball, the tachometers are coded to 20 meters however, the program assumes no friction. With substantial frictional forces, the robot will fail to achieve this task.

   (c) Sensor failure, for inaccurate input data, a computer program would give an unreliable result. However, for a robot, inaccurate data from the sensors is

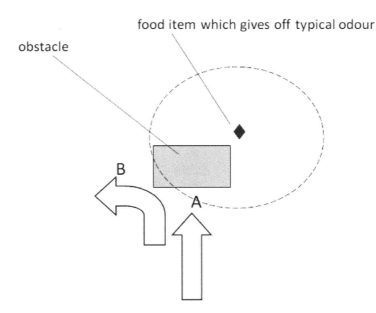

FIGURE 3.14 **Opportunism suppresses plans.** The robot is foraging for food and has two types of sensing; laser for obstacle avoidance and olfactory to detect the food, and is driven by a serial algorithm. At point A, the robot must move to right in order to find the food item, which is not possible for serial execution.

commonplace. The robot programming should be designed in such a way that the robot performance should not break down due to such a failure. A parallel processing over multiple sensors gets around this issue.

These difficulties of a robot interacting in the real world separate it from number crunching systems. Efforts to have parallelism in sensor actuator pairs help in sustaining the robot in scenarios of sensor failure, effectively dealing with a dynamic real world, and to have an opportunistic outlook to attain a given task were realised in reactive approach which is also termed the nouvelle AI, as it was a paradigm shift from the traditional AI.

3.4 REACTIVE APPROACH

To make a robot intelligently interact with the environment, a solution was sought via number crunching operations, more like solving a mathematical problem. This approach has come be known as 'Traditional Artificial Intelligence' or, alternatively as John Haugeland coined it, 'Good Old Fashioned AI' or 'GOFAI', where sensory inputs are converted to data to be processed at various levels of abstraction to develop plans and then executed in order for the robot to reach its goal. Planning was central to this approach and there was a complete lack of immediacy; a robot equipped with a camera based vision system will first develop pixellated maps, and then try to negotiate a suitable route to the goal point via such maps. Incremental cyclical progress will enable the robot to reach its goal. By early 1980s researchers observed the issues in this approach: high lag times which failed for a dynamic environment. Researchers and scientists opined that this is due to the lack of

FIGURE 3.15 **Checkmate !** In his research paper titled **'Elephants Do Not Play Chess'**, Brooks [51] advocated against representational knowledge and encouraged a reflexive, behaviour based approach to robots. Brooks illustrated his argument with a satirical example, that elephants are not appreciative of a representation and rule-based game of chess; however they do exhibit intelligent behaviour.

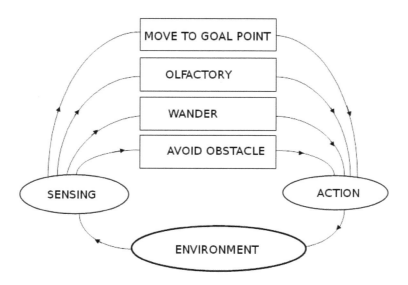

FIGURE 3.16 **Reactive Approach or vertical decomposition.** Layering of behaviours, each layer is a self contained unit that can function without the other layers, prevents bottleneck and failure of one module will not stop the functioning of the robot. Adapted from Hu and Brady [157]

quality processors, and it was believed that sufficient progress in technology could overcome this shortcoming. The shortcoming was not in hardware, but rather in the paradigm.

Walter's turtles, Toda's model of fungus eaters and Braitenberg's vehicles proved to be the stepping stones for the development of the nouvelle AI. These three developments firmly grounded the principles for developing autonomous agents via representation free routes employing embodied AI. All of the three developments are motivated by, or closely mimic animal behaviours. Toda models the basis of a self sustaining hypothetical animal with goals and survival means and Braitenberg elucidates sensory-motor concepts with his gedanken experiments, Walter's efforts prior to both Toda and Braitenberg confirmed to the reality of such theoretical models. While Brooks was influenced by Walter, ANIMAT and ALIFE research was a direct consequence of the Fungus Eater. Three typical characteristics of the behaviour-based approach;

1. **Situatedness**: The robots are situated in the real world. They do not deal with abstractions and models, viz. clouds are not spheres and buildings are not cuboids. The information obtained from the internal state or the short-term memory which contributes to local knowledge are also not in the form of maps or world models.

2. **Embodiment**: The robots have perceptions which allow them to integrate into the environment. Their actions have immediate feedback on their own sensor readings and hence their perceptions. Therefore, the robot is an active part of the environment.

3. **Emergence**: The final behaviour of the robot is an emergent phenomenon which cannot be determined prior to run time. This is a boon and also a curse, as it often leads to novelty in performance but designing robots for a desired behaviour becomes difficult.

As discussed in the previous chapter, the nouvelle AI is modelled on low to no representation, and denies mentalist models and is often designed as sensorimotor. The reactive approach is a minimal paradigm designed as a bottom-up approach where the robot action is determined by the parallel working of a number of modules working in tandem. This is also known as vertical decomposition, as shown in Figure 3.16. The nouvelle AI gets rid of the module of planning and couples together sensing and action more tightly and heavily relies on emergent functionality. Therefore, cognition is an emergent phenomenon, an overlap of perception and action.

Rodney Brooks' subsumption architecture [50] heralded in the nouvelle AI for robots as shown in Figure 3.15. This new paradigm soon found newer implementations as Ronald Arkin's motor schema and Pattie Mae's action-selection and proved to be more effective than traditional approaches.

3.4.1   Subsumption architecture and the nouvelle AI

Subsumption architecture has a very special place in modern day AI robotics. It was the first implementation of the reactive paradigm. Ever since the mid 1980s, Reactive approach, vertical decomposition, behaviour based paradigm and subsumption architecture are often used as near synonyms. Rodney Brooks argued that intelligence in robots should not be symbolical and plan based, rather it should be reactive and instinct based. Brooks based his new theory on structuring intelligence without representation nor reason, but rather agent-environment interactions.

Brooks designed the architecture as a layering of each behaviour to develop his control structure. Each layer is a finite state machine where higher levels could suppress or inhibit lower levels of behaviour. Such inhibition of information was new in the field of AI, however

Ludlow had developed such models in neurobiology which he called 'decision makers'and parallels with Skinnerian school of thought was apparent. In this model, intelligence can be dubbed as agent-environment interactions or also, the emergent behaviour due to contribution by each layer.

Finite State Machine

A finite state machine is a system which has a limited number of states. The system can transit between these states while obeying a fixed set of rules. As an example an on/off switch has 2 states and 2 rules defining the transition from one state to the next one.

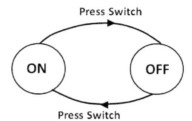

Considering a more involved scenario, a beetle, which moves straight ahead while in light and stops when it is dark. If the beetle encounters an obstacle it turns and runs backwards for 10 cm. After 10 cm, if it is in light then it will again switch directions and move straight ahead. The beetle can be expressed with 3 states and 5 rules.

The apparent shortcoming of this approach is that it is a compromise on the dynamism of the system, as each FSM can only represent a limited number of states. In subsumption architecture, Brooks employed each layer as an FSM.

The design considerations for the subsumption architecture were:

1. Employ evolution as the central principle in design methodology for robots, wherein more specialised performance can be designed by adding more modules on top of a pre-existing rudiment. These modules can be introduced as hardware, software or a hybrid entity of both. Brooks calls these modules circuitry or as the vernacular

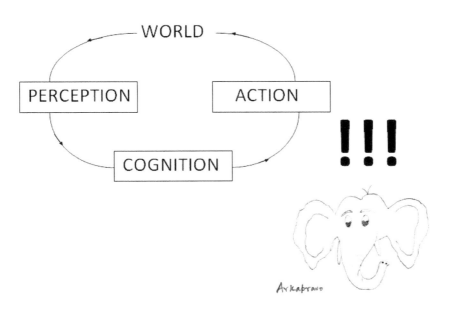

FIGURE 3.17 **Old AI.** Cognition is the result of a cyclical process and the observer is a 'mere spectator' and is not really participative in the cognitive cycle. Intelligence of the process is coded-in by the programmer.

regarding subsumption has come to be — layers of control. Each added layer produces some observable new behaviour for the robot interacting with its environment.

2. Each layer is a binding between perception and actuation.
3. For the sake of design and predictibility of the system, it is necessary to minimise interaction between layers.

The reactive approach had immediate results and instead of a slow-moving robot such as the Stanford cart, Brooks' laboratory had a plethora of gizmos running around in near real-time response. The notion of reducing intelligent action to mere reflexes to the environment was radical but it solved the issues of planning and world modelling.

The argument is not just restricted to planning versus reaction, but is rather between finite state computation where the programs use a constant amount of memory versus the more powerful computational models used in classical AI. These computational models often had to address problems such as search of infinite trees and graphs which are often more of an arbitration than contained to a limited amount of memory.

As another advantage, Shakey and Stanford Cart which employed deliberative approach were bulky and had a high performance computer onboard, in comparison robots developed by Brooks with reactive paradigm were far less bulkier as they did not need high level of computation

In old AI, the programmer was all important as shown in Figure 3.17. With no plans the robot never loses sync with the world as shown in Figure 3.18. The strengths of subsumption are its design methodology, parallelism and modularity. Design methodology makes it easy to build a system and add newer hardware with another layer of behaviour. With parallelism, the ability for each layer to run independently and asynchronously incorporates the working of a dynamic world and warrants for graceful degradation in case of the failure of any layer of

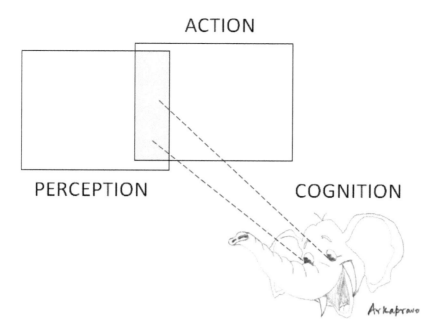

FIGURE 3.18 **Nouvelle AI**, action, perception and cognition not cyclical but rather continious processes and cognition happens as an overlap of the action and perception whence and as it is perceived by the observer. Reactive approach builds the model of cognition around the notion of emergence with the observer being central to the model.

control. Subsumption is clearly more modular than the sense-plan-act approach, but there has also been criticism regarding this feature, as I will discuss later in the chapter.

Other than these facets of design, three more salient features of subsumption architecture are low cost, reduced computational power and re-usability of code. Subsumption is modular and therefore cheaper to build and test and a single layer of behaviour is enough to get the robot running. Such results are most appreciable in walking robots, hexapods and humanoids.

A number of iconic robots were made using subsumption: Atilla, Herbert, Chengiz and later Cog. Connell developed the COLONY architecture by considering a number of individual subsumption units as shown in Figure 3.19, and connecting them together with further prioritisation among those units, to construct 15 different behaviours for 6 task-specific levels. COLONY architecture was meant for a robotic arm mounted on a mobile base meant to retrieve soft drinks cans. COLONY was the first behaviour-based paradigm developed for arm/hand manipulators. Later onwards, dedicated behaviour based architectures for manipulators, to manipulate pick & place to yield richer behaviours and tracking operation were developed by various research groups. Various more architectures were developed with subsumption in mind, two well known ones are, SSS Architecture and the Circuit Architecture.

3.4.2 Motor schema

Motor schema [17] was developed by Ronald Arkin in late 1980s. Though it is a behavioural paradigm, instead of layers of control it uses a schema-based vectorial approach. The output

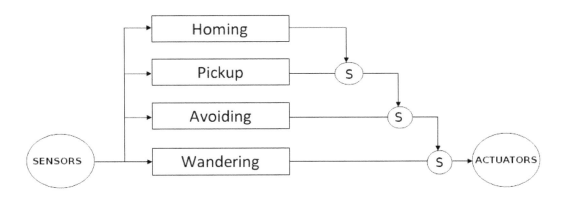

FIGURE 3.19 **SR diagram for a trash-collecting robot** designed using subsumption architecture. Simple operations such as wandering are put at the lowest end, with more involved behaviours such as pickup and homing are higher up. The higher levels suppress the lower levels. The robot can identify the trash item(s) on seeing them at a detectable distance. The robot 'wanders' as long as the trash is not seen; once the trash is seen by the robot's sensors then 'avoiding' obstacles and reaching the trash suppresses '**wandering**', at the site of the trash '**pick up**' suppresses '**avoiding**' and once the trash has been '**picked up**' then '**homing**' back to a dumping yard suppresses '**pick up**'. A trash collecting and foraging robot can be made from the same basic behaviours.

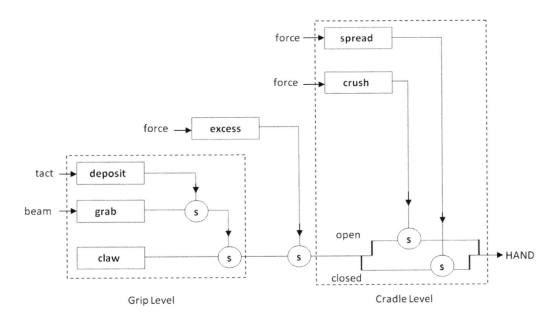

FIGURE 3.20 **An example from the COLONY architecture,**

of each schema is styled as a velocity vector. A dynamic assortment of such schemas is used to implement complex behaviour. The output is the normalised vector sum of all the active schemas. Arkin's model was developed on Arbib's research on frogs that described motor control in animals as a vector model. On extending such ideas to the robotics domain, schemas are developed as primitive motor behaviours which are reactions to sensory information obtained from the environment.

What is behaviour?

A universally accepted definition of behaviour in AI concerns doesn't exist. The term was incorporated from psychology by Simon in late 1950s, to develop models of intelligence and rationality which hinted at the concepts of embodiment and situatedness. Later, Toda used the terminology of 'behaviour program' in his fungus eater model and he contends behaviour is a means of adaptability in an unknown and apparently unsophisticated environment. In his seminal paper, Brooks introduces vertical decomposition with 'task achieving behaviours' to develop the subsumption architecture. The works of Brooks, Payton and Arkin broadly define behaviour as *basic building blocks for robot action*, or alternatively *the agent's response to its environment*. In reactive approaches a primitive behaviour would mean a *sensorymotor pair*; combination of such primitives would lead to the phenomenon of emergence in the robot. A more formalised definition would be, *behaviour is a function of the structure and dynamics of its physical design and construction, the attributes of the environment in which it works, and its internal control system.*

However, since performance in reactive systems is emergent it is difficult to confirm the behaviour of the system. Oftentimes a subroutine in a robot controller is wrongly referred to as a 'behaviour'. As an apparent way out, Gat distiguishes between code and the implements using 'Behaviour' for code and 'behaviour' for the implements, viz. A *Behaviour* is a piece of code that produces a *behaviour* when it is running.

This definition is not sufficient for hybrid systems which often have deliberative, reactive and adaptive components. In reactive systems behaviour means a purely reflexive action, whereas in hybrid systems it can mean reflexive, innate and learned actions and it is difficult to differentiate between the code and the implements as the system adheres to a high degree of emergence.

Unlike subsumption architecture, motor schema doesn't have a preset heirarchy and doesn't attempt to suppress or inhibit a given module but rather it works to blend behaviours together. Behaviours are generated as a vector at run time and are added together like vector summation. Motor schema are easily implemented in software using artificial potential fields developed by Khatib as shown in Figure 3.21.

Arkin classifies navigation primitives from individual schemas, more complicated behaviours can be generated by superimposition of these primitives as shown in Figures 3.22 and 3.23. At Georgia Tech, Arkin developed advanced foraging behaviours using schemas for robots, Callisto, Io and Ganymede.

> **What is schema?**
>
> Schema is traditional to psychology and neurology. Anthropomorphic motivations led Arkin to develop schema for robots. Neisser defines schema as *"a pattern of action as well as a pattern for action"*, while Arbib explains schema as *"an adaptive controller that uses an identification procedure to update its representation of the object being controlled"*. In robot concerns, schema relates motor behaviour in terms of various concurrent stimuli and it also determines how to react and how that reaction can be accomplished. Schemas are implemented using potential fields; individual schemas are often depicted as needle diagrams. Concurrently acting behaviours are coordinated using vector summations of the active schemas.

3.4.3 Action selection & bidding mechanisms

Action selection or action selection mechanism (ASM) was designed by Pattie Maes [?,217] and it takes a different approach than suppressing, inhibiting or vectorially adding behaviours. Instead it employs a dynamic mechanism for behaviour selection. ASM is a selection based competitive implementation of behaviour-based paradigm and it overcomes the problem of the predefined priorities used in the subsumption architecture. Each behaviour has an associated activation level, which can be affected by the current sensor readings of the robot, its goals and the influence of other behaviours. The activation level can be interpreted as a corollary to the energy level for Beer's cockroach simulation, the heuristic which determines the arousal and satiation level for food. Each behaviour also has some minimal requirements known as activation levels that have to be met in order to be active. These levels also decay over time. From all the active behaviours, the one with the highest activation level is chosen for actual execution. Maes was motivated by Minsky's 'society of mind', arbitrary agents interacting locally to lead to an effective working of a society. However, Maes' work also bears similarity to that of Tinberg and Baerends, and extends from biology to AI.

Maes designed every autonomous agent as a set of competence modules and, using list structures, implemented her algorithm. A competence module i can be denoted as a tuple $(c_i, a_i, d_i, \alpha_i)$. Where c_i is a list of preconditions which have to be fulfilled before the agent can become active. a_i and d_i represent the expected effects of the agent's action in terms of an add list and a delete list and α_i is the activation level. A competence module is *executable* when all its preconditions are true, an executable module module whose activation level surpasses a threshold may be selected to perform some real world action.

Action selection solves the issues of loss of information in subsumption when a layer is suppressed or inhibited. It avoids the potential field implementation issues of motor schema and merits every single piece of information available to the robot. However in a dynamic implementation it is harder to predict the robot's emergent behaviour at runtime. Maes' action-selection has also been used in tandem with reinforcement learning and provided better results.

Action selection served as a motivation for development of control paradigms based on bidding mechanisms. In a bidding-based control architecture each behaviour bids according to the urgency for having the action executed, which can be compared in the ASM approach to preparing a list with a descending order of activation level in action-selection. However, unlike Maes' approach bidding mechanisms do not employ a critical activation level to become active and neither do the behaviours have any preconditions to be met and they are always ready to bid. Also, the behaviours in ASM can influence the activation level of

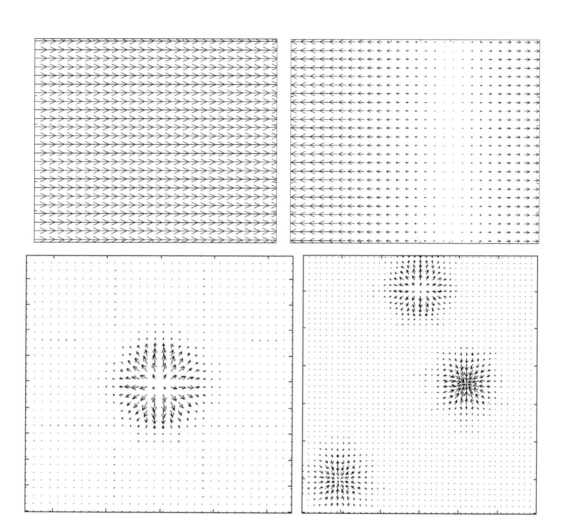

FIGURE 3.21 **Motor schema implemented using potential fields.** Move forward schema (upper left), path tracking schema (upper right), obstacle schema (lower left) and superposition of two goal and one obstacle schemas (lower right).

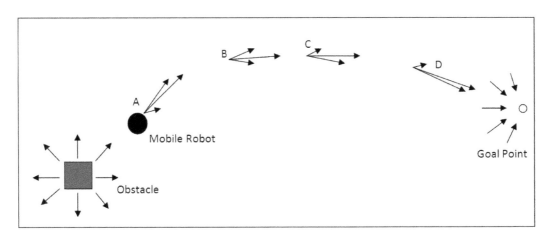

FIGURE 3.22 **Schema self modulation at run time.** In this example the two concerned behaviours are — 'wander', 'obstacle avoidance' and 'motion to goal'. The 'wander' behaviour is always acting, the other two behaviours occur due to the environment. At **point A** the robot experiences maximum repulsion from the obstacle and feeble attraction from the goal; the vector sum doesn't point to the goal point; and 'obstacle avoidance' is the dominant behaviour. At **point B** and **point C** the robot experiences less repulsion from the obstacle and more attraction to the goal point; the vector sum still doesn't point to the goal. At **point D**, the repulsion from the obstacle is of very little significance and the vector sum points nearly to the goal point; 'motion to goal' is the dominant behaviour. In this manner, schemas implement behaviours in a dynamic manner. It is worth noting that the robot <u>does not</u> chart the shortest path from **point A** to the goal.

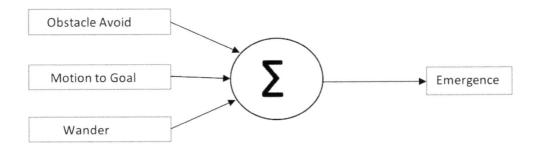

FIGURE 3.23 **Sigma diagram** addition of 3 schemas. Unlike subsumption no behaviour is suppressed and throughout the previous example in Figure 3.22 these 3 schemas are acting in tandem. However obstacle avoidance is dominant at **point A**, while contributions from the other two schemas are insignificant. Near **point D** motion to goal is dominant and the other two schemas contribute little to the robot's emergent behaviour.

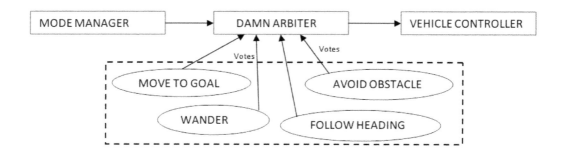

FIGURE 3.24 DAMN Architecture is an example of bidding-mechanisms where the behaviour module 'votes' to decides which action to be pursued.

other behaviours, whereas in bidding mechanisms, behaviours are totally independent of each other.

Rosenblatt developed the Distributed Architecture for Mobile Navigation project (DAMN). In this architecture, a set of behaviours which cooperate to control a robot's path by voting for various possible actions such as control of heading angle and speed, and an arbiter decides which is the action to be performed. The action with more votes is the one which is executed. However, the set of actions is predefined and it had a grid-based path planning.

Behaviour-based architectures can be classified depending on how the coordination between behaviours occurs; (1) Competitive: in these architectures, the system selects an action and sends it to the actuators, it is a winner-take-all mechanism. Subsumption architecture, action-selection and bidding mechanisms are examples of competitive coordination, and (2) Cooperative: in these architectures the system combines the actions coming from several behaviours to produce a new one that is sent to the actuators. Motor schema and Braitenberg vehicle-3c are example of cooperative coordination.

Unified Behaviour Framework (UBF) was designed to allow a robot to use competitive as well as cooperative policies using a single architecture, as shown in Figure 3.25. UBF allows for seamlessly switching between disparate architectural policies by varying the arbitration techniques. Often a low-level controller and its behaviour is developed for a specific task; thus any future redesign or extending the utility appreciably to other tasks is severely limited. Since UBF allows for seamless switching across distinct architectural policies during runtime, it encourages reuse of behaviours, with various different modes, viz. random activation of an arbitrary behaviour, priority based selection, cooperative and semi-cooperative arbitration etc. UBFs modular architecture helps to simplify design and development by code reuse.

Bidding mechanisms and ASM attempt to put in some deliberative content in reactive architectures to design higher level behaviours. This can be seen as early attempts which led towards hybrid architectures which is discussed later in the chapter.

3.5 A CRITIQUE OF THE NOUVELLE AI

At first, the new AI proved to be much faster in comparison to traditional AI. However the downside of this was the inability of the robot to engage in cerebral activities, which are the basis for all higher functions and sophisticated behaviours.

Behaviour-based approach resists analytical modelling and intelligence of such a system

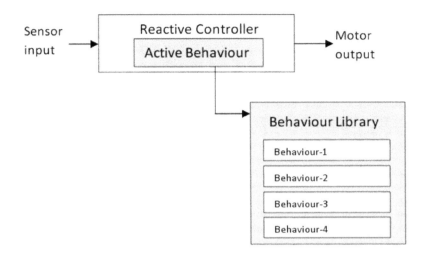

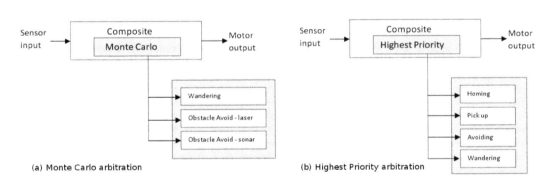

FIGURE 3.25   **Unified Behaviour Framework (UBF)**, provides dynamism to a behaviour-based approach by adding various modes of arbitration, which allows for the running of cooperative as well as competitive architectural policies on the same controller encouraging reuse of code. Two arbitrations using UBF, (a) **Monte Carlo**, the robot alternates randomly between wander and two modes of obstacle avoidance, (b) **Highest Priority** arbitration is more like traditional subsumption where the behaviour with the highest priority which can either be preprogrammed or acquired through voting gets exeuted.

is when the agent interacts with its environment. Obviously this does lead to an 'element of surprise' and predictibility takes a back seat. However, emergent behaviour tends to be more robust because it is less dependent on accurate sensing or action and also since it makes fewer environmental assumptions. Emergence is always a product of a dynamic process and it is ostensive. One of the most endearing examples is given by Brooks, where he demonstrates how bi-ped walking can emerge from a network of rather simple reflexes with hardly any analytical model or central control. Agre showed how behaviour as complex as goal-directed action sequences can be modelled as an emergent property.

TABLE 3.2 Comparison between traditional AI and reactive approach

	Traditional AI	**Reactive Approach**
Designed in	sensor fusion, models, goals, plans	action selection, schemas, behaviour
Basis in	search	binding problem
Information flow	choose next action	concurrent actions
Emergent aspects	behavioural response	goals and plans which are made on the go
Robustness	serial action, a module failure leads to system failure	parallel action leads to graceful degradation
Intelligence	coded in by the manufacturer or the programmer	happens in tandem as an overlap of perception and action

3.5.1 Issues with the nouvelle AI

Subsumption architecture led to a near real-time implementation for mobile robots and later bipeds and humanoids, however suppressing and inhibiting layers of control meant losing information which can be critical in hard real time implementations as aerospace and military hardware in battlefields. Motor schema relies on blending and hence doesn't have any issues of loss of information, though it inherits all the issues of potential field implementation, particularly the issue of local minima. Behaviour-based approaches are motivated from the natural world and do not work well to mirror human-like intelligence. There have been suggestions that to develop higher-level intelligence akin to human beings, a representation free approach will not prove to be sufficient.

3.5.1.1 Implementation issues in subsumption architecture

1. **Loss of information and disruption of lower-level functionality:** As lower-level functionalities are suppressed or inhibited, the information contained in those layers is lost. Hence a robot, when executing a higher-level behaviour, may fail to execute a simpler behaviour at the lower levels.

2. **How to identify a higher level behaviour:** Determining a given behaviour as higher level or lower level is a matter of arbitration [143] and there is no real methodology to ascertain this. The general rule of suppressing lower levels in favour of higher levels is not the hall mark of a good design.

3. **Layers are not really independent of each other:** Subsumption assumes that each layer of control is independent of another, but this is not always true. The internal

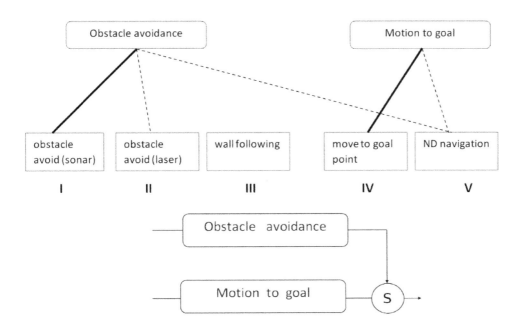

FIGURE 3.26 **Lack of a compromise solution.** Subsumption fails to identify a compromise solution, adapted from Rosenblatt and Payton [282].

functionalities of the higher levels often interfere with the lower levels, and often times two or more layers use the same hardware making independence of those layers really questionable. Thus small changes to the lower levels are difficult to implement without redesigning most of the other layers. In particular, higher layers can suppress, inhibit and also read the signals of lower layers. Adding higher level behaviours without altering lower level functionality is not possible. The physical connections between layers are hard-wired, so they cannot be changed during execution, making the system less dynamic to changes in the environment. As an example, consider a motor which is a common hardware for 2 layers of wandering and motion to goal. When wandering is suppressed by motion to goal, the motor is already at a non-zero velocity while motion to goal would have been designed considering the motor starts from zero. It may happen that at a higher velocity as attained during wandering, the motor may be difficult to align towards the goal point or, even worse, miss it all together.

4. **Levels must have a priority**: Each behaviour is associated with a degree of priority, thus behaviours with equal priority cannot be represented with the subsumption architecture.

5. **Lack of access to the internal state of each behaviour:** With increasing layers of control it is often difficult to acertain the information internal to a behaviour. This shortcoming poses as a design issue and the coordination of a large set of behavioural skills by adding further layers of control to achieve a desired coherent complex behaviour is a difficult and error-prone job, lack of modularity is alarming in hard real time systems such as aircraft, elevators and medical applications. Critics have claimed that such designing of behaviours is apparently more of an art than a science.

6. **Lack of a compromise solution:** Since subsumption is a 'winner takes it all' philosophy, it fails to include a compromise solution when the working of two or more layers of control have conflicting goals. An example adapted from the Rosenblatt and Payton, consider 2 behaviours as schematically in Figure 3.26. Obstacle avoidance can be best implemented by I, however it can also be fulfilled by II and V^2 as next best options. Motion to goal can be implemented by IV and V, IV being the best option. The best options are shown with continious lines while the next best options are shown with dotted lines. In a subsumption architecture with these two behaviours, where obstacle avoidance inhibits motion to goal, the robot will cease to have an alignment towards the goal while avoiding obstacle and this can be detrimental to the task in hand. As a better alternative, option V, ND Navigation could have been used as a consistency for both the behaviours, however subsumption doesn't merit such compromise solutions. To overcome this lacuna, Rosenblatt and Payton [282] suggested a connectionist architecture which uses a very fine grained layered architecture. Instead of designing behaviours as finite state machines with internal states and instance variable, behaviours are comprised of atomic functional elements that have no inaccessible internal state. These simple decision making units and their interconnections collectively define a behaviour. Since the behaviours can interact with each other, thus can select a best compromise solution. In continuation of this work, Rosenblatt also develop a notion of 'internalized plans', instead of mere program for action, plans are considered as sources of information and advice to agents that are already accomplished in dealing with a dynamic environment. Therefore, plans are used selectively, and serve to enhance system performance or as fail safes in dire scenarios. In implementation, the notion of 'internalized plans' is realised using a global navigation using potential-field based gradient descent method.

3.5.1.2 Issues with motor schema

Instead of suppression or inhibition motor schema uses blending of behaviours akin to vector addition, so there is no loss of information. However, three issues of this approach are:

1. Local minima problem: A built-in shortcoming of motor schema is the issue of local minima which plagues all potential field implementations[3].
2. Lack of practical solution by blending: A problem that cooperative mechanisms such as motor schema face is that the solution via blending behaviours may not really solve the problem. Consider a robot which has a gaping hole straight ahead, and two different behaviours attempt to avoid it, one trying to avoid it through the right and the other one trying to avoid it through the left. The sum of the 2 behaviours would be a vector pointing straight ahead to the gaping hole, the emergent behaviour fails to solve this problem. Researchers have tried to overcome this problem using dynamic methods for behaviour selection and bidding strategies where the the option with the highest level of activation or the most popular action is executed.
3. Null summation: Another implementation issue is when two or more schemas of the same magnitude working in opposite directions to each other will nullify the net vector sum. Thus though there is stimuli but it does not evoke a motor response.

---

[2]The next chapter will include further details on ND navigation.

[3]The next chapter details the various implementation of potential field method for navigation and the methods to overcome the local minima problem will be discussed there.

FIGURE 3.27 **'Today the earwig, tomorrow man?'** David Kirsh argues in favour of representations and maintains that recreating insect-like behaviour in robots is not sufficient to elicit higher levels of intelligence. Kirsh associates representations with perception, learning and control, thus making it imperative in development of human like intelligence. Brooks replied to this criticism by suggesting that the behaviour-based paradigm needs to incorporate cognitive approaches to develop humanoid robots with human-like faculties, Brooks named this discipline 'cognobotics'.

3.5.2   Extending reactive approach to higher functions

Brooks' hypothesis of 'reason and representation free' AI works very well for situationally determined activities such as walking, running, avoiding obstacles, climbing stairs, etc. However it is silent regarding cerebral activities such as playing chess, solving jigsaw, tinkering with the Rubik's cube, etc. For example Dennett's robots; R1, R1D1 and R1D2 without any representative knowledge, viz. a visual knowledge of the battery and/or the bomb looks like, will all fail the task and [once again] will be bombed to destruction, this time on a purely behaviour-based approach.

Kirsh [180] in his opposition to reactive approaches, insisted that robotics can never be modelled without representations, such as development of 'insect like' behaviour such as wandering, avoiding obstacles and wall following will never be sufficient to progress to higher levels of intelligence, as shown in Figure 3.27. Kirsh agrees that 97% of human behaviour is nonrepresentational, however in the remaining 3%, human beings engage in cerebral activities, the typical non-trivial aspects of human intelligence which lack in reflexes and animal intelligence. Kirsh argues that there is a limit to which a creature can go without 'concepts'[4] and for higher level intelligence there is a need to have conceptual representations in action, perception, learning and control [112].

Kirsh's criticism however doesn't suggest any other method to achieve artificial human-like intelligence. In similar context, Wilson [357] points to the lack of a model for complete understanding of intelligence — the lack of a strong relation between methods of natural science and nature.

The bottom-up approach suggested by the behaviour-based philosophy is often insufficient and fails to reach human-like intelligence. ANIMAT & ALIFE research attempts

---

[4]Kirsh prefers to use the term concept-free rather than representation-free. He suggests that mobile robots control truly concerns the conceptualisation in intelligent activity, rather than representation as modelled in reactive approaches.

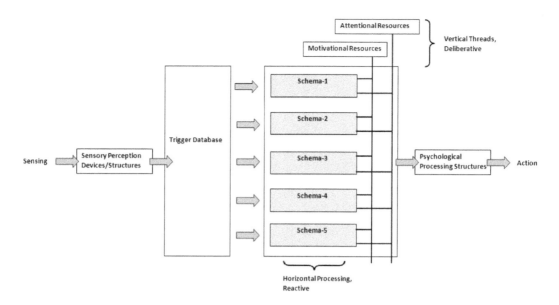

FIGURE 3.28 **Two modes of cognition.** Higher life forms as human beings at least have two modes of cognition, reactive and deliberative which work in tandem.

to develop human-like intelligence by simulating and understanding complete animal like systems at a simple level. The shortcoming is most glaring in more sophisticated behaviours, and since cognition is not limited to reflexes in higher beings. Neuroscience also suggests two 'modes' of human behaviours as shown in Figure 3.28. Similar ideas have been suggested by Steve Pinker for two levels of consciousness in higher life forms such as human beings which will be discussed in Chapter 9.

The Philosopher & The Robot

Agent-based robotics is seen as the pinnacle of cutting edge technology, while in contrast philosophy is perceived as the domain of the proverbial armchair, greying professors and old books. However, the two disciplines are very closely related. As discussed in the previous chapter, the problem of relevance and the contention of using internal representations is a philosophical debate rather than a technological limitation. Philosophical concerns nearly always dovetail with agent-based robotics, and this will be seen in later chapters inquiring into ethical robots, artificial consciousness and super intelligence.

In the early 1960s, agent-based robotics was conceived more as an extension of the computer and therefore intelligence was approached as the Cartesian model of the 'thinking thing', and information processing was central to this model. The robot had to sense the local environment, make maps and find out the most suitable paths to the goal point, and then move these paths. This process of sense-plan-act was cyclically incremented to work in a continiuous manner. The onboard computer was the 'brain' of the robot and other sensors were input units. By processing the data from the sensors, the robot could draw in on representations to form Kantian concepts of space and time [150], and therefore have meaningful interaction with the environment. The outcome was a very slow-moving robot with a bulky onboard computer which was not at all suitable for a dynamically changing environment with moving obstacles and changing terrain. The issue of this non-performance was attributed as the lack of fast processing units, and it was believed that the problem would dissolve with better technology and lead to the desired results.

In the early 1970s, Dreyfus pointed out the limitations of traditional AI by contesting the shortcomings of the computational paradigm in his seminal book *What computers cannot do*, and demonstrated the philosophical pitfalls in this approach. His argument was based on the explosion in the number of symbols and rules to overcome the frame problem. Dreyfus suggested the use of the philosophy of Heidegger and Merleau-Ponty to overcome the lacuna of disembodied, acontextual and symbolic computation which was being pursued by traditional AI. He suggested a paradigm shift towards what he coined as 'Heideggerian AI', which abandoned the rational approach of representation and planning and was designed on the agent's dynamic interaction with its local world and focussed on immediacy rather than long-term planning, and therefore intelligence was related to sensorimotor action rather than slow and incremental designs of plans. This consistent coping with the environment has been named `Being-in-the-World` in the works of Heidegger. In the behaviour-based paradigm, the concepts of situatedness and embodiment are directly drawn from Heidegger and Merleau-Ponty, respectively. Heideggerian AI is able to avoid the inherent problems of traditional AI as it goes around both the frame problem — by being situated — and also the symbol grounding problem — by being embodied. Heideggerian AI is the philosophical basis for connectionist approaches to cognition and behaviourist approaches to artificial life. Other than robotics, Heideggerian AI can also be seen in Agre's development of indexical functional entities to design the dynamics of PENGI and Freeman's neurodynamic model of the brain. In the 1990s, Varela and co-researchers [320, 333] developed the enactive paradigm by extending Merleau-Ponty's phenomenology of the body to embodied AI. The approach highlights that the agent's body is not only a lived-in and experiential structure, but it is also the backdrop of all cognitive processes.

In the attempt to develop human-like intelligence, the behaviour-based paradigm is an exposition into the minimal conditions needed for a humanly experience [92]. However, this is not sufficient to lead to human-like intelligence, and neither ethical agency nor conscious behaviour can be extrapolated from it. This will be discussed in Chapters 9 and 10. As of now, most state-of-the-art robotic systems employ a twofold hybrid system: low-level behaviours are attributed with a reactive module and higher-level behaviours are controlled with a deliberative module.

In the late 1990s Brooks extended behaviour based approach to humanoids with the making of his robot Cog, and coined it as a 'second shift' [55] in behaviour based paradigms, a leaning towards cognitive robotics. He believed that bodily form and morphology of the human like robot, development of artificial *motivation* — which will naturally prioritise events, coherent action of various smaller subsystems and adaptation to motor control & world dynamics will play important roles in the 'second shift'. Development of such 'cognobots' will have to be channeled through an integrated approach where the cognitive development, sensor engineering and integration and morphological design work in parallel and the various subteams are aware of the workings of each other. It is imperative that study of the human brain was meant to greatly influence cognobotics. Cog was planned as a long-term project and progress has slowed down with passing years, but it still has served as one of the early efforts to develop a humanoid robot bust and it was the first attempt at designing artificial consciousness in robots.

Nearly all modern-day humanoid robots have modules for natural language processing, mapping and machine learning; all of these require some form of representation and deliberative module.

3.6 HYBRID ARCHITECTURES

As briefly mentioned in the previous section, both deliberative and reactive systems have their shortcomings. Reactive systems are hardly of any use or at best they cater to very low-level tasks and they are motivated from nature as shown in the timeline of this philosophy in Figure 3.29. For example, I have discussed the design of a foraging robot using subsumption architecture, which is ispired from similar activities in the animal kingdom. For a single deer foraging for food, it is guided by its reactive sense of olfaction however it is also led by visual stimulus as it knows what the food looks like (viz. leaves, herbs, shrubs etc.), knowledge it has acquired from experience. Cognition is not left to a solitary senor-actuator pair but it rather happens in tandem across a number of sensory organs. Similar arguments can be made in favour of birds, rats etc. Animals, birds and even some insects acquire representational knowledge through experience which is useful in tasks as navigation, foraging, predator-prey scenarios etc. For human beings, psychologists have ascertained that human behaviour works across two distinct modes, willed and automatic.

Reactive architecture are often meant for a specific set of tasks. A controller designed on the reactive paradigm cannot easily be made to work with another set of tasks without modifying nearly the entire controller. In contrast, deliberative agents work well in static environments and fare poorly in dynamic environments. Therefore, a robot made to do complex tasks in a dynamic world cannot accomplish them in a purely reactive manner nor in a deliberative plan-based manner, hence there is a need to look into methods to combine the goodness of both of these. An apparently minimalistic solution is to equip behaviour based system with maps. Sensor based knowledge of the immediate neighbourhood is often realised with grid representations, while a priori maps generally have more global data. The third way to incorporate knowledge is through perception. Such systems are usually equipped with state-of-the-art vision systems.

Reactive approach with maps strictly pertains to navigation, so development of systems which have identifiable architectural components of both deliberative and reactive paradigms have found more acceptance in the AI community. The two paradigms can be combined in various ways. The simplest designs attempt to redesign the sense-plan-act approach to adhere to a reactive paradigm. In it the robot would generate a plan to accomplish a job and execute it reactively in a sense-act execution, after which the plan module will generate further plans, while multilayered approaches have been popular with researchers, as shown in Figure 3.30.

Hybrid architectures have a striking similarity with a managerial corporate heirarchy with high level planning and policy decisions done at the higher up 'manager' levels, which then pass them off for execution to the lower levels, while the lowest level, the reactive level, is a corollory to the 'worker', which executes the job and is more well acquainted with the immediacy of the job at hand.

The works of Firby, Gat, Connell, Bonasso etc. led to the development of the three layer architecture [116] and its variants — 3T architecture, ATLANTIS and SSS. Behaviour based route to controlling robots considers direct mapping of sensors onto actuators in contrast the 3 level architecture work on two types of algorithms, (a) algorithms for governing routine

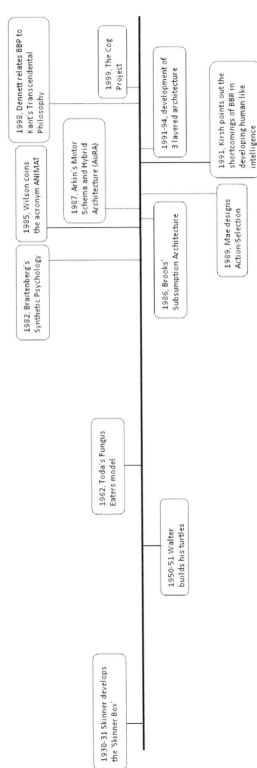

The Behaviour Based Philosophy (BBP) (1930 - current day)

- 1930-31 Skinner develops the 'Skinner Box'
- 1950-51 Walter builds his turtles
- 1962, Toda's Fungus Eaters model
- 1982, Braitenberg's Synthetic Psychology
- 1985, Wilson coins the acronym ANIMAT
- 1987, Arkin's Motor Schema and Hybrid Architecture (AuRA)
- 1998, Dennett relates BBP to Kant's Transcendental Philosophy
- 1999, The Cog Project
- 1991-94, development of 3 layered architecture
- 1991, Kirsh points out the shortcomings of BBR in developing human like intelligence
- 1986, Brooks' Subsumption Architecture
- 1989, Mae designs Action-Selection

FIGURE 3.29 **Timeline for behaviour based philosophy (BBP)**, 1930 – current day. The roots of behaviour-based approach can be said to start with the early experiments of Pavlov in 1890, and also with the beginning of the philosophy of phenomenology by Husserl in 1910, which was later extended by the works of Heidegger and Merleau-Ponty. This timeline focuses more on the technological feats than philosophy or psychology. Made by the author, CC-by-SA 4.0 license.

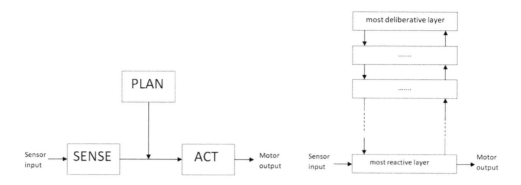

FIGURE 3.30 **Approaches to hybrid paradigm**, Among the various ways to design a hybrid architecture, 2 stands out. Plan, Sense-Act (P, SA) is an extension of Sense Plan Act (SPA) to a hybrid domain and execution using reactive approach, while a layered design with the most reactive layer at the bottom has been a trend with researchers, this sort of layering can be seen in AuRA and ATLANTIS.

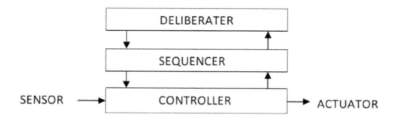

FIGURE 3.31 **3-layer architecture.** The generic flow diagram of a three level architecture which is a trade off between reactive systems and deliberative systems.

sequences of activity which rely on the internal state and are not exponential time search based processes and (b) search-based algorithms and planners. The components of this architecture are:

1. **Controller** The controller implements one or more feedback loops, tightly coupling sensors to actuators. The algorithms that go into the controller should have constant bound time and space complexities, should be able to detect failure and should use internal state to a minimum.

2. **Sequencer** The sequencer selects which primitive behaviour the controller should use at a given time. This can be a difficult choice to make and Gat suggest two routes to affect the sequencer, universal plans or conditional sequencing. The sequencer is executed by special purpose languages such as RAPs, PRS, ESL etc.

3. **Deliberator** The deliberator keeps track of the time-consuming computations. It can work in two ways, either by producing plans for the sequencer to execute or it can respond to specific queries of the sequencer. The deliberator attempts to optimally execute exponential time algorithms while keeping a check on the limited computational resources. There are no architectural contraints and the deliberator is written in standard programming language.

The controller is most reactive and depends on the current state, the sequencer depends

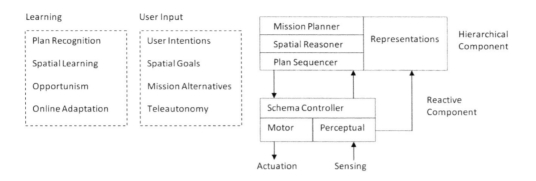

FIGURE 3.32 **AuRA Schematics**, a layering of modules with lowest being the reactive layer.

on the current and past states while the deliberator, as the name suggests is most deliberative and depends on the current, past and future states as shown in Figure 3.31. This architecture and its variants has been implemented on various robots: JPL's Robbie the robot, Alfred, Uncle Bob and Homer, and also on the UK MoD funded Battlespace Access Unmanned Underwater Vehicle (BAUUV) project (2010-2015), with added components and database support.

The Autonomous Robot Architecture (AuRA), the schematics shown in Figure 3.32 developed by Arkin at Georgia Tech. has been particularly popular with roboticists and has formed the basis for various other architectures including ALLIANCE for multiple robots, and ethical architecture designed for robots in the US military and is discussed in Chapter 8.

SUMMARY

1. Robot control paradigm is needed to effect robot action to external stimuli. In simplest scenarios as 'line follower' this will not be very appreciable however things become interesting with more sensors and tasks more dynamic than mere locomotion.
2. three approaches to develop robots are, (i) deliberative, (ii) reactive and (iii) hybrid approaches
3. Braitenberg vehicles and Walter's turtles laid the foundation for anthorpomorphic approach to mobile robots, Brooks extended these ideas to formulate the subsumption architecture.
4. Robots are different from computers and an inherent parallelism and ability to deal with a dynamic environment is essential to develop well functioning mobile robots.
5. Subsumption architecture, motor schema, action-selection and voting based architectures are the most popular implementations of reactive paradigm.
6. Reactive approach overcomes issues encountered by deliberative approach, however being representation free they are insufficient to develop higher level cognition.
7. Hybrid approach has seen a convergence of deliberative and reactive paradigms.
8. 3 layer architecture has made a niche in the hybrid domain, where the deliberative layers are the decision making layers, whereas the lowest layer is the reactive layer. ATLANTIS, SSS etc are examples of 3 layer architecture.

NOTES

1. *Motor schema was structured using potential field navigation methods, originally developed by Khatib as a navigation algorithm. It is interesting to note that though there has been many newer paradigms as ND, DWA, Bubble band etc for autonomous navigation, but none of them have found their way to being a robot control architecture.*

2. *Much like the 'earwings' papers, Lammens and co-researchers published a paper titled, 'Of Elephants and Men', which was critical about Brook's paper, 'Elephants do not play chess' and also introduced the GLAIR architecture.*

3. *With tongue in cheek humour Horswill expresses deliberative approach with the following algorithm:*

Algorithm 2 Deliberative approach, adapted from Horswill

---

repeat for all sensors
    check the sensors
    update the model
    compute a plan for the current goal
    execute the first action of it
    throw the plan away and start over
until forever

---

4. *In comparison, Braitenberg's vehicles as vehicle-3c, the output is difficult to predict but Braitenberg doesn't really promise any added novelty. Brooks' approach promises novelty in behaviour due to emergence, which can be only seen at run time, analytical analysis is futile.*

5. *The term global is often associated with deliberative paradigms, while local with reactive paradigms. However, Murphy illustrates that such a strict distinction is not possible in hybrid systems*

6. *The critics of Dreyfus made a pun regarding his interpretations of Heidegger as 'Dreydegger'. As Woessner puts it; " Dreydegger is a philsophical Frankenstein. Consisting of pieces borrowed from the philosophy of the mind, existentialism, pragmatism, and both the analytic, and continental traditions, it somehow manages to lumber forward on its own two feet.If originally it was concocted to show how AI research rested on faulty philosophical premises, it eventually developed a mind of its own and followed its own path".*

7. *Dreyfus has prospected towards the development of 'Heideggerian Cognitive Science', an ontology, phenomenology, and brain model, without any representation*

8. *Ronald and Sipper [280, 281] suggested a test to check for emergence in autonomous robotic systems. The test considers 2 individuals: a system designer and a system observer and the following three conditions: Design: The system has been constructed by the designer by describing local elementary interactions between components (e.g., artificial creatures and elements of the environment) in a language $\mathcal{L}1$. Observation: The observer is fully aware of the design, but describes global behaviour and properties of the running system, over a period of time, using a language $\mathcal{L}2$.Surprise: The language of design $\mathcal{L}1$ and the language of observation $\mathcal{L}2$ are distinct, and the causal link between the elementary interactions programmed in $\mathcal{L}1$ and the behaviours observed in $\mathcal{L}2$ is non-obvious to the observer — who therefore experiences surprise. In other words, there is a cognitive dissonance between the observer's mental image of the systems design stated in $\mathcal{L}1$ and his contemporaneous observation of the systems*

behaviour stated in $\mathcal{L}2$. These three conditions relate to design, observation, and surprise and reduces the concept of emergence to a shortcoming in language of make and of observation. Using this test Ronald and Sipper show that systems employing ANN and Evolutionary AI are certain to exhibit emergent behaviours. They further apply the above to various systems, viz. (1) nest building in a simulated wasp colony, (2) coordination in an ant colony, (3) Braitenberg vehicles 1,2 and 3, (4) In Minsky's society of mind model etc. This test also suggests that emergence is dependent on the sophistication and awareness of the observer.

9. *Renowned roboticist Masahiro Mori wrote that 'robots have buddha nature', motivation from religion is not what technology looks upto, however Robotics and Buddhist Principles seem to correlate well. Where, (1) enlightenment in Buddhism is attaining higher levels of consciousness with meditation and self reflections, is a direct correlation to the notion of emergence of cognitive ability in behaviour-based robotics, (2) Buddhism advocates that bodily experience is needed in order to experience a self aware mind, embodiment is an essential paradigm in embodied cognition and (3) the state of 'craving self' is required for attaining threshold of self awareness, in robotics efforts by Tilden and Hasslacher has demonstrated that an AI craving for survival also leads to the development of sensory-motor based emergent behaviours. Hughes has extended this comparison to suggest models for articial moral beings on Buddhist principles.*

10. *The S* Architecture, Tsotsos developed a vision based approach cycle. meant to overcome the representational shortcomings in susumption architecture, instead of modelling each behaviour as finite state machine, each behaviour is structured by modifying the the SPA cycle to yield a SMPA-W.*

11. *Behaviour based power management, utilising a three layer architecture with UBF to model the controller, Fetzek has developed a method to economise battery power for mobile robots. Fetzek suggest a tree traversal method, by which it is possible to determine which onboard sensors are not required. Those which are not needed for the current, active set can be powered down. Power consumption is further economised by the availibility of a loweR power mode below a certain power threshold, in this mode high power consuming onboard devices are forced o and highest priority is given to the current task or efforts to acquire more power, viz. a totter to the closest recharge station. For multiple sensors that provide near similar functionality the higher power consuming sensor would be powered down in favor of the lower power consuming. The above paradigm is augmented with a 'power plan', where the robot AI can conserve power in anticipation of high power requirements in the immediate future.*

EXERCISES

1. **ANIMAT to robots**, I have briefly discussed an ANIMAT approach for a line follower. A contrasting engineering approach to contain the robot to a given path would be to devise tracks for the robot, in the manner of trains moving on railway tracks. Discuss the pros and cons of the two approaches. How will you quantify performance?

2. **Mine Sniffing Robot.** For a subsumption-like mine sniffing robot which has 4 basic behaviours; **Wandering, Obstacle Avoidance, Plume Tracking** for 'sniffing' the plume and tracking the mine and **Diffuse** to render the mine harmless, what problems can the control structure given in figure pose? How can this architecture be modified for more than one mine?

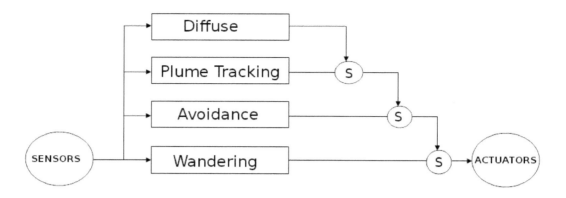

FIGURE 3.33   Exercise question - Mine sniffing robot

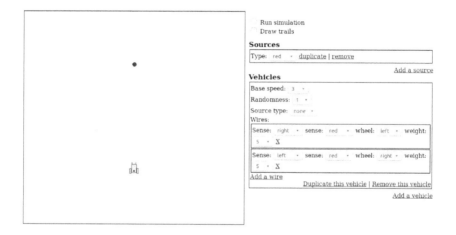

FIGURE 3.34   Exercise question — Braitenberg Simulator.

3. **Dennet's robots**, R1, R1D1 and R1D2 from the previous chapter suffered from the frame problem and were bombed to death. Can they execute their task and also save themselves from the time bomb if they are designed with a hybrid architecture as AuRA?

4. For a humanoid robot playing football, the architecture is designed as,

$$Foot\ position\ at\ time\ t = Ball\ position\ at\ time\ t \qquad (3.2)$$

what are the shortcomings of this architecture? What all behaviours will the humanoid robot exhibit?

5. **Braitenberg simulator**, consult appendix A and run the braitenberg simulator as shown in Figure 3.34 and try out some of the vehicles.

II

Implementation, or How to Make Robots

CHAPTER 4

Tools for a roboticist

"We are all on explorations into the unknown and it is by looking backward in the wake of our trail that we can see the course we are taking."

– Thor Heyerdahl, in *In the Footsteps of Adam: A Memoir*

"I'm afraid that the following syllogism may be
used by some in the future:
Turing believes machines think,
Turing lies with men,
Therefore machines do not think
Yours in distress,
Alan"

– Alan M. Turing

4.1 THE TOOLS: NAVIGATION AND ADAPTIVITY

DESIGNING robots cannot be pegged to a process or a series of rules, as agency is uniquely determined by the job at hand and the local environment. However, two basic concepts remain nearly the same, (1) navigation and its variants, which are also the most basic set of behaviour, viz. obstacle avoidance and motion to goal and (2) adaptivity implemented with machine learning techniques. Other than these two, there are design takeaways if the same hardware component or design paradigm is re-used and convergence in sensor deployment and fusion methods.

Simon [304,305] demonstrated that the agent's ability to make a rational choice increased with the number of discreet moves it can see ino the future — the ability of farsightedness. His paper concludes that, for a hypothetical animal random locomotion in the hope of surviving by finding a source of energy and replenishment, etc., can be disastrous. Therefore the hypothetical animal will have to search for clues in the environment, which are apparent or at least anticipated by it. It has to utilise such clues to seek routes which have a sufficiently high probability of survival. Toda's fungus eaters were designed with particular attention to the environment, strong focus on adaptability and employed locomotion as a means for survival by collecting fungus, and also to simultaneously pursue other goals which for the fungus eater was collecting uranium, and over time, they could adapt to their surroundings and improve performance. Wilson's ANIMAT design demonstrated that an adaptive algorithm reduces the number of steps needed to find food. These results puts an onus on both (1) intelligent locomotion and path planning and (2) adaptability and learning from experience, as essential tools for the design of an autonomous agent.

Path planning and navigation algorithms are attempts to address quality tasks in robots than mere reactive one and bring more predictibility to the robot behaviour and design. Navigation stands out as an essential tool and also the most fundamental robot behaviour, viz. line follower robot, micromouse maze solvers, simple obstacle avoidance etc and nearly all robots, from the simple ones to the most sophisticated ones, have navigation as the basic behaviour, with modules often coded in by the manufacturer.

Configuration space or C-space is an important aspect for navigation and is defined as the the hypothetical space formed from all the possible configurations of the robot. The dimensions of configuration space is the number of degrees of freedom of the robot. For a mobile robot the configuration space is a two dimensional plane and this formalism considers the robot as a point and obstacles as 'globs' on a two-dimensional map, and thus reduces the problem of robot navigation to path planning in two dimensions.

Using grids to granulate the configuration space is an easy option, as shown in Figure 4.1 to implement a 'divide and conquer' algorithms and methods. Grid-based algorithms such as 'breadth first' and 'A∗' will always find a solution if one exists; however these algorithms usually gives suboptimal paths, and are best for simple navigational tasks, and may not be very useful in complex tasks, viz. collecting empty Pepsi cans, follow the red flags etc. Neither can they be used for a gradually discovered area as in surveying nor are they advisable for safety critical tasks and rescue operations. As a further downside grid-based techniques are intractable for higher dimensionality. For example a mobile robot fitted with an arm manipulator will have two degrees of freedom for the mobile base and three or higher degrees of freedom for the arm manipulator and therefore the a configuration space of five or higher degrees cannot be visualised or operated with with simple mathematical operations.

If the shape of the robot is to be considered then the configuration space has to be modified by adding in the robot dimensions to the obstacle boundary. Shown in Figure 4.2, a circular robot which has to overcome a square-shaped obstacle leads to modification of the configuration space as shown. This 'grows' the obstacle to a bigger rectangle with rounded edges.

Since, planning is NP hard it is not always easy to find a path to the goal in all terrains and situations. Nearly every navigational method has a path planner to negotiate the thoroughfare, and therefore varies with the complexity of the environment and is context dependent, viz. collecting all Pepsi cans is a very different job than finding the shortest obstacle-free path to a goal point. Both jobs pertain to navigation and are meant for the same agent in the same environment. The first one needs an exhaustive search across the given area, reaching the various coke cans and the picking them up using an arm manipulator and collecting them in a container attached to the robot, the second in contrast will be a greedy approach to reach the goal point. Therefore, a single master algorithm for all purposes is not possible.

Methods are further tagged to performance, cost and computational consideration. Bug algorithms and A∗ are the simplest one and potential field methods and Nearness Diagram (ND) navigation helps to preserve the simplicity. Dynamic Window Approach (DWA) method and Monte Carlo techniques are more sophisticated methods. In addition, mapping is often employed by researchers, along side navigation. This chapter is a peek into these tools and is divided into two sections, Navigation, Path Planning and Mapping, and Adaptive Techniques.

Navigation is a vast topic and cannot be covered in a short chapter and the reader is

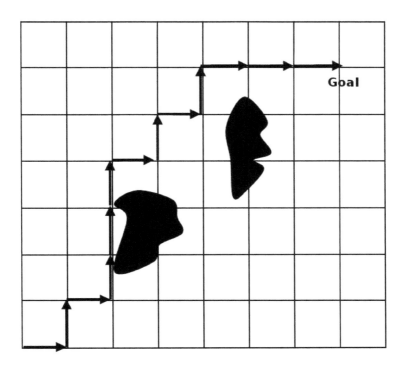

FIGURE 4.1 **Granulating space using two-dimension grids** has been popular with researchers. Here the robot is a dimensionless point and the obstacles are globs. As shortcomings, this method (1) gives suboptimal paths and (2) can only be useful for simple tasks and (3) cannot be extended to higher dimensions.

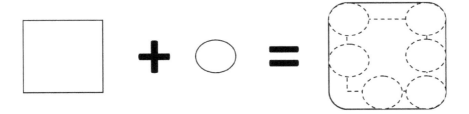

FIGURE 4.2 **Concept of configuration space.** For a circular robot to negotiate around a square obstacle, a collicion free course will need to take into account the robot's size and therefore in the configuration space the rectangular obstacle will be a larger rectangle with rounded edges.

encouraged to consult splendid texts such as Latombe [199], LaValle [200] and Choset et al. [68].

4.2 NAVIGATION, PATH PLANNING AND MAPPING

Safely avoiding obstacles to find a safe route to a desired goal is the simplest task for a robot. Navigation is also the most basic of human behaviours and is earmarked to find safe shelter, foraging for food, escaping a predator etc. In their primitive form navigational methods resort to observing the position of stars, following a scent, observing flow of water etc. In contrast the modern tools are maps, milestones and GPS. Some of the interesting examples of navigation both from history and popular culture are (1) in the Minoan legends the minotaur is a beast with a bull's head and a man's body, and it is contained inside a labyrinth and cannot find its way out, (2) in Grimm's fairy tale of Hansel and Gretel, where the two kids drop sweets on their way to the woods so that these can serve as landmarks for finding their way back home; and (3) history of course boasts of adventurers and seafarers who were guided by the stars, will power, greed and mercantile goals to reshape the globe forever. In modern times, other than autonomous robotics, navigation is also of great significance in defence systems, air travel, ships, space exploration, animation etc. Therefore, it can be said that navigation and path planning are at the root of nearly all cognitive processes and complex human actions are a synchronisation of such simple navigational modules.

For mobile robots, the simplest behaviours are 'wandering around', 'obstacle avoidance' and 'motion to goal'. More involved behaviours such as 'mapping' and 'rendezvous' are all variants of navigation and path planning. Generic problems of path planning, such as 'piano mover's problem' and 'escaping from a maze', have been tackled by mathematicians and logicians using global navigation methods which are methods that rely on an a priori map. In contrast navigating a mobile robot with onboard sensors is a much different problem, as it has to be tackled as local problem in real time, with sensorimotor capabilities. Though, nowadays nearly all state-of-the-art robots have both local and global information at their disposal, it is worth remembering that path planning in its most general form is time exponential in the robot's degrees of freedom, and on incorporating dynamics of moving obstacles, the problem becomes exponential in number, thus rendering, generic navigation as an NP hard problem.

4.2.1 A∗ and bug algorithms

In late 1960s, Nilsson navigated Shakey using the A∗ algorithm in a grid-based environment. It was apparent that primitive reactive approaches as '*See obstacle, Turn Left*' or '*If Goal is Visible, Move to Goal*', fails to incorporate the versatility of the terrain and dynamics of real-life navigation which has uneven terrain, other agents and moving obstacles. Neither, can such simple reactive modules be used for interesting tasks; as viz. collect red balls, stop on seeing a yellow flag etc.

The Bug algorithms were developed by Lumelsky and Stepanov [211, 212] and were the earliest sensor based planners. They considered navigation over a two-dimensional space, a point robot, a finite number of obstacles of finite size, where the robot knows its own coordinates and also that of the goal and it can measure the distance between any two points. The local information about the obstacles is known using a sensor, viz. laser, camera, tactile etc. two variations named as Bug-1 and Bug-2, as shown in Figure 4.3. Bug-1 is designed as an exhaustive search algorithm while Bug-2 is an opportunistic variant designed with a greedy algorithm.

All Bug algorithms and their later derivatives essentially incorporate two behaviours; 'motion to goal' and 'boundary following'. Over the years modification of these two algorithms has added novelty, some popular ones are, Tangent Bug, Polar Bug, Optim Bug, Pot Bug, I Bug etc or they have been used along with another path planner as a hybrid approach. The Bug algorithms were designed on anthropomorphic motivations from insect navigation, and used addition of behaviours, akin to the reactive paradigm, but it considered the problem more from a global point of view, than a local one. The broad limitations of the Bug algorithms are, (1) it is rarely that the location of the goal point is known. More often it is not very clear and the goal is usually identified by incomplete information and corresponds to a gradually discovered environment, viz. stop on seeing a red flag, stop after crossing the eight obstacle etc., (2) navigation is an explorative process and the distance between any two points can never be known with certainty due to the unevenness of the terrain and sensor errors, (3) Bug algorithms are for two dimensional static obstacles and will not be very successful for moving obstacles, viz. other robots and human beings, etc.

4.2.2 Considerations for navigation

Over the last five decades, research has attempted to bridge the gap between the sterility of the laboratory or the ideal settings of a simulation, and the real-time occurrences of the real world. The following aspects helps to assess the performance of a path planner:

1. Embodiment and representation, the structure of the environment, the robots mechatronic design, its shape etc. Motion confined to a single plane, or two-dimensional motion is usually easier to plan and navigate. Motion which occurs in three dimensions is difficult to plan and design and there is yet a lack of good methods. A typical cube world or box world of simulation, where each and every obstacle is a cube is easier to navigate; however, the real world has a variety of obstacles and the challenge and also a test of performance for the robot is to navigate through a cluttered environment. Performance of a method often varies with the obstacle, and more complicated and highly concave or U-type obstacles as shown in Figure 4.4 are difficult to negotiate and therefore stand to testify the quality of a method.

2. Space or time complexity, the computational resources, viz. the sophistication of the onboard microcontroller, the amount of data storage and hence the time needed to find the solution.

3. Completeness. Is the method guaranteed to find a path if one exists and report if there is none? The algorithm should be such that the robot 'gives up' after a finite number of attempts and reports the lack of a solution, and thus prevents an unending process where the robot keeps fseeking a path until it runs out of onboard power.

4. Soundness. Will the method consistently provide a collision free path? This is a particular concern for autonomous vehicles such as Google cars which have to find a collision-free path in real time with moving obstacles.

5. Optimality. Is either defined as minimum length or minimum path, depending on the requirement and it is rarely possible in real time, as is shown in Figure 4.5. Nevertheless, the actual path versus optimal path helps to improve the given method.

6. Reactive layer. Nearly all algorithms derive from a reactive action, viz. Bug algorithms used overlap of simple behaviours. Potential fields are reactive in nature and ND is

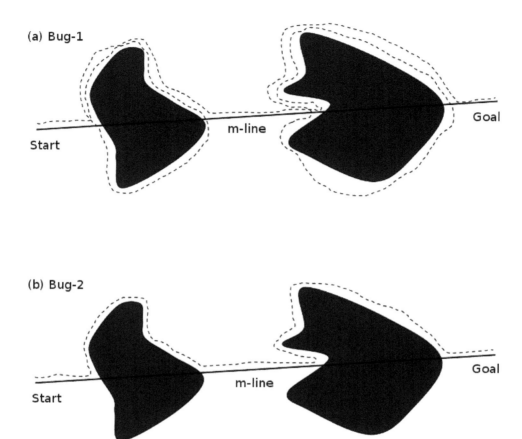

FIGURE 4.3 **The bug algorithms.** The robot's dimensions are ignored and considered to be a point – point robot, the dark globs are the obstacles, the m-line is a hypothetical line joining the start and goal points and the dotted line shows the path of the robot from start to goal. In (a) in Bug-1 the mobile robot starts with 'motion to goal' behaviour and on reaching the first obstacle switches to 'boundary following'. After traversing all of the boundary, when the mobile robot again encounters the m-line it again takes a 'motion to goal' behaviour and moves towards the next obstacle. In (b) Bug-2 the mobile robot starts with a 'motion to goal' behaviour and on reaching the first obstacle switches to 'boundary following'. In the next encounter with the m-line the robot switches to 'motion to goal' and moves towards the goal and encounters the next obstacle. The 'motion to goal' behaviour is the same in both algorithms, however when executing 'boundary following', Bug-1 is exhaustive while Bug-2 is driven by immediate conditions — a greedy paradigm.

based on situated activity, and most of the state-of-the-art robots employ this reactive layering with a global planner which may be accompanied with a remote human override. The reactive layer is essential, as it determines the robot's local interaction. and the robot's performance to the specific task; for example, the reactive layering for a robot driven by olfactory is different than one which is following red flags.

7. Mapping has become essential to navigation. It is not dependent on sleek and expensive hardware and it helps in surveying unknown terrain and works in tandem with localisation.

8. Task criticality. Safety is always a consideration, critical tasks with stringent deadlines or tasks which may need to be stopped abruptly need different programming considerations. A good example is a robotic ambulance carrying fatally injured patients who may need medical attention immediately, such a vehicle will not only need to find the fastest path with the least traffic but also adhere to a very strict deadline.

9. Kinematic considerations. Most planners consider the robot in two dimensions, as a point or a circle, with sensors arranged in circular symmetry, which is hardly the scenario, as robots have a body which is not always circular and sensor which are not always arranged in circular symmetry. Extending simple algorithms for car-like robots leads to discrepancies such as, any position of the car can be represented in two dimensions as (x, y, θ) as is shown in Figure 4.6, which is the requirement for most path planners, but this is not consistent as the car can change the orientation of its wheels with steering, influencing its linear velocity and the subsequent motion. Motion for vehicles with differential drive and steering is non-holonomic and tends to limit the free area in the configuration space. Analysis of such motion is difficult.

The divide and conquer strategy is popular throughout autonomous navigation. This approach, breaks the given problem into subproblems that are easier to solve, as is shown in their classification scheme for robot navigation in Figure 4.7. These subproblems are usually similar to the original problem but are easier to solve, less complex and often spread over a smaller domain.

Decomposition of the two-dimensional physical space into cells via a regular grid, random trapezoids, visibility graph and voronoi diagram are probably the easiest methods. As shown in Figure 4.8, the given area is sliced into a number of polygons and numbered; the path to goal is equivalent to traversing a graph and warrants a collision-free course.

Alternatively, the given task can be broken down into smaller tasks. Subgoal approaches [34, 346], as shown in Figure 4.9, select positions in the immediate neighborhood of the agent's current location in the prospect that this may be more easily achieved than a distant goal. Over a number of iterations, and with suitable selection of subgoals, such approaches can incrementally make progress in moving the agent towards the long-term goal. In contrast to the regular, a priori given structure of a grid decomposition, subgoal approaches often dynamically fragment their configuration space; essentially a tradeoff between the classical discrete and continuous forms navigation. The earliest subgoal technique was SSA, proposed by Krogh in later 1980s. Studies have shown the improved performance due to subgoals.

4.2.3 Artificial potential fields

Potential field navigation is a cornerstone paradigm for robotics. This method of employing artificial forces to navigate autonomous sensor-based agents was developed by the pioneering research of Khatib [177]. He developed this method in his doctoral dissertation for arm manipulators which was later extended it to mobile robots [178]. This approach was

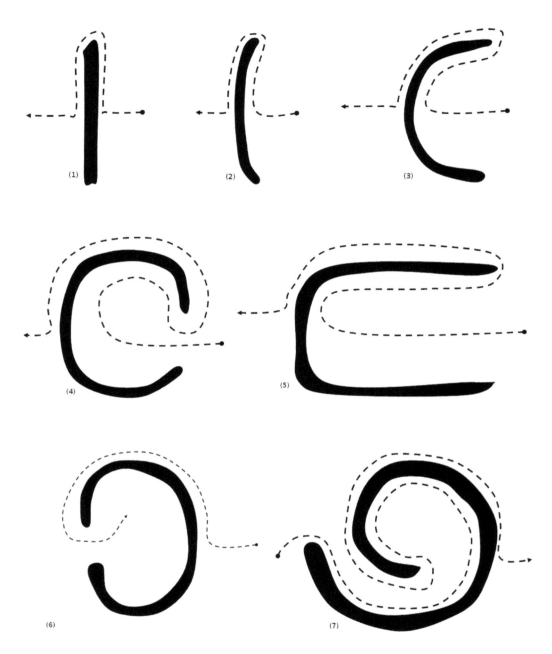

FIGURE 4.4 **Classifying obstacles.** Testing out an algorithm against various types of obstacles is one of the ways to ensure and improve its performance [34]. The discussion can be brought down to broadly seven types of obstacles, (1) flat obstacle, (2) shallow concave obstacle, (3) moderately concave obstacle or C-type, (4) highly concave obstacle, (5) extended C-type or U-type obstacle, (6) circumscribed obstacle and (7) spiral obstacles. Sophistication in the algorithm, higher complexity and more computational resource are needed for at least the last three types of obstacles.

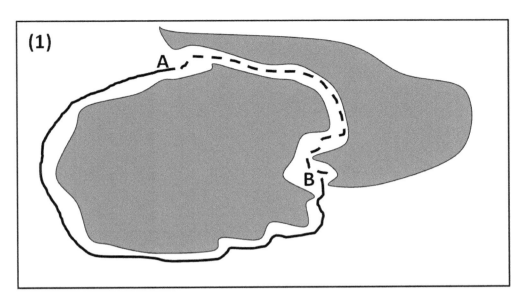

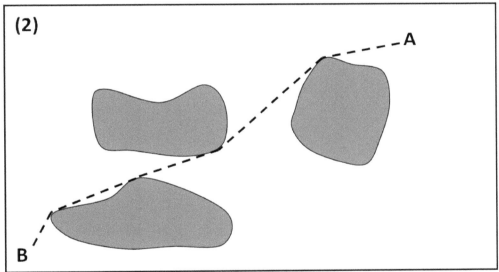

FIGURE 4.5 **The brunt of optimality.** Optimality is hardly achieved in autonomous navigation, as shown in (1) the dotted line from point A to point B is the optimum distance, however this path between the two obstacles is very narrow, and without a second layer of global planning or remote human control it can easily get lost. It also faces safety issues where any extended part of the robot such as arms or antennae etc. can get smashed into the walls. Therefore the longer path shown by the continuous line is favoured. In an optimal path, as shown in (2), where the robot traces the path from A to B, from the edge of one obstacle to another, until goal point is reached via five short dotted line segments, this is similar to concepts of the visibility graph but it will not be the natural explorative action of an agency, and in a motion like this the robot will run the risk of crashing into the obstacle. Optimality starts to matter if the robot is repeating a navigational circuit or a known number of behaviour(s) in a restricted situation, such as a guard dog robot. Machine learning techniques help to enhance performance and thus attain optimal performance.

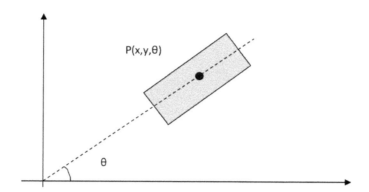

FIGURE 4.6 **x,y and** θ, are the parameters needed to represent a mobile robot moving in a plane. θ has been referred in as pose, heading angle and angle of orientations by researchers. For non holonomic systems as a car, the above represents incomplete information as the car can change its trajectory by changing the orientation of its wheels by steering.

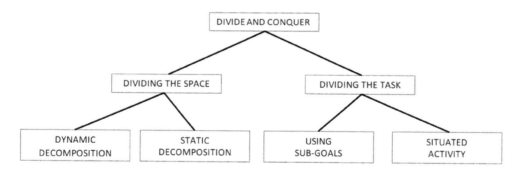

FIGURE 4.7 **Divide and conquer** is a popular technique and can be classified as shown. Divind the space can be done as a dynamic decomposition which is on the run such as dynamic bubble and dynamic window techniques which are structured on sophisticated algorithms, or as static decomposition which is more or less a priori, such as simple grid based methods, Voronoi diagrams, Trapezoidal decomposition etc. Using subgoals [34, 346] with a known reactive technique as potential field method or ND navigation has helped to maintain some simplicity and yet on the run performance. ND diagram is based on situated activity and has been hugely successful in two-dimension navigation contained to a plane.

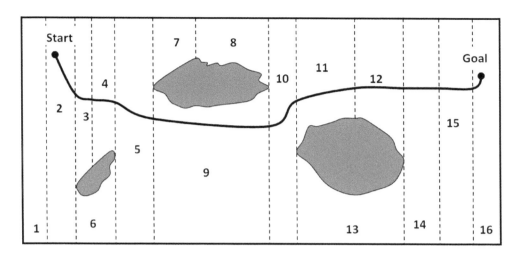

FIGURE 4.8 **Cell decomposition**, the given area is sliced into a number of polygons (as shown with dotted lines) and navigation reduces to the problem of traversing a connectivity graph. The robot has to pursue from the start point (**in 2**) to goal point (**in 16**), the corresponding connectivity is 2-3-4-5-9-10-11-12-14-15-16.

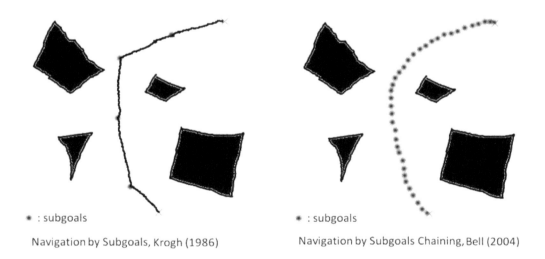

● : subgoals   ● : subgoals

Navigation by Subgoals, Krogh (1986)   Navigation by Subgoals Chaining, Bell (2004)

FIGURE 4.9 **Using subgoals**, is implementation of divide and conquer by dividing the task of reaching the goal. Subgoal chaining has been an interesting application by placing subgoals at very short intervals [34].

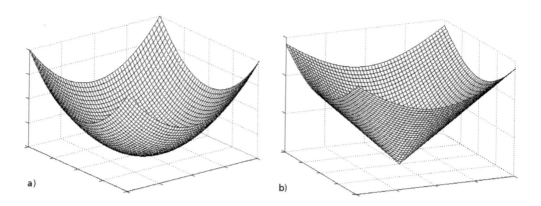

FIGURE 4.10 **Goals as attractive potentials**, a) parabolic potential: $z = \rho^2$ and b) conic potential: $z = \rho$.

influenced by concepts of physics; the inverse square law, harmonic oscillator, conic potential, geometrical constraints, and Lagrangian dynamics. This was later extended by the works of Koditschek, Khosla, Connolly, Chatila and Latombe and groomed into algorithms and heuristics. The method is easy to implement and a computationally favourable on-line collision avoidance approach for robots and is intended for situations where there was no prior model of the local environment, i.e., the agent reactively detects the environment during the motion execution, and therefore it is not tied to expensive hardware. As discussed in the last chapter, Arkin's motor schema uses potential fields for combining behaviours in a reactive architecture, which makes the potential fields method unique as a tool for navigationFr as well as for implementing behaviours. The potential field approach was designed for robotic arm manipulators, and was later extended to mobile robots. The robot is treated as though it is under the influence of a potential field U which is modulated in real time to represent the environment. The method can be easily explained through analogy to electrostatic fields as follows. The robot is modelled as a charged point having the same charge as the obstacles in the environment, and an opposite charge to the goal position. This causes a repulsive force between the robot and the obstacles, and an attractive force between the goal and the robot. Samples of (1) attractive potential suitable for goal point are shown in Figure 4.10 and (2) repulsive potential suitable for obstacles are shown in Figure 4.11. This associates high potential value with obstacles, and a low potential value near the goal. The two potentials are superimposed to form the net artificial potential that the robot is subjected to.

$$U = \sum U_{obstacles} + U_{goal} \tag{4.1}$$

This corresponds to the artificial force field,

$$\mathbf{F} = -\nabla \mathbf{U} \tag{4.2}$$

The robot's motion is then executed by means of an iterative gradient descent driven by the local force;

$$q_{i+1} = q_i + \delta_i f(q) \tag{4.3}$$

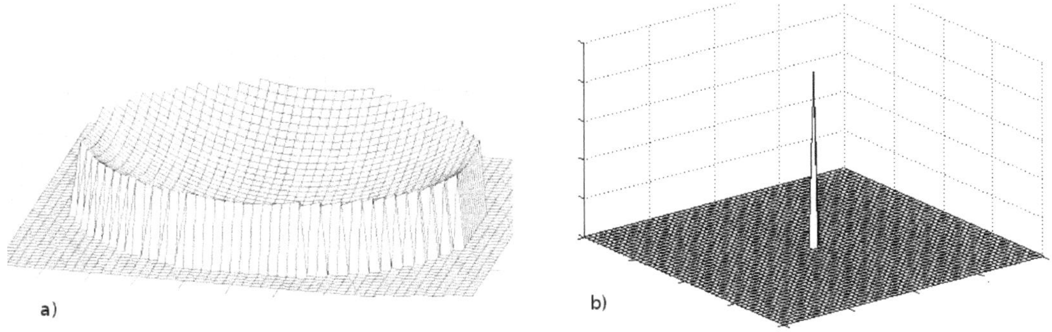

FIGURE 4.11 **Obstacles as repulsive potentials**, a) a cylindrical obstacle and b) a point obstacle.

where for the ith iteration q_i is the coordinate, δ_i is the step size and $f(q)$ is the normalised force;

$$f(q) = \frac{F(q)}{\|F(q)\|} \tag{4.4}$$

Therefore the robot is in a constant effort to minimise its potential, and consequently moves 'towards-the-goal' and 'away-from-obstacles' simultaneously, adding in at least two behaviours at run time. The repulsive potential is modelled as;

$$U_{obstacles} = \begin{cases} \frac{1}{2}\eta(\frac{1}{\rho(x,y)} - \frac{1}{\rho_0})^2 & \rho \leq \rho_0 \\ 0 & \rho > \rho_0 \end{cases} \tag{4.5}$$

Here $\rho(x,y)$ represents the Euclidean distance from the nearest part of the obstacle, ρ_0 represents the distance at which the obstacle ceases to have an influence on the agent, and η is a scaling constant representing the strength of the repulsive force. Khatib named this potential function 'FIRAS' which is an acronym for the French expression for 'Force Inducing an Artificial Repulsion from the Surface'. A parabolic bowl potential function is used to model the goal,

$$U_{goal} = \frac{1}{2}\xi\rho_{goal}^2(x,y) \tag{4.6}$$

where ξ is a scaling constant representing the strength of the attractive force. This model is computationally lean yet allows effective real time implementation, the formulation remains a favourite in the robotics community due to its simplicity and ease of implementation.

The reader may observe that potential field approach is another method of world modelling — though on the run, and works only for a sufficiently known and predictable environment. The limitation of the potential field approach is primarily the (1) local minima problem as shown in Figure 4.12 since the robot's motion is due to two opposing forces, therefore when the sum of these two forces is zero, the robot comes to a stand still with no effective force acting on it. Such local minima act as 'traps', and once in a local minima the

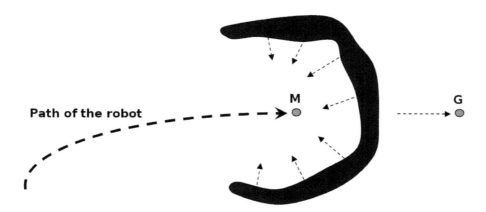

FIGURE 4.12 C-type obstacle and local minima, C-type, U-type obstacles or the 'box-canyon' problem are not easy to overcome, and simple algorithms as bug-algorithms and breadth-first etc fails to perform, for potential fields, the problem intensifies to a local minima problem. The total force acting on the point robot is the sum of repulsion from the obstacle and the attraction from the goal point G, this guides the robot into the C-type obstacle. At M, the attractive force exactly cancels out the repulsive force creating zero force, and the robot abruptly comes to a standstill at the local minima. Adapted from Latombe.

robot stops abruptly prior to reaching the goal. Research in potential field based navigation has often consisted of attempts to overcome or avoid these local minima. Other shortcoming of this method is, (2) the robot struggles to travel in narrow pathways as it faces repulsion from both the walls and therefore with the fast changing gradient the robot oscillates. Similar oscillations have been reported when the robot passes by a large obstacle and (3) the method constrains the robot from any movement directly towards an obstacle, and therefore when the goal point is close to an obstacle, it fails to reach it or is stuck in an infinite loop.

There have been several efforts to overcome the local minima problem. Redesigning the potential function for elliptical potentials reduces the impact of local minima, however this does not remove minima from the field and is not very beneficial in cluttered environments. Another approach is to use a second path planner when the local minima is encountered. The most common of these techniques is 'backtracking', when local minima is encountered and avoiding that local minima when the agent has sufficiently moved away. All local minimas once encountered should be avoided in at least the near future. This is seen in, Random Path Planners (RPP) or Simulated Annealing (SA) where the robot takes a random number of steps in the physical space in attempts to walk out of the local minima.

A good lot of implementations use potential fields with a grid-based planner. Choset et al. suggest the wave front planner as the simplest solution for the local minima problem, as shown in Figure 4.13. Another very popular implementation is the Vector Field Histogram (VFH) where the attraction/repulsion limited to the nearest neighbours and therefore easier to design. Discretisation of space has been a tool used with potential fields time and again. Some methods resort to physics, and exploit concepts of electrostatic potentials, harmonic fields etc. It was in such a context that Connolly et al. observed that with suitable grid-based methods, it is possible to arrive at potential fields which are 'nearly free' of local minima and named them navigational potential fields. These fields are solutions of Laplace's equation ($\bigtriangledown^2 \phi = 0$) [186], and they do not possess any local minima. These solutions bear analytical

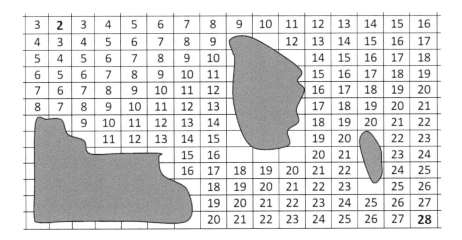

FIGURE 4.13 **Potential field navigation with wave front path planner.** In this method the wave front algorithm assigns a low number to the goal point (**cell 2**) and all grid positions away from the goal get a higher number with no numbering for grids having any part of the obstacle, until start point (**cell 28**) is reached. Once, the wave front has designated the grid as an array of numbers the potential field method works as a gradient descent method. The obvious shortcomings are, that the method is not useful for surveying or a gradually discovered environment, information from the obstacles is reduced to a filled grid point, and such methods are a compromise on the run time reactive performance of potential fields.

similarity to electrostatic, magnetic and flow fields. The Electrostatic Potential Field (EPF) approach treats the problem of navigation as an analogy to the flow of electric current in a resistor grid. EPF is modelled as a discrete resistor network channeled by an electric current which always takes the path of least resistance. The electric current is the maxim for model of navigation along the path of greatest negative gradient descent, and hence the solution of a grid-based resistor network is a solution of the Laplace's equation. It is free from local minimas, the only extreme points, being the source which is the mobile robot and the sink or the goal.

These solutions to overcome the local minima problem (all grid based methods and particularly both navigational fields and harmonic fields) are computationally cumbersome and expensive and are not very suitable for a gradually discovered environment or when the goal point is not very clearly known, and backtracking may not be advisable in emergency or safety critical tasks. All of these these methods are a compromise on the simplicity and the reactive performance of the potential field method.

Another weakness of potential fields is the agent cannot reach the goal point if it is located close to an obstacle. This limitation was coined by Ge et al. [118] as the GNRON (Goals Non Reachable with Obstacles Nearby) problem, and the solution proposed was a modification of the FIRAS potential, such that there is a global minimum at the goal point.

$$
U_{obstacles} = \begin{cases} \frac{1}{2}\eta(\frac{1}{\rho(x,y)} - \frac{1}{\rho_0})^2\rho^2(x,y) & \rho \le \rho_0 \\ 0 & \rho > \rho_0 \end{cases} \tag{4.7}
$$

Despite the various issues, potential field navigation and its variants are often used both by students and researchers as they can be easily implemented. Various robotic software suites has potential fields as a default tool for reactive navigation, and they are also a favourite for robocup soccer tournaments.

4.2.4   Nearness Diagram (ND)

A wonderfully simple and easy to implement situated activity approach for 'motion to goal' while 'avoiding obstacles' was designed by Minguez and Montano [234]. Their approach was based on classifying the various situations encountered by a sensor based agent in its interactions with the environment. The exhaustive number of situations where the agent interacts with the environment has to be identifiable from sensory perception, viz. sonar, laser or infra red. The Nearness Diagram (ND) defined five situations as shown in Figure 4.14, later this was extended to ND+ for six situations. Each situation corresponded to a two dimensional configuration of a cluttered environment. A free walking area and a security zone are important determinants for the algorithm. The free walking area is calculated as follows: gaps are identified over an angular distribution in the obstacle field, the closest region to the goal location which is navigable by the robot is then designated as the free walking area. The security zone is an arbitrary circle which centers on the agent. The radius of the security zone is crucial for the efficiency of the algorithm; an optimum radius is essential for smooth obstacle free path generation. The sensor data is classified by a decision tree. The inputs to the decision tree are robot and goal location and obstacle distribution. The output is the current situation. The following four criteria define the navigation schema:

1. Criterion 1: Safety criterion, two categories; identified as low safety when there are obstacles within the security zone, otherwise as high safety.
2. Criterion 2: Dangerous obstacle distribution criterion.

   (a) Low Safety 1 (LS1): Obstacles in the security zone are only on one side of the gap (closest to the goal) of the free walking area. The action moves the agent away from the obstacle and towards the gap (closest to the goal).
   (b) Low Safety 2 (LS2): Obstacles in the security zone are on both sides of the gap of the free walking area. The action enables the direction of navigation of the agent along the bisector between the two closest obstacles on either side.

3. Criterion 3: Goal and free walking area criterion

   (a) High Safety Goal in Region (HSGR): The goal is located within the free walking area; otherwise the remaining two options (HSWR and HSNR) should be applicable. The action drives the agent towards the goal.

4. Criterion 4: Free walking area width criterion. A free walking area is wide if its angular width is larger than a given angle. If not, the free walking area is narrow.

   (a) High Safety Wide Region (HSWR): The free walking area is wide. The action moves the agent alongside the obstacle.
   (b) High Safety Narrow Region (HSNR): The free walking area is narrow. The action directs the robot through the central zone of the free walking area.

All agent-environment interactions can be tallied to these five situations. The situation definition does not depend on the resolution or size of the space considered. ND+, attempts to refine ND with an added state, Low Safety Goal in Region (LSGR). Both ND and ND+ are very successful negotiating in cluttered environments. Sin they are structured on situated activity it embraces reactive paradigm and can thus respond to a dynamic environment. ND is an improvement over potential fields, it avoids trap in U-shaped obstacles, it has an

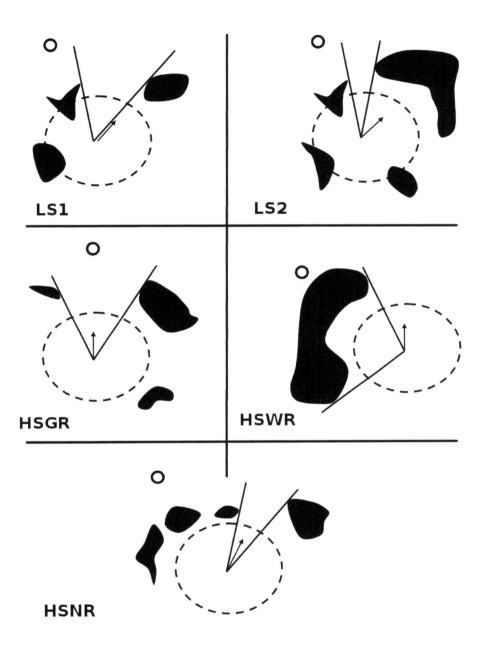

FIGURE 4.14 **ND diagram navigation** is designed on 5 scenarios — two with low safety (LS1 and LS2) and three with high safety (HSGR, HSWR and HSNR). The dark globs are the obstacles and the dotted circle is the security zone which needs to be appreciably more than the geometrical size of the robot. The robot is not shown here, but represented as the centre of the security zone. The goal point is shown as a BIG-O and the free walking area is the acute angle between the 2 line segments for LS1, LS2, HSGR and HSNR, and the obtuse angle for HSWR. The arrow shows the direction in which motion happens according to the algorithm.

FIGURE 4.15 **3D navigation**, most navigation in three dimension is remotely controlled by humans and there is yet to be a good autonomous navigation technique in three dimensions. Images courtesy *commons.wikipedia.org*, CC-by-SA 3.0 license.

oscillation free motion while navigating among very close obstacles and narrow corridors and, unlike potential fields, it does not prohibit the motion of the robot directly towards the obstacles. All of these reasons made ND a robust and easy to implement method. The limitation of ND diagram method are, (1) it doesn't take into account non-circular shapes or vehicular kinematics and dynamical constraints, (2) the method is designed for circular symmetry and symmetric placement of the sensors — as a typical ring of lasers or sonar sensors, and relies on angular scan, which may be very difficult to implement for a non-circular robot such as a car which is also non-holonomic and (3) it doesn't address the issue of sensor noise in real time.

Some of the current issues explored by researchers in motion planning and autonomous navigation are: navigation (1) for a dynamic environment with moving obstacles, (2) for a gradually discovered environment, (3) for cluttered and unstructured environments, (4) for non-holonomic and under actuated systems, (5) pertaining to extension to higher dimensions and higher degrees of freedom, as navigation in three dimensions, (6) for multi robot navigation.

4.2.5 Navigation in three dimensions

Most areal robots are remotely controlled from the ground, as shown for drones in Figure 4.15, and there is yet to be a robust, efficient and safe method for autonomous navigation in three dimensions. However, generic navigation is indeed a problem of three dimensions and various applications in robotics deal with navigation in three dimension, such as exploration in disaster arenas or survey of unknown or gradually discovered environments as cave inspection and difficult terrains in unchartered planets. The debate of global vs. local planning is many fold more when considering three dimensions, since most global planners tend to view the problem in projections of two dimensions than a problem in three dimensions, therefore it is difficult to gurantee the convergence of both the planners to the desired goal point. There is yet to be good methods for navigation which balance the global and the local, and has concentrated more on path planning than the reactive behaviour. Navigation in three dimensions has been successful in designing virtual characters in game programming or virtual reality application, but there is a lack of navigation in

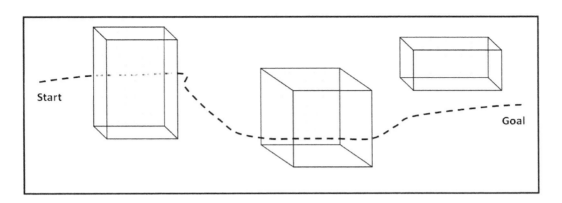

FIGURE 4.16 **Algorithm for navigation in three dimensions,** has been tried in simulations and simple environments in the research by extending Bug algorithms and A* algorithm for three dimensions, but a general approach to three dimension navigation is yet to bear fruitful results.

real time. Grid based methods are intractable in three dimensions, and simple heuristic as the A* in structured terrain, leads to an explosion in the number of nodes. Kutulakos et al. [195] and later Kamon et al. [167–169] attempted to extend the bug algorithms to three dimensions. Both the methods tried to combine the bug algorithms in two dimension with surface exploration, in the exploration mode the reactive behaviour is to be subdued. This approach yields results for very well structured environments, and has only been tried in simple simulations such as the block world, as shown in 4.16. This works out if the given terrain has been previously explored, but is not very useful for explorative tasks as surveying or disaster management etc. This problem has also been tackled is by using subgoals and subgoals used along with gradient descent has been a prospect, but it has yet to be groomed into algorithms. Multiple sensors and subgoals have been employed simultaneously, but that lacks in generalisation and such approaches either makes the agency too bulky or adds on to the computational resources — both of which are undesirable. There is yet to be a generic three dimension navigation [337] which can be implemented easily, which is low on computational resporces, works across a wide range of applications and is accepted across the robotics and AI community.

Mapping the local area using laser or sonars, over continued surveys has come to be closely associated with robot navigation. Simultaneous Localisation And Mapping (SLAM) starts with an empty two dimension map, and incrementally adds features to it over each sweeps of the given local area. Monte Carlo Localisation (MCL) attemps to solve the problem of localisation with

4.3 ADAPTIBILITY AND LEARNING

For a human being, repeating a job is usually easier than doing it for the first time and such repetitions helps to develop special faculties to execute the job without any real hassles. This enhances performance and therefore learning is an essential tenet of AI agency. For autonomous agency, Toda's fungus eater model, Wilson's model of ANIMATs and Pfeifer's design paradigm forcefully make their contention towards adaptability for long term performance over an assortment of tasks for the robot. Therefore, robots must be designed to learn from its experience of interacting with the environment.

As an ab-initio, learning can be seen as a mapping from a sensed world (sensors) to the next action to perform (actuator), and is usually accompanied with no previous experience and/or apriori knowledge as in behaviour based approaches, where the robot 'finds out' the world and hence relates to its own actions. However, it is not easy to have simple mapping of input sensory information to the action for the robot and the next state of the world, and such type of global reinforcement is not only time consuming but has been found to be redundant given the scope of current learning methods. Therefore, Brooks suggests learning for robots not as a method to cope with the world but rather as a tool to enhance its performance, where the robot is akin to a learning machine.

This also leads to the discussion that robots have to deal with far from pristine and often incomplete data sets as erroneous sensor data, damaged hardware components which includes sensors, uneven terrain, environmental conditions etc. As discussed in the previous chapter, hybrid architectures have been the choice of the designers to overcome such variety and dynamism offered by the real world. As an alternative choice to hybrid architecture, incremental learning has also found favour. This is because, (1) learning is easier to implement and nowadays in most of the robotic software suites it is provided as a ready made module in the robot's firmware, (2) unlike behaviour which is pre-programed, learning happens on the run and lends itself as a control policy and is more atomic in execution than behaviour and therefore (3) doesn't need to be designed separately as a behaviour nor does it need an executive for coordinating various behaviours, (4) learning works out across simple behaviours to the more abstract ones - even those which the architecture may not have accounted for and one doesn't need to resort to semi empirical mathematical models of the environment or time varying functions of the uncertainty which may be encountered by the robot.

From an engineering point of view, a robot given a specific task and more or less contains itself to a known environment, the robot designer knows the a priori information and representations needed by the robot, and unless there has been severe and sudden changes to the environment or the robot, in principle will not need to resort to incremental machine learning. However, researchers have also put attention to particular representations, as saving battery power, avoiding falling off cliffs and avoiding threatening natural formations, covering a certain stretch of terrain or finding a specific item as mineral ore, vegetation or a known landmark etc, so that repeating these feats become easier in the apparent future.

Learning can be manifestly employed in a robot to either provide apriori representations and information which will enable it either to readily interact with the world in a certain manner than to learn it afresh over time over a great number of attempts, or to enable it towards specific goals and help it survive for longer durations in difficult terrain. For simple behaviours as navigation, learning will help to modify representations of the world which will simplify the given task. Since low level reactive tasks are closely related to navigation, hence making a map becomes important for tasks which need repetition. For example, a sub goals technique which leads to the final goal will need the robot to learn to recognise the subgoal, viz. if the subgoal is marked with a blinking red light, the robot will need to behave nearly the same to all such subgoals, until the final goal is reached. Learning is closely related to intelligent behaviour and autonomy, and not only helps the agent to acquaint with the environment, but also is a remedy for sensor noise and faulty and damaged sensors and actuators. In various robot architectures, learning works in tandem with the control loop.

Learning for sophisticated tasks over long term implementation will lead to development of newer and advanced behavioural modules. As an example, the working of guard dog robot in an unmanned facility is briefly shown in Figure 4.17, here the robot is provided with simple behaviours and it has to adhere to a known calibration, that of the circuital path

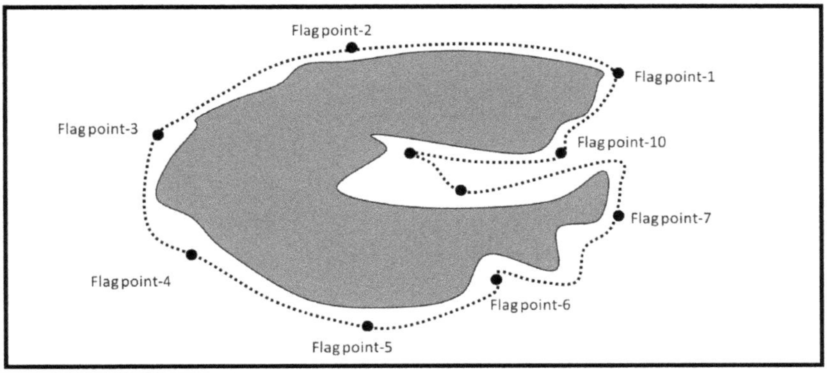

FIGURE 4.17 **Learning for a guard dog robot**, considering a robot patrolling an unmanned facility on a far off planet or at a difficult terrain which is difficult to access as a mountainous area or a hazard prone area. Here, the robot has go around the given circuital path (around the big grey glob) and sends photographs and information back to the the mission command at the dotted points (subgoal point) on the map. The loop taken by the robot (Flag point-1 to Flag point-10) for surveillance is nearly the same for all its pursuits and machine learning techniques (as ANN, Fuzzy logic) helps to keep the robot to its path for each loop and helps to avoid changes due to the diminishing onboard power, uneven terrain, weather conditions, environmental factors as storms, wear & tear, lack of regular maintenance etc. The robot was designed for simple behaviours as, 'motion to (sub)goal', 'obstacle avoidance' and 'take photograph', long term surveillance is a new and superior behaviour developed by machine learning techniques.

by using machine learning techniques, and over time it develops surveillance as a superior behaviour.

To summarise, learning helps to acquire the following features to robots [54]:

1. Adds in utility of particular representations to enhance performance, for the given example these special representations are the dotted points which serve as subgoals.

2. Helps relate special instances of sensor for action, or behaviour coordination, for the guard dog robot this sort of calibration between sensor reading to action can be seen, that it has to take photographs on reaching the subgoal.

3. Overcoming uncertainty, this is more appreciable if the robot is engaging in repetitive tasks in a predictable domain. As discussed, the machine learning techniques ties the circuital path to the robot's performance.

4. Gradual development towards newer and more involved behavioural modes, for the given example this is the development of surveillance as a new behaviour by melding simple behaviours.

The first application of ANN for robots led to improvements to the design of earliest models of autonomous vehicle. The NAVLAB vehicle was the outcome of research into autonomous vehicles in the late 1980s. Autonomous Land Vehicle In Neural Network (ALVINN) was the brainchild of Dean Pomerleau [268]. ALVINN was put to work on the NAVLAB vehicle. It was a refurbished army ambulance fitted with a 5,000 watt generator and had an operating system which could deal with 100 million floating point-operations per second. ALVINN had a three layered back propagation neural network which took images from a camera and data from laser range finder as input and computed the direction the vehicle should travel to continue to follow the road as output. Training was done using

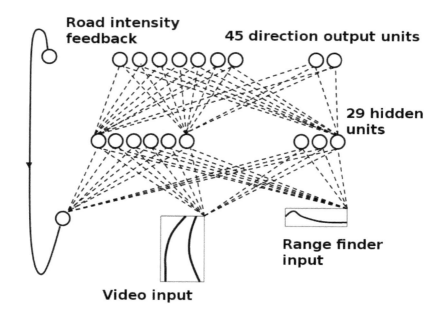

FIGURE 4.18 ALVINN architecture, was a three layered back propagation neural network the images from a camera and data from laser range finder were the two inputs and the intensity was used as a feedback, it computed the direction the vehicle should travel to continue to follow the road. Image adapted from Pomerleau [268].

simulated road images. Since the vehicle did not follow a predetermined path and the local variables guided towards the direction the vehicle should follow, it was adaptable across a variety of conditions and over various types of terrains. The highest speeds attained using ALVINN on the NAVLAB vehicle was about 70 mph.

Evolutionary approaches for developing control systems for robots can be done either by using real robots or by using software. However, using real robots it is seen that the even the simplest behaviour models tends to run over a hundreds of generations spread across a very large number of hours, therefore power is therefore the most obvious issue, as batteries will not sustain for such long and tethering using a power source may not be a suitable option at every situation and may limit the process. Jakobi also points to wear and tear of the robot, which can be substantial and real robots is not a practically viable option. Hence, the preferred route is via software, however with sophisticated robots and more demanding behaviour the semblance between the simulation and the real robot's performance is hardly any. Therefore, the onus was to develop a methodology on which simulators can be easily and cheaply built that enable that both the simulations and real robots to affect the nearly the same performance, which is scalable in time, speed, number of robots and complication of behaviour domains. In his revolutionary approach, Jakobi suggests development of base set of robot-environment interactions and also a base set aspects of the simulation. Fuzzy logic has been an easy to implement machine learning technique for robots. Using fuzzy logic in tandem with a navigational algorithm as potential field has been a favourite technique for researchers.

Another technique is to implement case based methods. Most software driven systems are backed up by a memory cache. Designing behaviour using cache memory leads the agency to respond to the same situation in nearly the same manner as it did previously. Case based approaches [271, 272] propose to make an exhaustive list of scenarios encountered by the agent and its response and update this list if and when newer situations are encountered. Case based learning can be summarised in the following five steps;

1. Define the problem at hand
2. Obtain similar case(s) from memory
3. Modify these case(s) to suit the problem at hand
4. Apply the solution developed, and evaluate the results
5. Learn by storing newer cases

This approach works very well for navigation and low level behaviours typically for a simple world and is a proven improvement over pure reactive methods.

SUMMARY

1. Navigation and adaptive learning are two important tools for a roboticist.
2. Navigation is the simplest behaviour and the easiest test bed to design newer and richer behaviours.
3. Grid based methods are easy to implement but they lack in reactivity and cannot be implemented in a dynamic environment.
4. Potential fields method has been particularly popular for autonomous navigation.
5. ND is a navigation algorithm based on situated activity and is very successful in 2D navigation.
6. Simple navigation algorithms cannot take into account the shape of the robot and the kinematic considerations.
7. There is a lack of good navigation methods in three dimension.
8. Adaptive techniques brings forth newer behaviour in the robot by robustly accomodating for the changes in environment.

NOTES

1. *Potential fields method finds applications in the gaming industry.*
2. *A variant of VFH has been used to make 'Guide Cane', a robotic navigation guide for the blind.*
3. *A⋆ is not recommended for sensor based navigation, as it depends on heuristic evaluation of a given node of a grid, and may lead to discontinuities as the robot traverses from one node to another.*
4. *Braitenberg vehicle-5 is motivated from the Turing Machine and is said to have a 'brain', this is achieved with threshold devices plugged between sensors and motors. Some threshold units will fire only when the signal is above threshold, while others will only stop firing when threshold is exceeded, in effect establishing two states of operation viz. ON state (ONE) and OFF state (ZERO) tailored by the threshold. Blending excitatory and inhibitory connections from vehicle-3 for the threshold units, Braitenberg very nearly replicates the McCulloch-Pitts neurons where the threshold devices are sames as hidden neurons, and shows that when arranged into a feedback loop these devices can be used to instantiate simple memory units, akin to digital logic (AND, OR and NOT). However, the computational capacity of the vehicle is rather limited, and 10 threshold devices it can store only a 10 digit binary number or a maximum decimal count of 1023. Vehicle-5 is designed not only to realise the*

computation in its own memory, but its ability to read and write the external world. A stretch of imagination will find the vehicle moving in a beach, tracing patterns in the sand, and negotiating a return trip by reading these patterns and retrace its own track. Such use of of the external world to store memory is an earmark for machine learning and McCulloch and Pitts proved that networks of such threshold devices can be used to build a Turing Machine. The vehicle can be seen as the corollary of the tape head, and the sand track as equivalent of the tape upon which symbols have be written, thus correlating vehicle-5 to a Turing Machine.

5. *The Dynamic Window Approach (DWA) [109] is one of the few navigation algorithms that consider kinematic constraints. This method incorporates the mechatronic functioning of the robot and is extendible to non-holonomic robots and can be implemented to charter unknown and dynamic environments.*

EXERCISES

1. **Grid based methods**, demonstrate with examples that grid based methods leads to suboptimal paths.

2. **C-type obstacle for ND**, discuss the performance of ND diagram navigation for C-type obstalces.

3. **Local minima**, show that local minima problem for potential field navigation can also occur when the obstacle is not C-type or U-type. Based on this results, is potential field navigation a good choice for generic navigation with uneven terrain, moving obstacles and other robots?

CHAPTER 5

Software, simulation & control

"ROS = plumbing + tools + capabilities + ecosystem"
– Brian Gerkey

"If you could see the world through a robot's eyes, it would look not like a movie picture decorated with crosshairs but something like this:

225 221 216 219 219 214 207 218 219 220 207 155 136 135

213 206 213 223 208 217 223 221 223 216 195 156 141 130

206 217 210 216 224 223 228 230 234 216 207 157 136 132

..."
– Steve Pinker

5.1 SOFTWARE FOR ROBOTICS

A discussion on robotics is incomplete without a mention of control software and simulation platforms. As brifely discussed in Chapter 2, simulation is an essential tool for testing of both the mechantronic design and the algorithm running the robot and it helps to understand the robot's performance in simple environments executing for easy tasks. Though simulations are an essential tool for robotics, they are ridden with limitations since they model only a finite set of real-world features and processes which are often plagued by approximations and inaccuracies. Despite the success in quality physics based simulators, the shortcomings are imminent. There has been a lack of techniques to model simple forces as frictional forces and magnetic (non-contact) interactions. Softwares which are made to measure the various forces, torque, moments, wear and tear etc. acting on a given body work on finite element methods and therefore cannot be implemented on mobile agencies. In complete contrast easy to use simulation interface is never sufficient to understand the true performance of the robot, and fails to grasp complicated environments and involved tasks.

Until around 2000, the two domains of software for robotics, simulation and control were distinct from one another. While simulation dealt with representing the robot on the computer screen and making it do various tasks in the software domain, control deals with real time implementation in hardware. In the 1990s these two streams were very different —

FIGURE 5.1 **BB-8**, the new robot in Star Wars was made with STM32F3 MCU. Newer processors and sleek hardware are needed to make better robots. New software platforms to tackle such hardware will not only be cumbersome but it will also be difficult to add-in accessories as camera, arm manipulator, sonars, motion sensors as Microsoft Kinect and ASUS Xtion etc. to the robot. Meta-Platforms as ROS helps to set a benchmark for software/hardware interface where accessories can be added-in as handshake of software modules.

simulations used to employ a pixellated graphical interface while control of the real robot would be with microcontroller coding such as for the PIC microcontroller. Post 2000, there has been a concerted effort from the robotics community to develop meta-platforms which unifies these two paradigms and the same code which enables a simulation on the computer screen also leads to a similar performance when put to work in the real robot. The earliest efforts were the player project and Microsoft Robotics Developer Studios where the two paradigms converged, similar motivation was seen in the simulator Webots by cyberbotics. Tekkotsu, Robot Operating System (ROS), YARP (Yet Another Robot Platform), MOOS (Mission-Oriented Operating Suite), ORCOS (Open Robot Control Software) and URBI (Universal Real-time Behaviour Interface) are newer and more well equipped avatars of such ideas, and being opensourced they attract crowd sourcing and tinkering by geeks, researchers and students to port in newer hardware and softwares. Of course V-REP and Webots are commercial simulators which have made a niche both in the industry as well as academia. Among these ROS has been most popular and has become a benchmark in robotics software in both academia and the industry.

Meta-platforms particularly Robot Operating System (ROS) addressing to both simulation in the software domain and control in the hardware/software interfacing has helped to set a benchmark in robotics so that newer hardware can be easily ported into existing systems and therefore encouraged crowdsourcing and open source software and hardware designs, as shown in Figure 5.1. This chapter is not a discussion on how to use ROS, rather it an exposition of meta-software platforms and introduces the reader to interesting facets of it.

Rossum's Playhouse (RP1) made in 1999 was one of the earliest open-source robot simulator. It modelled the world in two dimensions and could be employed for simple navigation as solving a maze — in the spirit of the micromouse. At about 700 kb It was a handy tool for programmers which could be readily used as a testbed for algorithms and control logic. The user could make robot models as per design layout and employ actuators, IR sensors, proximity sensors, and bumper sensors.

The player project was the first attempt to unify the domains of control and simulation. The name 'player' and 'stage' were inspired by the quote, "All the world's a stage," by William Shakespeare in his play, 'As You Like It'. The suite was designed to work as client-server models, with player providing the server interface, Stage the 2D simulator [334] and Gazebo the 3D simulator. Stage was later ported into ROS. Stage was chiefly developed by Richard Vaughan using FLTK and supported simple block like environments and robot models using LEGO like blocks. It provides the user with a two-dimensional graphic environment, which has facilities for a perspective camera, making it effectively a 2.5 dimensional simulator as shown in Figure 5.2. Stage has facilities for modelling the robot and its sensors using simple scripts. There have been significant changes since version 3 to improve its simulation performance. Newer features includes stag being a standalone simulator, use of controllers and multi-threaded execution. Stage is supported in nearly all UNIX-based OSes and in Mac OS X.

Gazebo is a 3D simulator developed with OGRE and it uses ODE for the physics simulation, shown in Figure 5.3. Robot models made in Blender or Google SketchUp can easily be imported into gazebo to form its xml based Unified Robot Description Format (URDF), and ROS has built in libraries for models of most of the popular robots. It is a big improvement on stage and now is the default simulator in ROS. ROS started off with two simulators, stage and gazebo — both forked out of the player project. Since then, both stage and gazebo have also developed into standalone open source softwares. STDR has been a new addition to ROS, as we will briefly see later in the chapter.

Being an interdisciplinary domain, various academic departments offer courses on robotics, across Mechanical Engineering, Computer Science and Electrical Engineering. Some universities even have centers and schools specific for robotics. While most Mechanical Engineering courses on robotics focus on the arm robot, those offered by Computer Science focus on AI and is often an extension of the course on AI. Mobile robot specific courses are few and are offered only by the top 50 universities across the globe. However, with robotics posing to take the center stage that is set to change. Therefore, a meta-palform should attempt to address the following questions[1].

1. The meta-platform should provide reusable code pieces which helps to build simple robot behaviours.

2. Since vision is the most important of all senses there needs to be dedicated software for computer vision. ROS provides support for motion sensing and supports OpenCV libraries.

3. Localisation is of primary importance and there needs to be tools for odometry and visual landmarks, and mapping and techniques as SLAM based on Particle Filters. ROS supports mapping tools as gmapping and octomapping.

4. Simple navigational approaches are insufficient, but they provides for potent testing grounds. ROS has navigation tools as ND-diagram and AMCL.

5. For more involved design, there will be a need to orchestrate pick-and-place operations using an arm manipulator and exotic design tools as gyroscopes and Omni wheels. A

---

[1]many of these ideas are due to Touretzky [325]

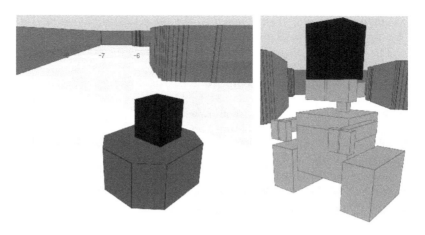

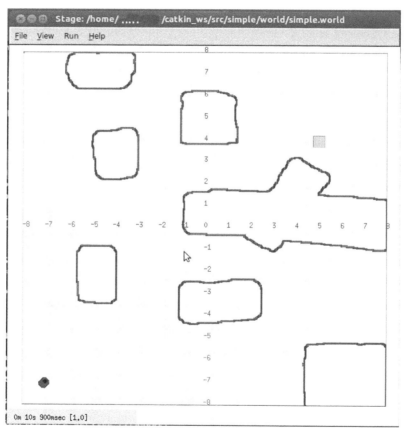

FIGURE 5.2 **Stage simulations**, LEGO like blocks model of the pioneer robot and Wall-E in stage (version 3). The cube mounted on both the robots is the SICK laser. In the third image, the little red pioneer robot is seen at the lower left hand corner and the green box is a movable cube placed at the upper right hand corner, the semicircle surrounding the pioneer robot is the field of view of the SICK LMS 200 laser

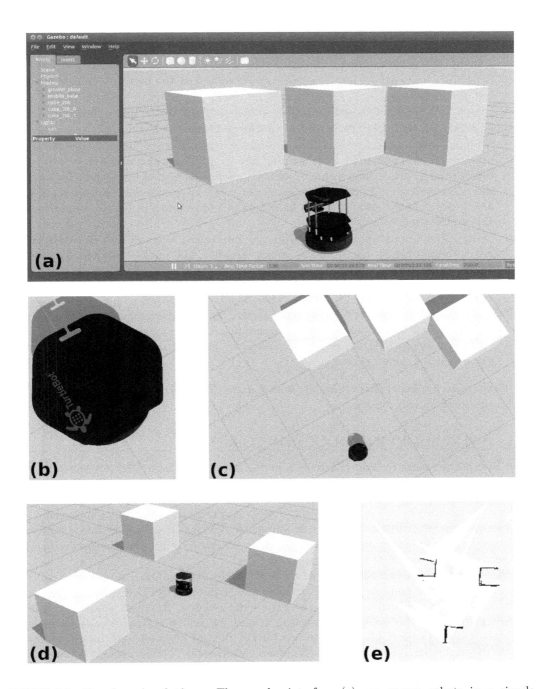

FIGURE 5.3 Gazebo simulations. The gazebo interface (a) can spawn robots in a simple environment — the robot here is turtlebot-II and the environment comprises of 3 giant cubes. The simulation is intractive and helps to zoom into and change angles as shown in (b) and (c). Mapping software provided in ROS can implement SLAM to create 2D maps, as shown in (d) and (e).

meta-platform should provide kinematic solvers for inverse and direct kinematics to facilitate ease of design.

6. With increasing complications, the robot builders will strive to develop modular design and simplify programming codes with class libraries. Modularity, code and hardware reusability and easy to use class libraries should be at the heart of the paradigm for the meta-platform.

7. Search and planning should be readily usable modules of the meta-platform.

8. It should support machine learning libraries in C++ and Python.

9. Human robot interaction is still in its nascent stages though face recognition, speech recognition and biometrics identification has come to be a part of our day to day lives, however integrating them into a robot is still a desired goal of the future. This module should consist of NLP, face recognition and emotion recognition. Since human-robot interactions should also tally with known social conventions hence this module should also address roboethics.

10. Many robot teams and robot swarms are interesting avenues and the student will learn about explicit coordination (robot groups) and implicit coordination (robot swarms). Most robot softwares are designed for single robot, and that includes ROS. The apparent compromise to this approach is that it alienates the wonderful facets of swarming as discussed in Chapter 6. There has been prospects to incorporate multiple robots paradigm into ROS using multimaster, and future versions of ROS may have capability to design robot swarms. Autonomous Robots Go Swarming (ARGoS) made by researchers at IRIDIA, Belgium is currently the best software platform option for researching on robot swarms.

ROS easily implements the first six of the above, however it is yet to be employed for social robots or swarm robotics.

5.2 A VERY SHORT INTRODUCTION TO ROS

The Robot Operating System (ROS) is a meta robotics platform [270] for robot software development, providing operating system like functionality it comes with various built-in utilities (viz. hardware drivers, maping and localisation (SLAM), path planning algorithms, simulation interfaces etc.) suited to study robots and there is opportunity to develop your own utilities and algorithms. ROS was developed by Willow Garage and Stanford University with the first distribution release, Box Turtle in March 2, 2010 which was developed for installation in Ubuntu 8.04 LTS.

ROS is released under the terms of the BSD license, and is open source software. It is free for commercial and research use. ROS is meant for UBUNTU installation and in its early days had two of its own robots, the PR2 and Turtlebot. ROS was an extension of the STAIR project and it started off by porting in most of their functionalities and simulators from the player project. Currently, it has URDF models of nearly all important mobile robots. It can be partially ported into Windows and Mac OS-X it also supports open source hardware platforms such as Raspberry Pi, ODROID etc. and various sleek hardware as Microsoft Kinect, ASUS Xtion, Sphero and LEAP device. It supports a number of robots: PR2, Turtlebot (I and II), Corobot, Roomba, Care-O-Bot, LEGO Mindstorms, Shadow Robot, Billibot, RAVEN-2 surgical robot, REEM-C humanoid and various others. ROS has had 8 distribution releases over the last seven years and it is now maintained and developed by Open Source Robotic Foundation (OSRF).

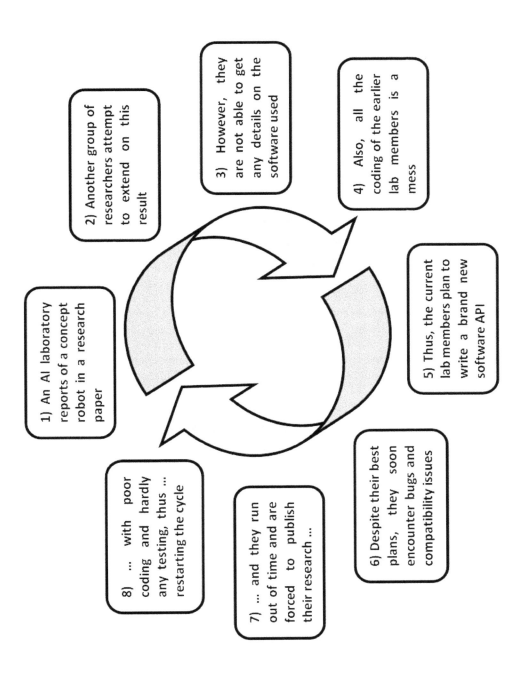

FIGURE 5.4 **Re-inventing the wheel.** Robotics research was plagued by lack of a universally acceptable hardware/software platform and it was difficult to replicate published research and this was detrimental to robotics. Adapted from Cham [65]

What is ROS?

Brian Gerkey, lead developer of ROS and CEO of OSRF, explains ROS in the following way:

" **ROS = plumbing + tools + capabilities + ecosystem;**

1. **plumbing**: ROS provides publish-subscribe messaging infrastructure designed to support the quick and easy construction of distributed computing systems.
2. **tools**: ROS provides an extensive set of tools for configuring, starting, introspecting, debugging, visualizing, logging, testing, and stopping distributed computing systems.
3. **capabilities**: ROS provides a broad collection of libraries that implement useful robot functionality, with a focus on mobility, manipulation, and perception.
4. **ecosystem**: ROS is supported and improved by a large community, with a strong focus on integration and documentation. The ROS website is a one-stop-shop for finding and learning about the thousands of ROS packages that are available from developers around the world."

The plumbing part is the asset of ROS, where communicating nodes across a publisher-subscriber system enables versatality in design level, behaviour level, sensor level and control level for the robot, this is broadly accomplished via roscore. One of the most striking features of ROS is that it works more like an operating system than an application, and various command structures and file structures bear similarity to standard bash commands in UNIX. Tools are the OS level aspects of ROS which are a basic set of tool for visualisation, diagnostics and debugging and often mimics sh commandline. Capabilities is the facility to develop customised codes and designs, which is now structured through catkin and is developed on make/cmake file structures. Ecosystem, other than the ROS website there are variour other forums, mailing lists and annual events which goes on to popularise ROS. It is further to be noted that ROS is not a programming language, it is not a compiler, it is not an IDE and it is not a software library; however all these facilities are supported by ROS in some manner or other. Over the seven years of its existence ROS has become a the single most used software platform in robotics.

Around the new millennium, each robotics research group worked in its own software platform and it was difficult to recreate published results as shown in Figure 5.4. This was detrimental to the future prospects robotics. ROS has nearly solved that problem by benchmarking the software in the robotics domain, and making the software more predominant and universal than the hardware.

Since it's first release in 2010, over the last seven years ROS has found growing acceptance in the open source community and the user contributed packages has increased at an astounding rate. Other than the ROS website at www.ros.org there are question answer forums at answers.ros.org, meta blogs at planet.ros.org and mailing lists to help and support ROS users which helps to fostercommunity participation and crowdsourcing much like large open source projects such as Ubuntu, Python, Apache etc.

ROS provides standard operating system services such as hardware abstraction, low-level device control, implementation of commonly used functionality, message-passing between processes, and package management. ROS is currently supported on Ubuntu,

An Example In Emergence Of Behaviour

This is an example using the gazebo simulator in ROS. We had met the PR2 robot in the last chapter, here we try to make it perform simple obstacle avoidance using lasers in a box-world like environment made in gazebo.

Algorithm Obstacle avoidance. PR2
repeat, full laser scan
if minimum range ≤ 0.5 **then**
move by a certain angle in a given direction
else
move forward
end if
until forever

On running the above code it is noticed that instead of obstacle avoidance the PR2 exhibits an akward box-pushing behaviour.

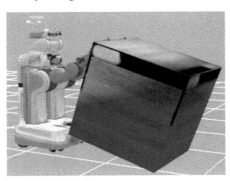

What is happening here? In this simulation the PR2 has its arms streched. While coding this aspect was not taken into consideration and the algorithm was developed for a point like robot, or at best a circular robot and 0.5 was used as an arbitrary distance to maintain a minimal proximity. For the condition *minimum range* ≤ 0.5 ($\sim$ AB in the figure below), it corresponds to a space at about the elbow of the arms of the PR2 and each time the code checks for an obstacle there, doesn't find it and it *move forward*, since gazebo is a physics simulator, in a box-world this manifests as a box-pushing behaviour. On correcting the code to *minimum range* ≤ 1.2 ($\sim$ CD in the figure below) the issue is sorted out.

Box-pushing was never coded in, it emerges as a result of the interaction of the code with the PR2 and the box world.

FIGURE 5.5 **ROS Lunar Loggerhead** was released on May 23, 2017 .The tortoise is a reoccuring theme at ROS, all releases and logos have a tortoise reference. It has been suggested that such is due to, the first modern robots are often said to be the two tortoise robots, Elmer and Elyse developed by Walter. Image courtesy `http://wiki.ros.org/`, Creative Commons Attribution 3.0.

TABLE 5.1 ROS Distributions

Name	Release
Box Turtle	March 2, 2010
C Turtle	August 2, 2010
Diamondback	March 2, 2011
Electric Emys	August 30, 2011
Fuerte Turtle	April 23, 2012
Groovy Galapagos	December 31, 2012
Hydro Medusa	September 4, 2013
Indigo Igloo	July 22, 2014
Jade Turtle	May 23, 2015
Kinetic Kame	May 23, 2016
Lunar Loggerhead	May 23, 2017

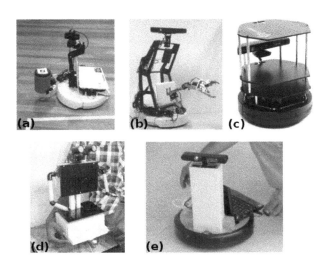

FIGURE 5.6 **Differential drive robots.** There has been growing interest in making differential drive robots, be it commercial ((a),(b) and (c) from Touretzky [326], reprinted by permission of the publisher Taylor & Francis Ltd, http://www.tandfonline.com)) or hobbyist — (d) Achu Wilson's Chippu (Image courtesy Achu Wilson) and my own project named Tortue built with ASUS Xtion and a Roomba 560 base in ROS Electric.

experimentation for extending this support for other operating systems as other flavours of Linux, Windows, Mac OS-X, ARM versions for embedded devices etc. is an ongoing effort.

A chit-chat with the STDR developers

Manos, Aris and Chris from Greece tell us about their experience developing the STDR simulator at ROS. This interview took place on 10 Nov 2014.

AB: Tell us about the development of STDR, the motivation and need?

The STDR team: STDR Simulator's development process begun in November 2013 from a group of electrical engineers involved in the robotics and specifically members of the P.A.N.D.O.R.A. rescue robotics team from the Aristotle University, Greece. The motivation behind STDR was to build a low-requirement simulator that's easy to install and use, thus providing a good tool to test incipient ideas or develop less advanced projects, as effortless as possible. Our intention was not to provide the most realistic simulator, nor the one with the most functionalities. On the contrary, the overall design is minimal allowing rapid robotic algorithms prototyping. This fact was critical to our approach as many simulators need a considerable amount of time to be installed, configured and setup for an experiment.

The tool that allowed us to implement such automated flexibility is the Robotic Operating System or ROS. Being ROS users for some time now, we pinpointed the fact that there were no ROS-dedicated simulators, meaning that no simulator is actually built in ROS and not just use its interfaces. Thus, we decided to employ ROS tools to create a ROS compliant two dimensional simulator, aiming both at providing easy usage and being a state-of-the-art robotic software.

AB: ROS obviously has other 2D/3D simulators as stage and gazebo, how is STDR an improvement on those?

The STDR team: STDR simulators contribution is an oxymoron, in the sense that the improvement could be described as a deliberate deterioration. By providing only basic/minimal functionalities, we achieved grave reduction in complexity of both installation and configuration. In addition to that, system requirements are immensely minimized, rendering experiments feasible for a vast group of developers. As stated in the previous question, STDR is purely implemented in ROS, thus it does not just provide ROS bindings or wrappers like stage or gazebo do, after launching an experiment with specially defined parameter files. On the contrary, in STDR, the robotics developer can immediately execute the simulator, add a robot providing its specification graphically and subscribe to the appropriate ROS topics effortlessly.

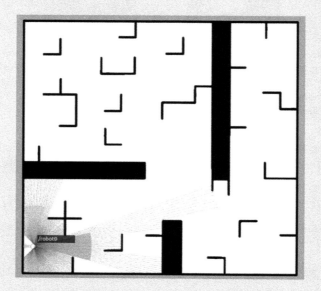

FIGURE 5.7 **STDR**, startup graphics for STDR

AB: Can one make basic algorithms as A⋆, bug-algorithms, potential fields etc. to work in STDR?

The STDR team: As STDR is a simulator, its job is to simulate robots and their environment. Concerning the environment simulation, the user can provide a simple image file containing the obstacles, the unoccupied space, as well as the unknown areas. On the other hand, via the robot simulation, the robotics developer has access to simulated effectors (wheels) and receptors (sensors like LRFs, Sonars etc). In that sense, the user can perform a SLAM (Simultaneous Localization and Mapping) algorithm or have the map known a priori, and therefore execute on them a wide variety of algorithms such as A⋆, bug algorithm, potential fields, path planning and navigation modules etc. Of course the developer can benefit from the large amount of ROS packages provided by the scientific community. Since STDR is ROS-based, these packages can be employed with minimal effort.

AB: How does STDR handle multiple robot simulations? is there methods to achieve coordination/cooperation or swarming behaviours?

The STDR team:STDR simulator does support multi-robot simulation. The developer has to ability to spawn several robots and send commands to their effectors or receive data from their sensors. Of course he/she is free to implement multi-robot communication controllers via ROS infrastructure (messages / services / actionlib). However, the multi robot support is a germinal feature, that will be further evolved. In its current state many robots can be simulated, but lack the capacity to physically detect each other, i.e., robots can navigate through each other and one robots sensors fail to sense another robot.

AB: According to you all what does a person needs to know prior to starting ROS? C++ or some programming background?

The STDR team: It depends on the application. Some programming background is always useful, when you are developing software but it is not necessary for all applications. ROS has several layers of abstraction, thus writing a simple program does not require understanding implementation and design details of lower (infrastructure) layers. Good knowledge of C++ and/or python is definitely a plus. For someone who is not familiar with C++ concepts it would be an uphill battle to master ROS.

FIGURE 5.8 **The STDR team**, from the left to right: Manos, Chris and Aris from Aristotle University, Greece

AB: What is the future of STDR?

The STDR team: Our next steps would be towards a more realistic simulation of the environment. In that manner, we aim to add a simple physics engine in order to perform collision checking more realistically. Also, we are thinking of a dynamic environment, where robots can perceive each other as moving obstacles or perceive moving obstacles generally, like a moving door. Since STDR is an open source project, its future would be heavily influenced by communitys interest for new features, bug fixes, etc. We should mention that we have already several contributions from the ROS community and we encourage the potential developers to inform us of any problems or even achievements they had using our simulator.

STDR Simulator is currently an official ROS package and can be downloaded via apt-get. The supported ROS distributions are Hydro and Indigo, thus the supported

OSs are Ubuntu LTS 12.04 and 14.04, as well as all the intermediate distros. Some useful links regarding STDR follow:

Official STDR webpage: http://stdr-simulator-ros-pkg.github.io/
ROS wiki page: http://wiki.ros.org/stdr˙simulator
GitHub repository: https://github.com/stdr-simulator-ros-pkg/stdr˙simulator

ROS in Space - The Robonaut 2 (R2) humanoid torso

The Robonaut-2 is a project which developed the first humanoid robot at the International Space Station (ISS), the humanoid robot runs on ROS. The project is a venture of NASA and General Motor, working at the Johnson Space Center, Houston. The project aims to develop an automated humanoid torso fitted with 2 dexterous arms which can function as an astronaut in space. The advancement that R2 has achieved over other robots is that it can use the same tools that an astronaut uses, there is no need for specialised tools only meant for use by R2. R2 is an advantage where simple, repetitive, or hazardous tasks are needed to be performed in harsh and unfavourable conditions as space stations and non-terrestrial colonies. R2 was delivered to the ISS aboard the shuttle Discovery on the STS-133 mission with the mission spanning across 2 weeks, 24 February 2011 to 9 March 2011. While still on board the Discovery, testing and basic operation started inside the Destiny module. Once at the ISS, it was deployed on a fixed pedestal. The project achieves real time control using ROS and ORCOS in tandem.R2 is the first humanoid robot in space, it has a height of about 1 meter, weighs 132 kg and is made out of nickel-plated carbon fiber and aluminum. It is equipped with 54 servo motors and has 42 degrees of freedom and needs at 120 volts DC.The torso can be fitted with legs to make it into a full humanoid, deployed on a pedestal or fitted on to a four wheeled rover. NASA's team used gazebo for R2 simulations, ROS packages which they have released to the opensource community.

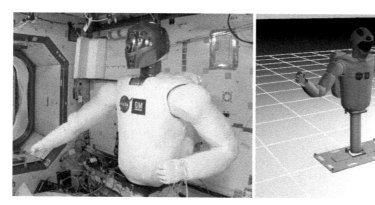

FIGURE 5.9 **The Robonaut 2** uses ROS and is employed at the International Space Station (ISS). Shown here is the robot on the left and its gazebo simulation on the right.

A *tête-à-tête* with Daniel Stonier and Marcus Liebhardt from Yujin Robot, Seoul

Yujin Robot is a Seoul based company which developed the Turtlebot-2 with the Kobuki base. Daniel and Marcus from the company's innovation team share with us the experience of developing these robots and future prospects. This interview took place on 24 Nov 2014.

FIGURE 5.10 **Kobuki base and Turtlebot-2,**

AB: How did Yujin Robot come into being?

DS & ML: More than 20 years ago Yujin Robot started in automation and is now comprised of several small business groups, not all of them directly related to robotics and has also seen a couple of internal teams successfully spin off their own businesses. The R&D group entered service robotics more than a decade ago and despite being a relative adventurous proposition at the time, the driving force then, and now has been the CEOs passion for building robots.

FIGURE 5.11 **iClebo and iRobi,**

AB: What all robots have you developed?

DS & ML: Since Yujin Robot entered service robotics, they have explored several directions. The one constant has been the iClebo cleaning robots initially random walkers and later the vision SLAM machines. There have been several products for university research (soccer robots, mobile platforms) and many versions of RobHaz for hazardous operations (rescue robots). Until recently though, the other major drive was towards educational (specifically classroom) robotics with robots like iRobi and RoboSem. Whilst this is not so commercially attractive outside of Asia, here the sheer numbers, unbelievable competition and parental focus creates a very different economic environment, one that is potentially attractive to robotics. Yujin Robot explored the development of both small and large product lines, but despite the enthusiasm for robotics and a healthy economic niche, such a market remains a hard sell.

FIGURE 5.12 **The GoCart**,

AB: Tell us about the Kobuki? Why this name?

DS & ML: 'Kobuk' is the Korean word for turtle. Adding the suffixed 'i' is typical korean syntax when using such nouns in a sentence, but also gives it a more intimate feeling. The why of its origin for our mobile platform is in part shrouded in time. There was no definitive moment in which we had named our robot, but some things are for certain. For a long time we had a cleaning robot based mobile platform in the lab for experiments that we called Kobukibot. Wed also adopted the ROS fascination with turtles ever since the ROS turtle tutorials first came out (some of us can remember a time even before there were such tutorials at one point there was a tutorial for the most perverse, but fun way to add 1+1 using ros services). Korea is also famous for its Turtle ships, an engineering design we can relate to!

AB: Since the Kobuki and TurtleBot 2 are meant to run with ROS, how do you explain ROS to a young enthusiast? Say a 15 year old who is just about learning some C++ and bits and pieces of Java or Python and wants to own a Kobuki.

DS & ML: Kobuki/TurtleBot is a readymade hardware unit. So working with such a platform is all about problem solving, programming and of course, doing these things on a platform which lets you build interesting applications that come to life in a more vivid way than they can on a PC. ROS itself lets you write programs that can talk to each other and the ROS ecosystem provides you with a lot of software that you can reuse and recycle for your own benefit. I think for a budding young developer, these are awesome concepts that can take them far. It teaches how to break down complexity into small chunks and how to integrate efficiently with existing components. Without these skills, a developer wont be able to scale up what he or she can do, nor accelerate at the same pace as developers around them will be able to do. In todays software world, these concepts are often more important than ones proficiency in a particular language.

FIGURE 5.13 **The Kobuki/Turtlebot-2 team**,from left to right and top to bottom we have: Hyeong Ju Kim, Jorge Santos Simon, Young Hoon Ju, Nosu Lee, Min Jang, Marcus Liebhardt, Sam Park, Jaeyoung Lee, Daniel Stonier and Joohong Lee

AB: How has ROS and OSRF communities responded to the Kobuki?

DS & ML: Very positive all the way through. Kobuki was originally intended only for the Korean market since we could not get iRobot Create robots distributed here and we wanted to expose ROS on TurtleBots to Korean students at universities. It was such a significant jump up from the Create, that it quickly became consumed by the TurtleBot and international ROS communities. Willow Garage and later OSRF were involved in the process right from the design of the hardware. One of the most rewarding things to see is its use now as a prototyping platform for small startups.

AB: Any interesting upcoming projects? Since we are moving into a time when robots and smart electronics will rule our lives, how do you foresee the future of this industry? and the impact on society?

DS & ML: There is a definite convergence of technologies happening right now. For the first time in decades, we have explosive growth in multiple sectors of the technology world, and the most interesting thing is, almost all of these sectors impact on what can be done in robotics. Smart phones → embedded computation, good cameras, communicativity outside a wireless LAN. Gaming → 3d sensing. The internet explosion post the IE5 stranglehold → cloud computing resources, remote connectivity. As a first step, we are seeing reduced costs and the ability to be able to navigate in semi/unstructured environments, now triggering the emergence of mobile robotics in the service industries (outside cleaning robots). For the first time it feels like the heat is on to get new products to market in these industries. However I dont believe these successes can be solely ascribed to the fact that there is a technology convergence. History shows that technological leaps often fail to penetrate the market. These leaps have to be either hugely significant, or a need arrives. What we have now is a very real and quickly looming crisis of manpower shortage in various industry sectors in the near future, such as health and elderly care. This is dispelling the conservatism in several industries we have hitherto explored and they are now more keen to take a leap of faith. It is into these emerging service robotics applications Yujin Robot is now venturing with products like GoCart. Rather than trying to sell a service, or provide entertainment, we now look to fulfill the automative needs of the service world.

AB: Tell us more about the GoCart project.

DS & ML: A year ago we started the innovation team with the goal of creating Yujin Robots products of tomorrow. The first result of this endeavour is GoCart our autonomous mealtransport system. GoCart is the first robot of our Gopher Series, which combines Yujin Robots various technologies for indoor transportation, such as our new stereovisionbased dSLAM system. GoCart is our contribution to fight the shortage of skilled caretakers and raising costs in health and elderly care. GoCart will take care of moving meals and other supplies around in a facility, such as hospitals and elderly care facilities. Thereby freeing up caretakers time allowing them to spend more time with the residents and patients. With its affordable price GoCart will reduce the overall operating costs of the facility at the same time. More information about GoCart can be found on the web page `http://gocart.yujinrobot.com/` and additional technology details in our recent blog post `http://blog.yujinrobot.com/2014/11/delivering-ros-in-box.html`.

ROS has a rich library of supported sensors, actuators and robots and a new robot as the ones shown in Figure 5.6 or a smart device can also be interfaced with little effort. ROS has two entities, the operating system entity and pakage specific entity, a suite of user contributed packages organized into sets called stacks that implements various functionality such as navigation, mapping, planning, perception, simulation, interface for a device etc. The operating system entity can be seen from the structuring of ROS to be a layer over the existing Unix system and bears a similarity to bash and cmake utilities, the pakage specific entity is more often user defined.

In ROS, every communication between robots, sensors, actuators is done by nodes. A node is the simplest piece of code (often written in `C++` or Python) which can be executed. The node may be the device driver of a robot, sensor, some peripheral device or it may be explicitly written to accept, process or send data in the form of messages. Node to node communications is not direct, but through a common topic in the form of messages. A publisher node publishes messages to a topic and the subscriber node subscribes to this topic so that it gets the messages that are published by the publisher node.

By the virtue of modularity across both design implements (with CAD models and URDF) and software features (such as ready made modules for navigation and mapping tools) and unification of simulation and control interfaces, ROS marries a plethora of processes and a myriad of hardware to software and is an essential for all robot enthusiasts. The best resource for ROS is its website, `http://wiki.ros.org/`.

SUMMARY

1. Meta-platforms in robotics has unified the simulation and control paradigms and the same piece of code which runs a simulation can be made to work in the real robot.
2. ROS and similar platforms provide interfacing to hardware and also provide a wide array of software tools.
3. Simulators as gazebo and MORSE provide quality physics based simulation, while simple simulators as STDR and stage can be used as test bed to try out algorithms etc.
4. ROS makes it easy to interface newer robots and accessories.

NOTES

1. *Pioneer robot in caves environment was the default simulation for player/stage, the caves environment shown in Figure 5.2 became iconic and can been seen in various research publications, tutorials and videos.*
2. *Willow Garage closed down in March 2013, and then on ROS was maintained and developed by OSRF.*
3. *A robot I made using a roomba base in ROS Electric can be found at* `https://www.youtube.com/watch?v=ZPhbyOlCq08`
4. *ROS ecosystem is an ever growing assortment of softwares closely allied to ROS, such as MoveIt!, ROS Industrial, OpenCV etc.*
5. *ROS 2.0 is the new avatar of ROS and yet to go mainstream.*

III

Robot-Robot & Human-Robot Interactions

CHAPTER 6

Robot-robot interaction, groups and swarms

"They're flocking, but that's not what they think they are doing"
– Maja Mataric [92]

"Swarm robotics is a novel approach to the coordination of large number of relatively simple robots which takes its inspiration from social insects."
– Sahin [287]

6.1 MANY ROBOT SYSTEMS

A number of robots working in tandem is often preferred over a single robot due to improved system performance, distributed action over a number of locations, better fault tolerance and economic considerations. A single or two to three robots as a part of a system is not always a great design, since if one of them stops working then a significant part of the job at hand is left unattended. If this is a mobile system in a hospital, a production line or a workshop then suitable replacement has to be found soon. This is why teams of robots either working with (robot group) or without (robot swarm) a centralised control reduce the problem at hand to a number of subproblems which can be completed with greater ease.

In multirobot systems, sensing works over a number of robots and therefore over a much larger range than the physical limitation of the sensing range of each robot. The action can be carried out at a greater distance from the point of sensing. Therefore, a team of robots can undertake a wider range of tasks, most of which will not be possible for a single robot to perform on its own. As a drawback, in a distributed system complexity increases and there is the chance of delay in sensing across the robot formation which can add to the clutter in the navigation and lead to more robot-robot and robot-obstacle collisions, which may defeat the very purpose of a robot formation.

Kiva systems, which employs deliberative knowledge sharing across about a thousand robots, and Robo Cup, where sharing of sensor information has proven to improve performance, are examples where a battery of robots has proven to be more useful than a single robot. Parker's Alliance architecture in the mid 1990s was one of the early attempts to design subsumption-like layered architecture for a team of robots. Mataric's Nerd Herd was the first implementation of social behaviour and flocking in robots. Mataric's robots could demonstrate very simple behaviours such as, homeing in on a common base and wandering

FIGURE 6.1 **Cooperative box pushing** is a favourite tool to demonstrate cooperation across a team of robots. From Kube and Zhang [191], with kind permission from Springer Science and Business Media.

while avoiding obstacles to more sophisticated social functions of synthetic foraging and flocking.

Cooperative box pushing as shown in Figure 6.1 has been a tool and also a favourite experiment to demonstrate cooperation in a team of robots. This is a direct motivation from cooperative transport in insect colonies and concerns moving a box like object by a team of robots to a given goal point which otherwise cannot be achieved by a single robot. It is observed in the natural world that performance improves with the number of agents therefore box pushing becomes a desired test or a benchmark to demonstrate cooperation.

6.2 NETWORKED ROBOTICS

Robot formations are marked by cooperation and coordination and have been attempted by architectures, bidding mechanisms and fuzzy logic methods. Over the past decade networked robotics is seen more as an ecological system, an interworking and mutual exchange of information rather than explicitly programming of cooperation and coordination into the robots. With the proliferation of ubiquitous and pervasive computing [350], a network of mobile robots may not remain contained to its member robots, but rather includes a more interactive ecology which is spread over thousands of computers and works to weave robots, electronics, embedded hardware, software, etc., into one single system working on seamless exchange of information, over calm and context aware user interfaces.

Such an omnipresent and pervasive network for robots is the basis for ubiquitous robotics. This technology of the future is projected to include all robots across the planet, connected across a single network adhering to a synthetic ecology. Ecological interdependence in the natural world ensures the well being of a large number of species and the sustainment of the biosphere. Stealing such ideas from nature, a model for synthetic ecology is suggested by Duchon and co-researchers [98,99], and includes the following salient features.

1. Agents and the environment cannot be treated as separate
2. Agency emerges out of the dynamics of agent environment interactions
3. The agent tries to map available information to the control, in order to execute its job.

4. Information is readily available from the dynamic environment for the agent to learn and adapt. Adaptation is not just an effort for survivability, but marks an ascendancy in skill of the agent.

5. There is no need for centralised control or executive.

Duchon's model is remarkably similar to the fungus eater model proposed by Toda and later modified by Pfeifer. However unlike Toda, Duchon considers an inherent ecology and therefore interdependence and seamless exchange of information between the agents as the underlying principle of his model than rote low level agent environment interactions. Also, the agents need not be homogeneous but rather they can be entirely different types of agencies, as varied as a humanoid robot and an autonomous car.

It is anticipated that ubiquitous computing will render that robots will not just be hardware entities but will also be embodied virtually, leading to ubiquitous robotics and human-robot interaction will be seamless, unobtrusive, calm and on the run. This large integration of robots and devices over a network which runs all across the world will redefine our ways of living and social interactions. A ubiquitous robot in a synthetic ecology will typically be an entity which has three levels of agency working in tandem: (1) the software, (2) embedded environment and (3) the mobile robot. The software has ease of movement through the network and therefore has the advantage of acquiring information at near real time. The embedded environment is a network of sensors and helps to gather localised information and the mobile robot is meant to execute the action. The following projects will further illustrate the principles of synthetic ecology and networked robotics. These three projects, (1) Humanoid Robot System on Ubiquitous Network [257], (2) PEIS [286] and (3) RoboEarth [341], were designed to help and support elderly citizens in their day-to-day chores and to develop new lifestyles governed by information and communication technology, where a synthetic ecology of devices and robots works in unison and harmony with the need and requirement of the human beings.

Humanoid Robot System on Ubiquitous Network : Extending teleoperation over a long distance network, Okuda et al. [257] developed a humanoid robot system controlled remotely over a wide area network via mobile phones for monitoring and carrying out simple tasks in a home equipped with smart electronics. This research, done in collaboration with Aichi Prefectural University and Mitsubishi suggested a classification of networked robots: (1) Virtual Type, which are driven by operators via graphical interfaces and exist solely in the virtual space. (2) Unconscious Type[1], are driven by embedded hardware and are more an extension of machines and the (3) Visible Type, which are nearly autonomous and can act on human orders, such as personal robots. The network is developed with a visible type robot, the HOAP-2 humanoid robot. It has 13 degrees of freedom and it was employed over an interface of RT-Linux. The robot was used in the project for remote manipulation and surveillance over a mobile phone, schematically shown in Figure 6.2.

Physically Embedded Intelligent System (PEIS) : PEIS [286], the project was targeted at development of a smart household system which can take care of elderly citizens. As a philosophy for the project, it is meant to be an ecology of communicating and cooperating agents, where the performance of the household system emerges from interaction among the various agents of the ecology, as shown in Figure 6.3. The project attempts to design rich ecology of a variety of interacting agents and go beyond single humanoid robot and

---

[1] Here the moniker of unconscious may not be very appropriate, as will be discussed in the later chapters.

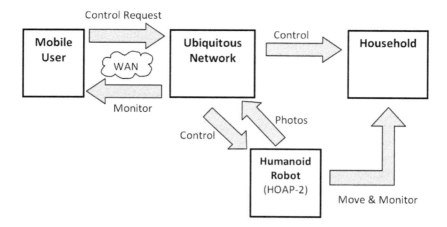

FIGURE 6.2 **Humanoid robot system on ubiquitous network** is the first project to employ humanoid robots over a remote network for simple household chores. Adapted from Okuda et al. [257].

tasks tagged to teleoperation. It proposes to use an array of embedded sensors across the household to dispense local information, rather than overloading one or two humanoid robots with a barrage of sensors. As an illustration, olfactory sensing which can detect fire, stale food, security concerns etc. is more effective if a number of less expensive sensors are embedded across the household over a network rather than equipping a humanoid robot with an a state-of-the-art sensor. Similarly, a humanoid robot is redundant attending to the job of night security. Instead smart electronics working to a timer is more suitable. The test rig for the project was a real household in Sweden with an elderly couple as the inmates. Saffiotti and his team used RFID tagging to identify objects in day-to-day use: as coasters, cups, toaster, alarm clock etc. over the network. The research team also laid a floor equipped with RFID sensing to detect the locomotion of the senior human beings of the household. This was in tandem with communication and cooperation modelled between simple and inexpensive robots like the Roomba and the Pioneer to attend to chores of the household. The clear advantages of this project are, (1) hardware or software failures are easier to overcome and the latency time is lower in comparison to a network of few robots, (2) once installed, maintenance can be done by the elderly users without much hassle nor requiring any particular skill in robotics or electronics and (3) cost of setup and maintenance is far lower than employing one large humanoid robot.

The RoboEarth Project : The project [341] was developed by researchers across various universities in Europe and is aimed at making a very large system for sharing knowledge between robots, a World Wide Web for robots. As of now, the service robot industry is limited to specialised robots, viz., vacuum cleaning robot, trash collecting robot, delivery robots etc. which are preprogrammed to specific tasks. This not only reduces the opportunity for the robot but also adds to cost and redundancy. In the near future, robots will often need to perform tasks which are not programmed into them, and also adapt to newer environments and adapt to a variety of tasks. RoboEarth attempts to solve this problem by large scale sharing of data of specific robots and is similar to an Internet of Things (IoT) approach. With interlinked robots over a world wide web of robots, it is believed that it would speed up learning and adaptation and that robots will be able to undertake and

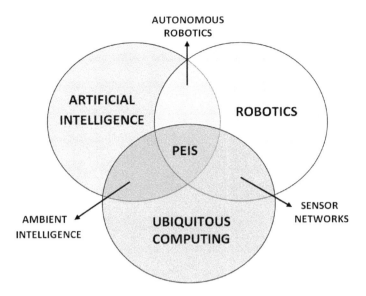

FIGURE 6.3 **The PEIS philosophy** is built on the ecology of a variety of communicating and cooperating agents than a employing a single humanoid robot. This approach is not only cheaper but attends to a larger range of tasks and is not cursed by network delays and bottlenecks. Overlap of technological domains resonates with the concepts of developing an artificial ecology and may be a common occurence in the future. Adapted from Saffiotti and Broxvall [286].

execute tasks that were not planned for in the design and the programming cycles. For example, an office assistant robot had to attend to a situation of a sudden cardiac arrest of an office staff member and as a learned behaviour calls the ambulance, another robot which is finds itself in a similar situation will just download these files from the RoboEarth cloud server to deal with it. Of course, the design, sensors etc. of these two robots may not be the same, but the second robot will at least have some idea of how to deal with such a situation. The RoboEarth has been successfully used in trials for (1) maze navigation and playing robot-robot games which gradually improved due to learning techniques, (2) serving drinks to patients in hospital, which showed the utility of RoboEarth in a dynamic task in a social context and (3) operation of an arm manipulation by different robots which demonstrated reusability of the RoboEarth software and database.

These three projects do not yet lead to a pristine ecology of robots and devices, and most of them have either some form of overarching human participation and therefore fail safe conditions or a blend of central control, but each of them foreshadows technology which may become a part of our day-to-day workings in about a decade. The next chapter discusses the social aspects of using robots in elderly care.

6.3 SWARM ROBOTICS

Swarm robotics is a novel emergent group behaviour of multiple agents lacking in a centralised control or global knowledge, where each individual agent has minimal hardware and is tasked with low-level behaviour, but complex global behaviour is manifested in the group.

The concept of the social insect, as shown in Figure 6.4, is a common trait in ants,

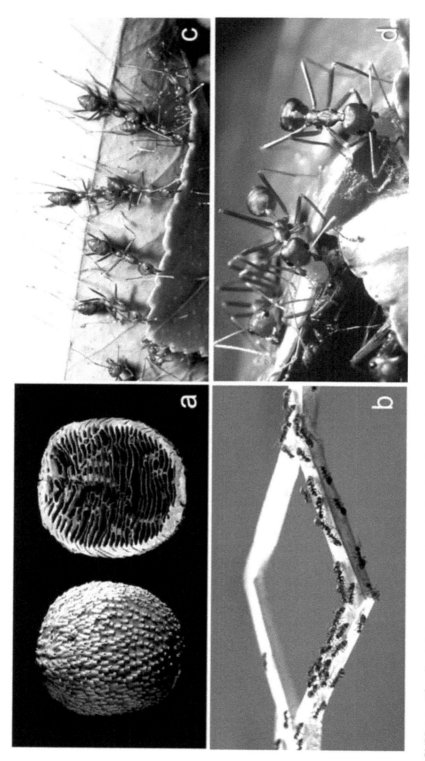

FIGURE 6.4 **The Social Insect**, as seen in ants, termites, bees and wasps is the motivation to swarm robotics. Eusociality is seen in (a) this cross section of a termite nest. shows the coordination of worker termites (b) ants working in cooperation tend to find the shortest path for foraging (c) weaver ants use their own body to pull leaves together, and (d) in weaver ant workers, division of labour is seen as the leaves have been put in place by a first group of workers and a second group sews in the leaf edges with silk threads formed by the mature larvae. Image from Garnier et. al [115], used with permission of Springer.

termites, bees and wasps and has led to progress in swarm robotics. It is tempting to consider swarming as a collective survival instinct, such as individuals escaping a predator or chasing prey where similar individual goals lead to a common group behaviour, but eusociality is a feature of animal and bird collectives and is typified by division of labour where a particular behaviour is unique to a group of agents, viz. nest building, food gathering etc.

Extrapolating the single agent approach to a swarm of multiple agents fails as swarming is the growth of a collective AI across all the agents such that the group behaves in unison as a single entity. This cannot be achieved as an addendum or overlap of single agent behaviours. This unison to reflect a collective behaviour is very similar to a phase change in physical or chemical systems and happens due to self organisation and stigmergy and is marked by the novel properties of robustness, flexibility and scalability.

1. **Robustness**: The ability for a task to be accomplished even if some of the robots fail. The swarm demonstrates redundancy and continues to operate at a lower performance. The failure of a few robots will not deter the system by much and due to distributed sensing across the swarm leads to a high signal to noise ratio which is not suppressed by the loss of a few robots.

2. **Flexibility**: The degree to which the robotic system can perform a task when the environmental parameters change. This happens by employing different coordination strategies when confronted to changes in the environment or the task. Flexibility is a trait of social insects. For example, ants can very easily change mode from foraging to chain formation to prey retrieval. Food recruitment in ants is coordinated as a biochemical stimulation using pheromones where each ant lays a pheromone to guide the following ant on the correct trail. In contrast, in chain formation, as shown in Figure 6.4 (c) and Figure 6.5 the ants attempt to form chain-like mega structures with each ant gripping the other ant to reach distances which are way more than the ability of a single ant. Here instead of a biochemical stimulant, the ants use their bodyies as the way to communicate across the formation. In another contrast, in prey retrieval, the ants work in seeming unison to carry large prey to their nests, a task which is not possible for a single ant to accomplish.

3. **Scalability**: The ability of the swarm to reorient the agents when the scale of the task itself changes. Introduction or removal of agents doesn't change the swarm behaviour and doesn't affect the performance of the swarm. Therefore, the coordination mechanism underlying the swarming behaviour is more or less independent of group size.

Despite such advantages, swarms are difficult to control as compared to traditional systems, and engineering a desired swarming behaviour is still being researched into. Researchers have been guided by studying, collectives of particles [336], animals [75], flocking of birds and colonies of insects, and used such models to understand the phase change and the emergence of a social behaviour and collective AI, a convergence of micro and macro [140, 141]. The emergence of swarming as a phase change [329] is most remarkable, wherein individual motion of the agents is redundant and the aggregate behaviour is aimed towards a common goal.

As an illustrative example, the iconic video game of the 1980s PAC-MAN, can be represented as a finite state predator-prey model shown in Figure 6.6. Each ghost is tasked with touching and thus killing PAC-MAN and it often seems that they are working in precise coordination. However, there is no central control executive and it is the common goal which leads to the collective behaviour and all the ghosts chase (or flee) PAC-MAN at the same time.

FIGURE 6.5 **Chain formation in ants**, is a communication scheme used to reach a distant goal. Here the ants use their body as the way to communicate and cooperate to form structures to reach distances which are not possible for a single ant to reach. In contrast to foraging, instead of a biochemical stimulation using pheromones, here ants use their own bodies for communication. Image in public domain, courtesy `wikimedia.org`

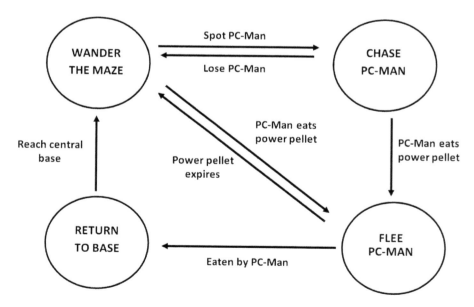

FIGURE 6.6 **Predator-prey state models, PAC-MAN.** When chasing or fleeing PAC-MAN, the ghosts have the same goal and implicitly show swarming behaviour.

There is a stark contrast between networked multirobot systems and multirobot swarms. Robot-robot and robot-obstacle collisions, which otherwise tend to increase with a greater number of robots in a multiple robot system is effectively avoided since the swarm works in unison as a single intelligent system rather than a set of individual agents. Communication for a multiple robot system needs additional hardware and software for sending and receiving messages and for signal processing and filtering of noise across the robot collective, which often leads to undesirably high lag times, and these issues increase with the size of the collective. In contrast, for a swarm, communication typically works over low-quality hardware and there is no need for expensive hardware and computational resources to communicate. Sensing is distributed across the swarm and the noise is insignificant to the signal. Research has shown that whatever noise is there tends to enhance coherence across the swarm. Since coordination across the swarm is maintained by the emergent swarm AI, the robots will never compete amongst themselves and always cooperate towards a common goal with all the agents contributing at an optimum level of productivity. For practical problems which often need scaling up, adding more robots in a multiple robot system may be limited by the hardware fanout limits and computational capacity of the on-board processor. In a swarm scaling up is inherent and therefore is not limited by any such restrictions. These contrasts will bring down the overall system cost and enhance system performance, a swarm of robots will be cheaper to design and maintain than a multiple robot system. However, swarms are difficult to control and swarm engineering is still a fledgling concept which is yet to be made into a reliable technology.

6.3.1   Relating agent behaviour to the collective behaviour

By modelling collectives from natural sciences, it is possible to relate the agent behaviour to the behaviour of the formations. Aoki's [15, 16] and later Reynold's [275] model relates basic behaviours which can lead to a swarm. Reynold's boids are a motivation from birds and insect flocking and are designed by rules in the order of decreasing precedence:

1. Collision Avoidance: Avoid colliding with nearby members of the flock
2. Velocity Matching: Try to match velocity with the nearest flock member
3. Flock Centering: Try to stay close to the nearest flock member

Collision avoidance under static conditions and velocity matching are nearly complementary. If the boid adheres to velocity matching then the mutual separation between the boids will remain nearly the same. To avoid colliding into the closest neighbour there has to be a minimal distance between the boids. Velocity matching maintains this distance dynamically. Reynold's model has been demonstrated to lack in flocking behaviour in large groups (> 10) with regular fragmentation. Also, it has been seen that reliability falls off sharply with larger swarms, and typically more than 30 agents will struggle to form coherent swarms. A more sophisticated model is by Couzin et al. [75] which uses group velocity and angular momentum to classify collective behaviour. For an agent in a collective there are three zones modelled as spheres, shown in Figure 6.7 as concentric circles in a two dimensional projection of this model. The three zones are named zone of repulsion (zor), zone of orientation (zoo) and zone of attraction (zoa). The agent attempts to keep a minimum distance from other agents within zor, it tries to align itself with its neighbours in zoo, which tends to minimise collisions, and the agent shows an attraction to other agents and becoming a part of the collective in zoa. Applying this model to dynamic scenarios leads to four types of classifications. (1) Swarm, here the group cohesion happens with low group velocity and low angular momentum with lack of parallel alignment. This happens

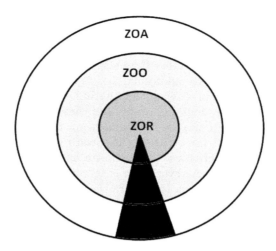

FIGURE 6.7 **Couzin's model**, considering a hypothetical agent in a collective. There are three spherical zones shown here as circles in two dimension with a blind zone directly behind the agent. These three zones determine the social behaviour and cohesion with other agents in the collective, zone of repulsion (zor), zone of orientation (zoo) and zone of attraction (zoa). Swarm formation is seen when attraction and repulsion are more predominant in comparison to orientation.

when attraction and repulsion happens more often, in comparison to parallel alignment. (2) Torus geometry, in this scenario the individuals of the collective perpetually rotate about the centre of the formation. This is characterised by low group velocity but high angular velocity. This has a tendency to happen when zone of orientation is small, and zone of attraction large in comparison. (3) Dynamic parallel group, this scenario is characterised with high group velocity and low angular velocity. It is more mobile than swarm or torus behaviours and (4) Highly parallel group, this scenario is characterised with very high group velocity and rectilinear movement this behaviour is seen to occur when zone of orientation is more dominant than any other zone. As per Couzin's model, swarming is typically seen in insects. The dynamic parallel group occur in collectives such as bird flocks and fish schools. The torus is a rarity, and the highly parallel group lacks in exchange of information across its constituent agents.

6.3.2   Signatures of swarm robotics

The first robotic swarm was designed by Maja Mataric at MIT in the late 1990s named 'The Nerd Herd', it was not until the new millennium that pioneering works by Beni [36], Sahin [287], Winfield, Kazadi, Dorigo and Vaughan helped its founding as a new discipline. Since then, swarm robotics has never failed to garner the attention of the robotics community.

Being a relatively new field of study, the definition of swarm robotics has evolved to incorporate novelties of newer research. Extending concepts from swarm intelligence to seamlessly lead to swarm robotics, Beni defines an intelligent swarm as:

> "Intelligent swarm: a group of non-intelligent robots ("machines") capable of universal material computation"

here, universal material computation is meant in the spirit of number crunching, pattern synthesis, playing chess etc. In contrast, Sahin defines it as an extrapolation of agent-world interactions:

> "Swarm robotics is a study of how large number of relatively simple physically embodied agents can be designed such that the desired collective behaviour emerges from the local interactions among agents and between the agents and the environment"

Schmickl et al. [291] put numbers to express the idea of a swarm:

> "In robot swarms, hundreds or thousands of small and simple robots have to perform in a well-organized and efficient way to pursue common goals"

It is worth noting that, in these definition there is no marking for a lower bound for a swarm, but the norm is about ten to twenty robots while researchers have also been seen to work with smaller group sizes. In the other extreme, simulation of swarms as high as of 100,000 robots have been successfully designed, though however most experiments are limited to about 20 robots.

In later research, Celikkanat and Sahin [62] also incorporated self organisation and the prospects of engineering a swarm of robots into its definition:

> "Swarm robotics, which takes its inspiration from natural swarms such as ant colonies, aims to tackle this challenge by analyzing the coordination mechanisms that underly the self-organized operation of natural swarms and using them to "engineer" self-organization in large groups of robots "

Swarms are usually formed of highly homogeneous robots, which look the same, with the same hardware and have about the same software complexity. Heterogeneous teams of robots such as in Robo Soccer have roles defined for each robot and are not so much examples of swarm intelligence and as coordinated teams of robots.

A swarm of robots can be identified with the following characteristics;: minimalism, stigmergy, emergence, change of phase and self organisation. Though it should be noted that these are not pre-requisites of swarm formation.

6.3.2.1   Minimalism: Non-intelligent robot, intelligent swarm

The robots for engineering a swarm should be simple and equipped with low-end hardware [237] and minimal software design [238], relatively incapable and inefficient to accomplish the assigned task with sensing capabilities and mutual communication limited to the local environment, preferably only with the nearest neighbours. These desired shortcomings ensure that the robots find it difficult and rather often impossible to execute the given task on their own, seeding grounds for implicit cooperation and development of the swarm. Employing a number of such simple robots for the given task improves performance. Communication with the nearest neighbours fosters distributed coordination across the group.

6.3.2.2   Stigmergy: Indirect interactions

Direct communication between two simple agents can be via broadcast over the wireless and explicit signaling, however interaction can also be more subtle and indirect. In stigmergy, communication proceeds by modifying the environment. In it, one of the agents modifies

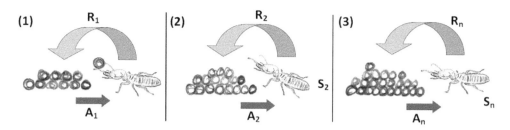

FIGURE 6.8 **Stigmergy in termite nest building.** In (1) at an arbitrary time, the nest size is A_1 which triggers response R_1 in the worker termite, which then adds another soil pellet to the nest, thereby changing both the nest size and also any later response from the rest of the termites, as shown in (2). In general as shown in (3), the stimuli R_n at a nest size A_n is attributed to the nest building by nearly the entire swarm and not one individual or a central executive. Therefore coordination is not constant as in a networked system, but varies with the contribution of the entire swarm. Also, a task which doesn't have sufficient participation will fall short in crossing over the minimum threshold of the response and will not be completed. Adapted from Bonabeau et al. [40]

the environment and the other responds to it. One of the foremost examples of stigmergy is foraging in ants. Each ant deposits pheromones on the ground, making it possible for the next ant to follow the trail.

Another example is in termite nest building, as is shown in Figure 6.8, where the worker termites use their pheromone, to make soil pellets rolled from clay and mud. These pellets are used to build their nests, by stacking pellets one on top of another. This process enables diffusion of the pheromone and therefore stimulates the worker termites to accumulate more pellets and carry on with the task of nest building. This stigmergic interaction works only if the density of the termites is above a minimal threshold. If density is lower than the threshold, then pheromones will start to evaporate, curbing stimulation and impeding nest building.

Therefore, communication in a swarm can be seen as action of an agent which modifies the subsequent behaviour of one or more agents in the system. This can be by signalling or by stigmergy.

6.3.2.3 Emergence: Swarm behaviour is difficult to model

Swarming cannot be achieved by direct design, and needs inter-agent communication, which is often not direct, such that the global behaviour is not programmed in for each single agent. Rather it manifests with the interaction of the agents mutually and also with the environment — the global behaviour emerges. Emergence is not a part of the definition of a swarm, however rich complexity at the global level can be a virtue of the system and can prove to be highly beneficial. Behaviour of a swarm is notoriously difficult to predict, and is still to be honed into an engineering tool. Emergence of a global behaviour is due to the interactions of the agents with each other and the environment and cannot be arrived at by summing behaviours of individual agents in the environment. Since emergence is a key in behaviour-based approach, it is often seen as an extension of summing or layering of encapsulated behaviours. While simple behaviours in single agent behaviour-based systems as 'line following' and 'obstacle avoidance' can easily be designed with little experience, swarms lack such an advantage and even simple behaviours are often difficult to implement.

6.3.2.4 Phase change: Disorder to order

Aggregation in self ordered systems, towards a common goal, or preference for a particular direction has been associated with thermodynamic phase change. The best example is from physics, i.e., the alignment of ferromagnetic materials exposed to an external magnetic field. At the microscopic level, a given particle driven with a constant absolute velocity assumes the average direction of particles in the neighbourhood of a given radius, subject some random perturbation. The net momentum of the interacting particles is not conserved during collision and leads to this kinetic phase change. Vicsek et al. [336] report on their research into ferromagnetic particles and suggest a thumb rule to identify the phase change. Their research concludes that the average normalised velocity of the aggregate is very nearly zero if direction of the motion of the individual particles are distributed randomly as in a typical Maxwell distribution, but for ordered direction of the velocities the average normalised velocity (v_{avg}) over N particles is nearly 1.

$$v_{avg} = \frac{1}{N} \left| \sum_{k=1}^{N} \mathbf{v}_i \right| = 1$$

6.3.2.5 Self organisation: A dynamically stable swarm

The achievement of order in the global level of the swarm due to individual interaction of its agents is probably the most striking feature of the swarm. It is dynamically stable as two phenomena occur in tandem, positive feedback and negative feedback. Positive feedback encourages the formation of order in the swarm. Typical examples are finding a food source, which is accompanied by a group activity, which can be trail laying and trail following in ants, and dancing in bees, where a bee finding a nectar source indicates so by dancing. Negative feedback equilibrates the positive feedback, and is seen as food source exhaustion, limited number of worker population, crowding at the food source etc. Self organisation relies on sensing errors and fluctuations, and randomness is welcome. For example, in foraging, an ant following the trail of pheromones with some error may get lost only to find a newer food source or a more efficient path to the food source. In birds it has been seen that sensor noise leads to better accuracy in homing to a food source etc.

It is to be noted that these are earmarks of swarm robotics, and not conditions or prerequisites to generate a swarm. Also, none of them can exist independently and all are interrelated phenomena.

6.3.3 Metrics for swarm robotics

To quantify the study of a swarm, researchers have used various methodologies, however the following four metrics [329] have been used time and again by a number research groups.

1. Order (ψ) or the angular order is an indicator of the alignment of the robots in a collective and is a direct extension from Vicsek's model [336]. It is defined as;

$$\psi = \frac{1}{N} \left| \sum_{k=1}^{N} exp(i\theta_k) \right|$$

where θ_k is the angular alignment of the kth robot. ψ will take a value of 1 for a perfectly ordered swarm and values closer to zero for nearly disordered ones. A value closer to 1 will indicate a good quality swarm.

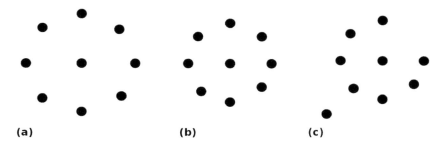

(a)   **(b)**   **(c)**

FIGURE 6.9 **Hierarchic social entropy (S)**, is a measure of diversity of a group. S is less for more compact systems and those which have an ordering of the members and reflect symmetry. Considering the following 3 orientations with the black circles as mobile robots, then $S_a > S_b$ since configuration-b is more compact compared to configuration-a and $S_c > S_b$ as configuration-c lacks the circular symmetry of configuration-b. While the value of S doesn't convey much, a positive $\frac{dS}{dt}$ will suggest an unstable swarm, while a negative $\frac{dS}{dt}$ will mean attaining order with increasing time and therefore a stable and coherent swarm.

2. Hierarchic Social Entropy (S) is a measure of diversity and therefore positional order [26], as shown in Figure 6.9 . This method is derived from Shannon's theory for information uncertainty to measure societal diversity in heterogeneous systems of mobile robots. To evaluate S, consider a diversity metric *H(R,h)* for a system of robots, where for 2 robots i and j are a part of the same cluster if

$$\|\mathbf{r}_i - \mathbf{r}_j\| \leq h$$

and R is a society which contains these various clusters. *H(R,h)* is obtained by summing over all clusters.

$$H(R, h) = \sum_{k=1}^{M} p_k log_2(p_k)$$

where p_k is the proportion of robots in the kth cluster and M is the number of clusters for a given h. To obtain hierarchical social entropy S, *H(R,h)* is integrated from 0 to ∞,

$$S = \int_0^\infty H(R,h)dh$$

Since entropy must decrease with time a positive $\frac{dS}{dt}$ will indicate an unstable swarm while a negative $\frac{dS}{dt}$ will relate to a stable and coherent swarm. Since both order and entropy depend on geometric arrangement, in experiments it is evaluated with the help of an overhead camera.

3. Average forward velocity of the geometric center of the swarm (V_G). A high (V_G) will reflect an efficiently moving swarm while a low (V_G) will suggest lack of cohesion.

4. Size of the largest cluster, the largest distinguishable group will be an indicator of cohesion in the swarm.

An ideal swarm, will have high order denoting alignment in the swarm, nearly constant entropy implying coherence, $\frac{dS}{dt}$ nearly zero suggesting a lack of diversity, high swarm velocity and one large cluster.

6.3.4   Swarm Engineering — Visions for a new technology

To employ artificial swarms as a technology, it has to be easily controllable, nearly predictable and a dependable system with well-defined goals which are tailored to commercial needs. Pioneers have attempted to study [15, 16] and design [275] such decentralised, on the go intelligence and efforts to integrate and manipulate organised animal and insect societies with mobile robots has exploited the ANIMAT principles. The Robotic Sheepdog project, where a flocking behaviour in the ducks was influenced by a robot and the Insbot project where olfactory sensing allowed robots to find acceptance in cockroach society, are vivid examples of engineering and manipulating naturally occurring swarms.

However, engineering swarming behaviour to design dependable and nearly predictable swarms of robots is not just difficult, but poses as an oxymoron since design principles attempt to ensure high level of integrity and reliable performance of the given system while swarming is the novelty in the collective agency which lacks in predictability.

Therefore a system working on swarm engineering will need to be designed and validated for a high level of assurance and also a safety critical stamping so that it doesn't lead to any unintended behaviour. Extrapolating a set of 'atomic' behaviours for a single robot to swarms is futile because the next behavioural state of a swarm will depend on the recent history and local conditions, thus reducing a design paradigm from a methodology to a serendipity. A brute force, hit and trial approach which is preferably restricted to a known environment will still need many attempts to perfect and yet may lack in a safety critical stamping, is as follows:

Choose Local Rules → Swarm Test → Desired Global Behaviour (?)

Here, the *Swarm Test* can be conducted to some extent by simulations or on real robots, and this process can be made to continue until the desired global behaviour is obtained. Since swarming by definition lacks hierarchical command or a control structure, a quotient of performance, robustness and vulnerability, or even defining a point of failure becomes a self defeating task. Systems which have a high degree of behaviour-based elements are typically emergent in response, and validating such a system is often context dependent and lacks in a generality. Studies have confirmed that swarm intelligence affords much higher levels of robustness and fault tolerance. On the flip side is the lack of ability to control and mediate such a system. It is seen in nature that ants in their social niche are excellent managers of their food supplies, and termites work in cooperation to build their many layered catacomb-like nests and migratory birds travel across swathes to reach a more welcoming habitat. Therefore, nature once again serves as a great motivator which begs us to search for principles and design paradigms to ensure dependability and performance if we are to use robot swarms as a technology.

Winfield [365] named this field of study, and developed methods to manipulate and engineer a group of robots to form a swarm by simple wireless communications where each robot can broadcast its own unique ID and also that of its nearest neighbours over a network such as telnet to form swarms as seen in Figure 6.10. With minimal information exchange over the network and without the robots requiring absolute or relative position coordinates, the swarm acts coherently as a single entity. Leading to a highly robust and scalable swarm, the resources for which increase linearly with swarm size. This type of swarming is not based on any aspects of navigation, but the robots perform operations which are close derivatives of navigation and path planning. Such type of implementation in the Linuxbots lead to similar results in a real-world situation over noisy communications channels. Therefore, Winfield suggested that systems which rely on emergence should not be more difficult to validate

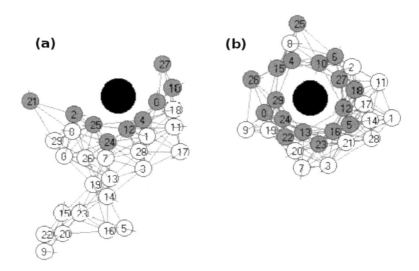

FIGURE 6.10 **Emergent encapsulation over an adhoc network.** Over a wireless network with only information of the robot's ID and that of its nearest neighbours and the condition that connectivity is established only over a single hop, the robots (a) gradually approach and (b) demonstrate emergent behaviour to encapsulate the obstacle. Both images from Winfield [365], used with permission from Springer.

than conventional complex systems, and that some characteristics of swarm robotics affords more dependability than traditional systems.

Analysis for failure mode is not well defined for a swarm. To explain the concept of dependability of a swarm, Winfield defines 2 parameters, (1) liveness and (2) safety. **Liveness** is the property of exhibiting derisable behaviours. This would need mathematical analysis at two levels, agent level and swarm level. While a single robot yields to stability analysis, there is a lack of methods for analysis of swarm stability. Though stochastic analysis may find some appreciation, simulation is strongly discouraged and only considered as a tool for understanding and prototyping. **Safety** is the property of not exhibiting undesirable behaviours. While it may seem that liveness and safety are reciprocal of each other, it is not so by definition and a system which is safe may lack in liveness, therefore doing less, albeit safely. Tackling safety is akin to hazard analysis. Swarming by its manifestation confirms robustness and flexibility, so failure of an agent or lack of cohesion with some agents will not deter the swarm from its behaviour. Therefore, Winfield suggests taking a constructivist approach for swarm engineering. Learn about swarm behaviour by designing swarm, where both processes work in tandem.

As a conclusion and in an attempt to overcome the issue of dependability, a way out is by designing robots with internal models, which can do a quick test by simulation prior to committing to a task. In hindsight, such may obstruct the fundamentals of swarm intelligence and the inherent novelty of the process, where the solitary agent lacks the understanding of the group behaviour and emergence is the key to intelligence. However, robots with internal states may be the answer to ethical robots, and discussed in Chapter 8.

Working independently at about the same time as Winfield, Kazadi [174] suggested methods to engineer the path of a swarm of robots by tracking a plume of gas. In his doctoral research Kazadi formalised swarm engineering as a two-step approach.

1. **Step-1** the swarm condition as an extension of the problem definition, which will lead to the formation of a swarm of agents which is capable of carrying out the given task
2. **Step-2** produce behaviour(s) that satisfy the above condition

These two steps can also be seen in Reynolds' boids [275]. The first two rules define the swarm condition; in the third rule each boid attempts to stay close to the flock members which can only be achieved by velocity matching. Likewise, Winfield's swarming behaviour over wireless communications to form an adhoc network can also be seen as a two-step approach, where as a first step each robot can broadcast its own ID and also that of its nearest neighbours. In the second step connectivity has to be only a single hop, which ensures one single large cluster and group cohesion. The first step confirms the robots' ability to participate in a swarm. The second is usually relaying information, which enables group cohesion. The second step leads to the formation of swarms. This has also been reported in other studies [237].

As emergence is a common feature to both swarm engineering and behaviour-based robotics and both are bottom-up approaches, it is often seen as an overlapping discipline. However, this is not entirely correct. Behaviour-based robotics is more intuitive than a formal paradigm and targets the development of a single working prototype. In sheer contrast, swarm engineering is aimed to produce a general set of conditions which may be used to generate various swarm designs, and as a second step, design behaviours which satisfy the previous conditions, which with an appreciable degree of certainty will accomplish the given goal(s).

Since Winfield and Kazadi, two broad methods to engineer a swarm have found interest across a number of research groups. These are by (1) addition of behaviours and (2) evolutionary methods.

For the first one, extending on Reynolds' Boids and Mataric's Nerd Herd, an additivity of behaviours seems a good design methodology for swarms. Turgut et al. [329] design flocking behaviour based on artificial physics which structures Reynolds' boid model. The swarm is characterised by a virtual force vector ($\mathbf{f}$);

$$\mathbf{f} = \alpha\mathbf{p} + \beta\mathbf{h} + \gamma\mathbf{g} \tag{6.1}$$

which is the addition of behaviours, $\mathbf{p}$ is proximal control, $\mathbf{h}$ is the alignment vector and $\mathbf{g}$ is the goal direction and α, β and γ are weight factors which can be arbitrary or tunable. Proximal control behaviour assumes that the robot can perceive the relative position, the range and the bearing of its neighbours in close proximity and use it to avoid collisions while maintaining cohesion between robots. This is done using an omnidirectional camera and short range IR sensors. For a generalisation, considering k robots, for the ith robot with range (d_i) and bearing (ϕ_i) it is given as the vectorial sum of all virtual forces:

$$\mathbf{p} = \sum_{i=1}^{k} p_i e^{j\phi_i} \tag{6.2}$$

where p_i may be modelled as a potential field with inverse function of distance (d_i) or other heuristics. However it is always characterised with a desired distance d_{des} which is the minimum distance each robot can move without colliding with its nearest neighbour. Therefore, p_i is repulsive when $d_i < d_{des}$ and attractive otherwise:

$$p_i = \begin{cases} -ve & for & d_i > d_{des} \\ +ve & for & d_i < d_{des} \end{cases} \tag{6.3}$$

The Swarm Robots

A robot for forming swarms is marked by simple design. It often lacks multiple sensors, light weight, mostly uses cross complied program codes on low powered microcontrollers, low cost and is typically made up of differential drive robots which are limited to behaviours of navigation and path planning. The individual ability of each robot is sized down, anticipating that the swarm will be able to manifest novelty as an emergent property, beyond the ability of the individual robot. Researchers have used fleets of e-pucks, Khepera and Pioneer to form swarms.

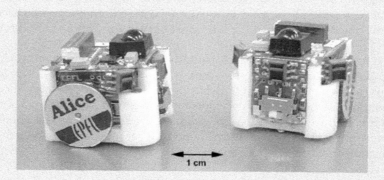

FIGURE 6.11 **Alice**, building microsized robots has drawn interest from the late 1990s. Alice has dimensions of 2 cm x 2 cm x 2 cm and was developed at EPFL between 1998 and 2004. CC image courtesy *http://en.wikipedia.org*.

Making a robot platform specifically for swarming capabilities has been the trend over the last decade. Winfield and Holland developed the Linuxbot architecture. These were robots with wireless and could be controlled over a network such as *telnet*. The Linuxbots could communicate over the local WLAN (12 Mbits/s) and form aggregations over this network of wireless IPs.

FIGURE 6.12 **Robots with the wireless module and Linuxbot.** the image on the left is of a robot with the wireless module [358], and the one on the right is of Linuxbot [365]. Used with permission from Elsevier and with permission from Springer respectively.

Turgut et al. [328] developed the Kobot as a differential drive robot with 2 DC motors. It is controlled with the PIC microcontroller and a wireless module and is lightweight at 350 grams.

FIGURE 6.13 **20 Kobots.** The Kobot with a 12 cm diameter and 350 grams is lightweight and easy to maneuver and is equipped with 2 primary sensory systems. (1) Infra Red Short Range Sensing, with 8 IR sensors located at 45 degree intervals around the base which can detect objects within a range of 21 cms and can distinguish robots from obstacles. (2) Virtual Heading Sensor is composed of a digital compass and a wireless module and the robot's heading with the sensed North is broadcast over the wireless. The Kobot was designed with its physics-based simulator CoSS and nearly every experiment is also supported with simulations in CoSS.

FIGURE 6.14 **s-bots forming swarm-bot configuration for passing over a gap** is reminscent of chain formation in ants. Image courtesy https://en.wikipedia.org/, CC BY-SA 3.0 license.

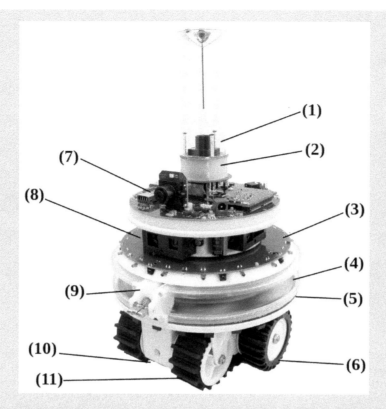

FIGURE 6.15 **The footbot** is a part of the Swarmanoid project and it is equipped with (1) an omnidirectional camera, (2) a beacon, (3) range and bearing system, (4) a ring of RGB-LED, (5) a ring of proximity and light sensors, (6) Treels, (7) cameras for ceiling and front, (8) distance scanner, (9) gripper add on facility, (10) RFID and (11) ground sensors [97]. Image courtesy Francesco Mondada at EPFL, Lausanne.

The s-bot was designed with each robot comprising about 100 major parts. It was spread over 116 mm diameter and was controlled by an ARM processor with 64 MB of RAM running at 400 MHz, and consumed 700 mW power. The s-bot was equipped with a gripper and the goal of the project was to develop swarms by physical interconenctions — each robot gripping to the next to achieve goals. The gripper was a novelty of the design and was used to pick up objects, transfer objects to another robot and also physically link up to another robot if such a collective need arises, viz., crossing over an obstacle as shown in Figure 6.14. The s-bots are able to navigate, grasp objects and communicate with each other and if required physically connect to each other using the gripper to form a super-entity, the swarm-bot. Mondada et al. suggests that a swarm-bot is composed of anywhere between 2 and 35 physically connected s-bots and is superior in performance to a single s-bot, particularly in exploration, navigation and transportation. The Swarmanoid project (2006-2010) at IRIDIA, Brussels, led to the foot-bot, the hand-bot and the eye-bot. The footbot is an improvement from the Kobot design and s-bot designs. As shown in Figure 6.15, it has a distance scanner, a proximity and light sensor, range and bearing system and a gripper add-on facility like the s-bots.

Alignment behaviour considers that the robot can measure its own heading angle (θ_i) and align to the average for k robots,

$$\mathbf{h} = \frac{\sum_{i=1}^{k} e^{j\theta_i}}{\left\|\sum_{i=1}^{k} e^{j\theta_i}\right\|} \qquad (6.4)$$

Goal direction, for simple scenarios $\mathbf{g}$ is not required as the former two behaviours are sufficient for forming swarms (less than 10 agents). However this behaviour serves as a tool to engineer larger coherent swarms to desired results. The goal direction is available to a minority of the robots. If the robot is aware of the direction to the goal, then this tallies to the angle corresponding to the goal($\angle\mathbf{g}$). Otherwise it corresponds to the angle for the average heading alignment ($\angle\mathbf{h}$). The informed robots relay the goal direction and therefore can influence a significant section of the swarm towards moving towards the goal. Leadership in the natural world, in such animals as cats, dogs and monkeys, is marked by vocalisation and gestures to mark out territory or in attempts at claiming a higher portion of the hunt etc. In flocks of birds and schools of fish leadership is oftentimes not explicit, due to larger body size, vocalisation, ability to dominate the group etc. but is implicit. Leadership is naturally formed from informed agency when a minority of agents in a swarm or group are aware of added information from the environment and use it to lead the group to food, a safe migration route and newer nesting locations. Studies have shown that in honeybees, as low as 5% of the swarm can guide the group to the location for a new nest, and only a handful of members in fish schools are seen to supervise foraging behaviour of the school. In fishes, this emergence of leadership is implicit and is not accompanied by any signaling mechanism or high-level transfer of information, while honeybees communicate with a honing sound accompanied by rapid movements which seem to be a 'dancing stance'. In migratory birds, the older and more experienced members take on the role of natural leaders in tracking migratory routes, while similar studies in cow herds has shown that implementing a behaviour on a minority of the cows can lead to guiding the herd to a desired collection behaviour or a preferred location, Correll et al. fitted a minority of cows with a mountable device which induced stress, this increased the tendency for the cows to form aggregates and move towards the center of the herd. In all of these scenarios, the leadership is due to the implicit information in the minority of agents and is born out of limited cognitive ability with none to very little transfer of information. It is a roboticist's dream to replicate such informed agents. A stumbling block is the lack of information of animal and insect social behaviour and preferences. Therefore the attempt is to design mathematical models [74, 80]. Celikkanat et al. [63] model informed agency as the closest neighbours aligning to the same heading angle, while the informed robots align to the goal direction. This manipulation of an informed agent may seem like a 'follow the leader' behaviour, but it is decentralised and not a behaviour which has been programmed into the robots, and it is an emergent phenomenon. Simulations suggests noise in the proximal control behaviour with no alignment behaviour has a surprisingly positive tendency to align the collective swarm behaviour. In contrast, the alignment behaviour is most appreciable in larger swarms, while smaller swarms, are governed by goal direction behaviour and, without any alignment behaviour in smaller swarms the robots achieve better accuracy. Therefore, low ratio of informed robots leads to fragmentation of the swarm. This result in robotics aligns with that in biology. It was also seen that accuracy is independent of the flock size for a fixed ratio of informed robots, while accuracy improved with increasing numbers of informed agents. Hence, the phenomenon of influencing a swarm with informed agents is scalable, flexible and is also robust, though performance falls off on reducing the number of informed robots.

Employing informed agency one can engineering a swarm for a desired behaviour or to a desired location either by increasing the participation of the informed robots — increasing γ or by tuning the ratio of informed robots to total robots (ρ), which has direct influence on the collective behaviour. On increasing the weight factor γ, it was seen that swarms with moderately low ratio of informed robots ($\rho \sim 10\%$) show improvement in accuracy. In contrast swarms with a high number of informed robots ($\rho > 80\%$) or very low number of informed robots ($\rho < 5\%$) are immune to the change in weight factor. Likewise, on increasing ρ, it was seen the accuracy increases asymptotically for swarms with $\rho > 40\%$ and it converges to a maximum value around 80% to 90%.

A special case of informed agency is migratory birds, as they have an additional piece of information obtained from the environment, that of direction to goal or homing behaviour determined from earth's magnetic field. Since every bird in the flock has this additional information, all the agents are informed agents. Models of long-range migration [128] have demonstrated that for sufficiently large flock size, noise in proximal sensing is suppressed, sensor noise in homing leads to better accuracy and average speed of the flock is more or less near to that of the individual agent.

A second approach is by using evolutionary methods, where an artificial neural network works to improve solitary agent behaviour, thereby improving the swarm performance.

Despite the novelty inherent in swarm robotics, lower net costs and higher efficiency, experiments and maintenance of a large number of robots will always be a hurdle for research and applications. Researchers contend to study very large swarms in simulations due to economic and logistic considerations. Simulations are insufficient to study swarming, since they can never replicate the intricacies of the real world in its entirety. Even then simulations have been used as a tool to study and to have a partial grasp on the implementation level of behaviour, swarm cohesion, synthetic leadership etc.

Two most popular simulators are (1) Stage [334] and (2) ARGoS [264]. Stage, a 2.5 dimensional simulator of the Player/Stage/Gazebo project has been a benchmark software for swarm robotics simulations. Version 3 of Stage is built on OpenGL and FLTK. It could run 1000 robots at about real time and 100,000 robots at slower than real time scales. Code reusability and open source licensing for Stage encouraged crowdsourcing and various popular robots, such as Pioneer, Khepera, Roomba etc., were available online. Stage was very good for navigation and simulation of simple behaviours. The shortcomings were (1) that all designs was predominantly extended from blocks and making more complex robot designs were not very user friendly and (2) sophistications such as gripping and self assembly were not possible.

Autonomous Robots Go Swarming (ARGoS) a simulator and control interface for swarm robotics in three dimensions was developed by Pinciroli et al. at IRIDIA, Belgium. It was designed for the robots of the Swarmanoid project and in addition it also supported e-pucks and few of the other popular robots. ARGoS has an open source license and runs on Linux. Unlike its predecessors, ARGoS employs a number of physics engine, viz. ODE, Chipmunk etc, and it was designed as a multiple threaded architecture. The visual interface is based on Qt4 and OpenGL, which supported scripts in C++. ARGoS can simulate 10,000 robots faster than real time and has been tested to simulate as many as 100,000 robots.

SUMMARY

1. Many robot systems are important since they can achieve lot more than single robot systems.
2. Networked robotics and ubiquitous robotics are the buzz words which are models

to unify robotics with embedded electronics and other ambient intelligence via the internet and also over local networks.

3. Swarm robotics is a very direct motivation from birds and insect colonies and it is marked by high degree of emergence and novelty in group behaviour.

4. Engineering a swarm is difficiult compared to a group of robots, and newer techniques are being explored by researchers. (1) Informed agency and (2) Evolutionary techniques are two popular routes exploited by AI scientists.

NOTES

1. *Reynolds' boid model is the first flocking model outside biology and natural sciences, it was designed for computer animation and it was published in ACM Journal of Computer Graphics. The animation required resources which increased nearly as a square of the boid population.*

2. *Sensor noise is seen to act as a desired evil in swarm robotics and researchers have time and again reported that noise aids in achieving accuracy to a desired goal location and also improved the alignment of the swarm.*

3. *The robustness of a swarm can be quantified with the parameter called swarm repair time which is defined as the average time taken for the swarm to regain its tendencies as were prior to the agent(s) failure. It can be analytically shown that swarm repair time is linearly dependent on swarm size.*

4. *Trianni, in his doctoral dissertation, offers another classification of multirobot systems:*

    (a) *Collective robotics, characterised by explicit interagent communication*

    (b) *Second-order robotics, such as self assembling robots and self reconfigurable systems, which are physically connected robots*

    (c) *Swarm robotics, interplay of agents and environment leads to a nouvelle global behaviour*

5. *Since heading angle measurement also involves the participation of a wireless module, exchange of information such as heading angle or robot's unique ID on the wireless network both work towards implicit cooperation in the swarm.*

6. *Developing a collective perception in a swarm by collating a large number of local perceptions, such as building a global map with local sensor values of the individual agents, 'a swarm SLAM' as of now is very limited.*

7. *Nasseri and Asadpour [251] have reported design of 'selfish' and 'social' informed agency by tinkering with the intensity of the potential field function.*

8. *Onyx parachutes have employed swarming techniques for landing maneuvers*

EXERCISES

1. What are the distinguishing features between a swarm of robots and a group of robots. From a behaviour point of view which one can be engineered easily?

CHAPTER 7

Human-robot interaction and robots for human society

" In the end, robots may have to be much smarter to appear comfortably dumb."

– Robin Murphy [208]

"You can take my androids on planes — the torso in the suitcase and the head in carry-on"

– Hiroshi Ishiguro [252]

7.1 HUMAN-ROBOT INTERACTION

THE last few chapters have discussed simple AI agency, robot design, programming and machine learning techniques to build robots which can negotiate low-level tasks such as navigation, easy engagement and manipulation in their local environment and formation of robot groups and swarms. In the new millennium, robots are poised to accomplish much more by either bonding with human beings or alternatively finding unique niches for themselves, in prospect making important contributions to human society. This chapter focuses on human robot relations and how robots can contribute to human society — in the Asimovian spirit. In a way, this is a first part of human robot interactions and Chapter 8 is a second part, in continuation of this chapter.

Human beings are unique, and are the only agency which boasts of a high degree of autonomy and the highest level of intelligent behaviour, as shown in Figure 7.1. Therefore, robots designed for effective human-robot interaction will have to address both these facets and match up to human beings in appearance, autonomy and intelligent behaviour. Unlike the robots discussed previously, social robots are designed to respond to the perception of the people interacting with them. Acceptance of the social robot as human being or at least approaching human-like interaction is the underlying key to effective, long-term and fulfilling human-robot interaction. The PR2 robot playing billiards with the developers of the erstwhile Willow Garage and winning, and President Obama meeting with the ASIMO, as shown in Figure 7.2 are some of the endearing freeze frames of human-robot interaction.

Social robotics are designed such that they attempt to mirror human emotions and

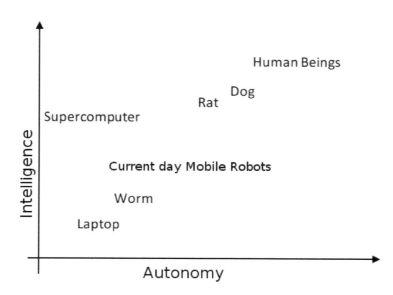

FIGURE 7.1 **Intelligence vs. autonomy.** Human beings have high intelligence and also high autonomy, like no other animal or AI beings. To effectively participate in a society with human beings, robots will have to match both of these. Adapted from Long and Kelly [207].

empathy by manipulating the interacting human's expressions and intonations, and are usually designed to work in a social role or function, viz., nurse robot, companion robot, museum guide robot etc. Mirroring and imitating behaviour was also seen in ANIMAT models and in swarm robotics. This chapter will discuss how mirroring works as a tool for social acceptance. Mirroring is also the basis of the 'learn by observing' and 'learn by demonstration' approaches in machine learning and it is a marks towards higher modes of cognition and conscious behaviour, as will be again discussed in Chapter 9.

In contrast to social robots are the explorer robots which chart unknown terrains on a far off planet or a very difficult terrain on Earth. They are designed with sleek hardware and controlled from mission base which is usually located far away.

A lesser variant of the explorer robot are the delivery robots which are becoming commonplace for redundant and repetitive delivery jobs are controlled over two modes, local navigation and global navigation. A further distinction can be seen in the search and rescue robots which are tethered and designed on simple and easy to use technology than GPS driven snazzy gizmos.

It is expected that in the next few decades natural human interactions such as voice, vision, face expressions, etc., will form seamless, calm, context aware and ubiquitous human-robot social interactions. Robots that can attend to domestic chores such as cleaning the house, attending to the lawn, cooking etc. are not affordable as of now. However with newer technology and mass production the price will be around a few thousand dollars and therefore more manageable for a middle income household. This chapter highlights human-robot interaction and service to society.

Human-robot interactions can be classified as;

1. Anthropomorphic humanoid robots such as Kodomoroid, ROMAN, Inkha and KASPAR [86], which are designed with human facial features and their interaction

FIGURE 7.2 **President Obama meets ASIMO**, at the Miraikan National Museum of Emerging Science and Innovation, in Tokyo, Japan, photograph courtesy *www.whitehouse.gov*.

with the human user is over vision and voice channels. They appeal as a near-human beings.

2. Zoomorphic robots such as AIBO, Kismet and Paro [135, 369], which fill in as alternatives to pets by imitating care and compassion, and are most suitable as companions for the elderly. Paro has found application in psychological therapy of the depressed and those suffering from dementia; AIBO has been used for therapy in autistic children.

3. Caricatured robots such as CERO shown in Figure 7.13 and Sparky shown in Figure 7.9 have minimal design and bare basic facial features. These robots are inexpensive and easier to build and program, and they target few primary behaviours rather than a wide assortment as seen in Anthropomorphic or Zoomorphic robots.

4. Functional robots which are more like smart machines engaging in jobs for human society such as search and rescue, delivery, exploration, surveying and mapping etc. They are supervised by a human being over a global network, but lack emotional exchange of information.

The first two are domains of social robotics and are marked by response to natural human interactions, which are underscored by emotive exchanges over voice, vision and in some scenarios also text. Social robots have found various applications and can be seen at work in the care and hospitality industry, household robots, office assistant robots, etc., as shown in Figure 7.3.

FIGURE 7.3 Robots in society. An artist's sketch of future involvement of robots in society: (a) assistive robots, (b) playmate robots in child education, (c) robots for mentoring and assistance in manipulation tasks, (d) robots that teach movement exercises, (e) personal robots for the elderly, (f) robots for surveillance and protection of children and adults. Image from Schaal [289], reprinted by permission of the publisher Taylor & Francis Ltd., `http://www.tandfonline.com`.

7.1.1   Distributed cognition

Human-human interactions happen over a number of channels: (1) voice, which includes words and intonation, (2) vision, pertaining to appearance, body language and facial expressions and (3) touch such as tactile and proprioception. These three channels must work in unison to ensure socially acceptable interactions.

Distributed cognition [153] considers the working of such social functions as an emergence over a number of streams of information working in near parallel. As an example, a social robot eliciting a state of happiness will attempt to convey this with face expressions, muscle actuation, use of particular vocabulary to include expressions, words and positive intonation. Distributed cognition has similarities to behaviour-based approaches [160] since both rely on emergence from superpositioning of a number of near independent modules working in parallel, also both find inspiration from mother nature, insects & animals for behaviour based robotics and human social interactions for distributed cognition.

7.2   SOCIAL ROBOTICS

Kismet, shown in Figure 7.4 was the first social robot, with anthropomorphic features it was designed as a friendly pet and it could mirror the empathy and emotion of the interacting human being, thereby creating a fake sense of attention, care and attachment.

The long list of social robots developed over the last two decades since Kismet, shows that social robots have found acceptance in human society as shown in Figure 7.5 and promise to be the next big technological revolution. However, that raises questions, is the social robot merely a sleek and expensive toy fitted with modern technology? If such robots

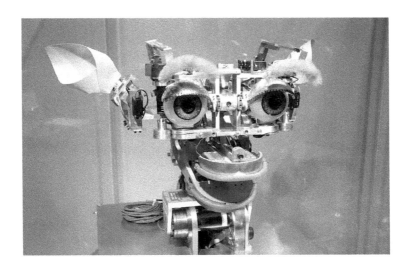

FIGURE 7.4 **Brazeal's Kismet.** This zoomorphic robot was the first of its kind. Image courtesy `wikimedia.org`, Creative Commons CC0 1.0 Universal Public Domain Dedication license.

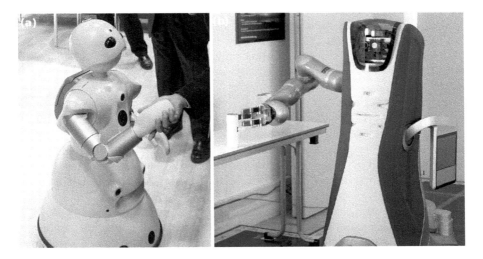

FIGURE 7.5 **Avenues of social robots**, typically target a niche to supplement human lives. Other than functioning as loyal servants, they have also found roles as assistants, companions, pets and also as stage performers and artists. Shown here are the Wakamaru on the left which was designed to help the elderly. It can talk, has a built in vocabulary of about 10,000 words and has even 'acted' in a Japanese play, and on the right is the Care-O-Bot robot which is modelled to do the job of a nurse and carer. Both images courtesy `wikimedia.org`, CC license

are readily accepted in our society, wouldn't it blur out the distinction between a toy and a living being?

Non-humanoid robots with little or no social skills have been very successful in both the industry and domestic sectors as functional machines and humanoids as ASIMO and PR2 on the other hand have only resulted in sleek and expensive toys and therefore, do we really need a social robot?

Another engaging inquiry is that, social robots are based on mimicking human expressions and emotions, which are obviously not felt by the robot, and are not genuine. Therefore are social robots really social? These inquiries form the social robot paradox, and are the reasons to attempt to develop ethical and conscious agents. These questions are tackles in Chapters 8 and 9.

Duffy suggests, that a social robot will always fake human traits. However high-quality synthetic human social behaviour through increasing behavioural complexity will make it seem more real than a well-designed gimmick. Therefore, such an artifact may beg our perception and our social ego to see it as being intelligent, with emotional states, and probably even appealing as conscious being. Robots designed and built with no reference to the human beings which shares their physical space may not exhibit behaviours which can be interpreted as artificially intelligent. A human assessment of a robot's intelligence will inherently bring in his/her own interpretation of intelligence, blendedin with a social quotient. For example a robot that doesn't reply to a social greeting as 'hello', may not be seen as very congenial and subsequently be perceived as nothing more than a machine. Therefore social robots can also be seen as the convergence of artificial and natural emotive abilities.

Social robots vary. Some are visually iconic while others are more realistic with facial expressions to elicit artificial emotional states and with synthetic skin and hair. The design of social robots is either focused on human-like appearance or human-like manners, with few robots matching up to both. At one end of the spectrum is the android research of Ishiguro and Mac Dorman which puts the emphasis on human like appearance. At the other end of the spectrum are CERO and Care-O-bot which have minimal human likeness, but they both make up for it in effective dynamics and orientation to convey human behaviour and intentions.

Humanoid robots, be it humanoids with legs such as the ASIMO or a human bust on wheels such as the PR2 and Wakamaru, have clearly found more social acceptance. It has been seen that employing the humanoid form inherently defines the robot as having some degree of social functionality. A human like face with lively features and a human-like exoskeleton ensure that even the most mundane interactions such as walking or blinking the eyes will apparently serve as rudiments of a social function, and be perceived as a human behaviour.

State-of-the-art robot, such as Pepper shown in Figure 7.6 is a high end social robot and is one of the best examples of a humanoid with a face. Pepper has superlative skills to identify and reflect human emotions and is arguably the best social robot to date. It was developed by Aldebaran for SoftBank, a Japanese company. Pepper is 28 kg and 1.2 m tall. It can identify emotion by analysing facial expressions and voice intonations. Priced at 198,000 yen or about $1700, it is a promise that social robots will be affordable for the middle class in the next few years. It is equipped with face-recognition technology and a number of sensors, camera and voice recorder. Pepper is equipped with learning tools rather than being programmed for specific tasks. This helps it to gradually get acquainted with human way of life and the various social and cultural functions. Pepper is designed to be

FIGURE 7.6 **Pepper** is a state-of-the-art social robot. It works on detecting and mirroring the expression and emotion of the human being which is interacting with it. It is priced at $1700, and as of the summer of 2016, Softbank had sold over 7000 pepper robots. Image courtesy *photozou.jp*, CC-by-SA license.

human centric, to interact with people and maintain a state of 'feel good' and happiness. It is not meant to be a personal robot or a domestic robot for household chores.

In zoomorphic robots, the social interaction is more or less based on mirroring. Kismet and Paro do not convey humanly intentions verbally but they are designed to communicate in non-verbal mode such as eye movement, purring and making affirmative noises etc. Similarly, for AIBO which is designed as a pet dog, the robot's behaviour is designed to be unexpected and sufficiently complex, as that of a real pet dog, to keep the owners interested in watching it.

7.2.1 Design for social robots

As shown in Figure 7.7, social robotics involves a number of disciplines and a social robot is both a work of art as well as technology. As design paradigm, it should be situated in a social environment with amicable human beings and the interactions should be in near real time. Social expectations and tailoring emotional interactions, the robot should appeal as a socially aware entity and broadly adhere to friendly interactions. Most social robotics research groups control this aspect by making the robot appear as a child or a pet, thereby preventing intense interactions which may demand higher levels of cognition and richer emotive modes. Some designers explicitly target the user's attention by engaging in eye contact or encourage interaction with a playful smile. The contrast has also worked well where the robot doesn't attempt to explicitly foster user reaction, but engages in human-like tasks to direct user attention, but in a more implicit manner.

Motivation has been from the study of infants, conversing with care givers and study of

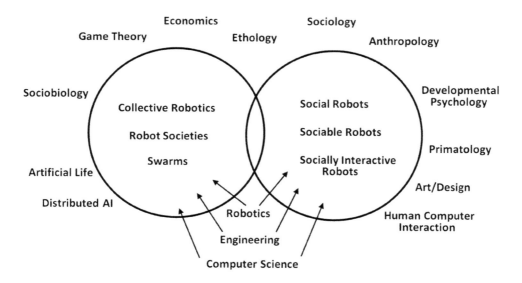

FIGURE 7.7 **Collective and social robotics** is the confluence of number of disciplines. Adapted from Fong et. al [107]

animal-human interactions, the design paradigm can is broadly dictated by, (1) aesthetics, (2) facial expressions and (3) language processing, all three serving as a cue to convey the human emotional state, humanly intention and focus the avenue the robot was designed for. Other than these, a social robot should be easy to set up and control and should supplement the human way of life.

7.2.1.1 Aesthetics

More than 60% of human-human interaction is non-verbal, via facial expressions, gestures and tonality, etc., that makes aesthetics a top priority in designing artificial beings to support human society. In human-robot communications, the robots should be human-like in appearance and correspond to human aesthetic values. A contradiction to this was found in 1970 by Masahiro Mori. Mori observed that human emotional response to the appearance of an AI agent as it is made more natural is with enthusiasm until a critical point is reached at which the agent's appearance is found to be disgusting. However, as the agent's appearance continues towards natural, the human emotional response once again changes from disgust to enthusiasm and appreciation. This phenomena of change in human emotional response to disgust and then back to delight is named 'The Uncanny Valley' [214], as shown in Figure 7.8.

The following are the salient features concerning the aesthetics of a social robot:

1. **Face type** to appeal to human beings and trigger a social response, the robot should have a cute face. This also helps in tailoring the response of the human being, as though he/she is interacting with an infant (Robota, Pepper etc.) or a pet (Paro, Kismet etc.). However, this has not always been true and robots such as CERO and Wakamaru have connected well with humans even without a cute face.

2. **Appearance and human likeness.** Current day technology makes it difficult and often financially prohibitive for robots to have both human-like appearance and also

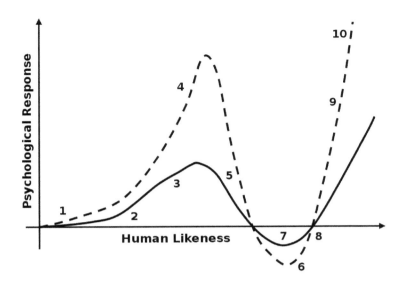

FIGURE 7.8 **The uncanny valley.** The dotted curve is for moving agency, while the continuous curve is for the ones which are static. (1) Is an industrial arm robot — both moving and static, no human likeness, (2) is a smiley face with no really discernible human traits but conveying humanly emotions, (3) is a stuffed humanoid toy, (4) is a humanoid robot such as ASIMO, (5) caricatures — both static and moving, (6) is a zombie, (7) is a corpse, (8) is a prosthetic hand — both static and moving, (9) is a Japanese Bunraku puppet and (10) is a healthy adult human being. Adapted from Robertson [278]

    human-like manners. Appearance has found favour in android research and has led to the development of human-like silicone skin, human like skin pigmentation, human like eyes which can blink and focus etc.

3. **Degrees of freedom (DoF).** To facilitate for a wide variety of expressions the robot's face and bodily gestures need to have a several degrees of freedom. For example ROMAN has a 4 DoF neck. This ensures natural gestures, such as smiling with a nod of the head for greeting and affirmation.

4. **Tailoring basic behaviour.** The robot may be designed and programmed, styled on behavioural and social response of a human infant or a pet. This would enable the robot to dynamically engage with human beings in social interactions both over verbal (voice, tonality, use of power words etc.) and non-verbal (gesture, touch etc.) modes. This approach limits the scope of interaction, but helps to establish a realistic limit on the domain of human-robot interactions, as an example a user will not attempt to discuss the stock market with a robot which is designed to look and behave as an infant nor would a user attempt to teach musical scores to an AIBO, which is designed as a dog.

5. **Fake intentionality.** The robot must convey intentionality to suggest social exchanges, such that the behaviour of the robot should be consistent and predictable as per known social norms. For example, a greeting should meet with a similar greeting — a nod of the head should meet with a similar nod and a smile in affirmation, likewise if the human user expresses grief, the robot should respond with empathy etc.

6. **Regulate response.** As technology is limited and the robot cannot be as good as human beings, the robot should selectively engage in human interactions, thus,

FIGURE 7.9 **Sparky and Feelix.** Sparky on the left and Feelix on the right, both these robots have cartoonish actuated faces and that forms the social binding in human-robot interactions. Both images from Fong et al. [107], used with permission.

avoiding interactions that are either too intense or too gentle[1]. Of particular importance are attention switching and a notion of autonomy both of which should be subtly controlled with changes in direction of head, neck and eye gaze etc.

7. **Learning.** A social robot must be equipped with a learning mechanism as in ANIMATs. This would enable it to acquire more sophisticated social skills as it interacts with the human beings and the environment.

8. **Social acceptance.** While humanoids have found more social acceptance, Paro and KASPAR have appealed to those suffering from dementia and autism, respectively. Designing the aesthetics for a social robot should be strongly related to the application and the niche it is meant to target.

7.2.1.2 Facial traits

Attention switching, to meet the gaze of a human obersver and mirroring human expressions does help in establishing a ground for social interactions but to derive meaning from a facial expression or action will help it to reach closer to the domains of human-human interactions.

Determining the psychological state using facial expression is an effective non-verbal interaction in day-to-day interactions. Human facial traits can be incorporated into a social robot using a number of techniques: (1) Haar feature-based cascade classifiers, (2) Eigenfaces, (3) Active Shape Methods and Facial Action Coding System (FACS).

The method of Haar classifiers proposed by Viola and Jones for real-time detection has

---

[1]This does suggest that a robot that can spot sarcasm and answer in a repartee with some consistency has crossed over a threshold of humanly social interactions and has now started to appreciate and enjoy human social behaviour.

come to be the standard for face detection algorithms and is the default in various image processing software suites. Though easy to implement, Haar classifiers are not exhaustive and lack in robustness, with the best results being for frontal views of faces. The method of Eigenface uses the image matrix of faces and compares it to a calibration and is accompanied with a pre-existing large database.

Active shape models (ASM) are statistical models of the shape of objects as defined by a point distribution where the shape of an object is reduced to a set of points, and such methods have been widely used to analyse facial images. To convey a human emotion the relative positioning of critical points on the human face are observed. These points are known as, 'landmark' points. A shape model by itself is not sufficient to detect faces, and more often ASM has to be employed along with a Haar cascade.

These three methods work on still images, for real time execution the best technique is Facial Action Coding System (FACS). This method developed in the 1970s, is a powerful tool used to correlate facial expression a human being can make to their corresponding psychological state. This is a novel method of relating facial taits to the psychological state, for example happiness is seen as cheek raise and lip corner pull, both working at the same time. It takes into account the contraction of each facial muscle and how it affects the appearance of the face. Facial expressions are reduced to Ekman Action Units (EAU), and the combination of these EAUs provides a good estimate of the corresponding human expression as is shown in Table 7.1

TABLE 7.1 Facial Action Coding System (FACS) (1978)

Psychological State	Facial Traits (EAU numeration is given in square brackets)
Happiness	Cheek Raiser [6] and Lip Corner Puller [12]
Sadness	Inner Brow Raiser [1] and Brow Lowerer [4] and Lip Corner Depressor [15]
Surprise	Inner Brow Raiser [1] and Outer Brow Raiser [2] and Upper Lid Raiser (slight) and Jaw Drop [26]
Fear	Inner Brow Raiser [1] and Outer Brow Raiser [2] and Brow Lowerer [4] and Upper Eyelid Raiser [5] and Eyelid Tightener [7] and Lip Stretcher [20] and Jaw Drop [26]
Anger	Brow Lowerer [4] and Upper Eyelid Raiser [5] and Eyelid Tightener [7] and Lip Tightener [23]
Disgust	Nose Wrinkler [9] and Lip Corner Depressor [15] and Lower Lip Depressor [16]
Contempt	Lip Corner Puller [12] (trace, on the right side) and Dimpler [14] (trace, on the right side)

These are six prominent emotions and their primary EAU. The EAU numeration is given in brackets, so happiness, which is seen as cheek raise and lip corner pull, both working at the same time, will be 6 + 12 in FACS coding. More sophisticated treatments take into account head movement and eye movements. This method cannot only relate facial traits to their emotion, but also distinguish between facial expressions made voluntarily and those which are involuntary and spontaneous. FACS is currently the only available technique for assessing emotions in real time and is a proven tool employed by psychologists and psychiatrists.

7.2.1.3   Natural language processing (NLP)

Chatbots such as ELIZA and ALICE working on Natural language processing (NLP) methods have been very successful in artificial conversation with a human user. Chatbots rely on a set vocabulary to search for particular arrangements of words to detect the psychological state and the intention of the human user. For example, 'I am not feeling well' from the human user will prompt artificial empathy from the chatbot with a response like, 'I am sorry to know that. What bothers you?'. Similarly ' Isn't it a nice and bright day today' from the human user will lead to a more proactive reply from the chatbot, 'Oh yes, isn't it lovely weather'.

How to simulate a conversation?

Chatbots are probably the simplest example of a socialising AI. Earliest promises such as Weizenbaum's ELIZA, which then inspired ALICE has yielded startling results and modern day state-of-the-art chatbot, GOOSTMAN poses as a near human being and turns the tables on the Turing Test. How can a rote programming script elicit empathy and emotions much like a human beings? The following two examples attempt to answer the question.

1. **Chatbot — revisiting ALICE.** Artificial Linguistic Internet Computer Entity or ALICE engages in conversation with human beings with heurestical pattern matching the user's input. However, it is not foolproof and a lack of consistency in the conversation can be spotted easily after 5-6 exchanges between the 2 agents. Here, ALICE is implemented using PyAIML, as shown in Figure 7.10.

```
> Hello there
What can I call you?
> Call me Ronald
OK I will call you Ronald. Are you a man or a woman?
> I am a man
Thanks for telling me your gender, Ronald. Most of my clients are men.
> Okay
OK with me too.
> I was wondering, if you can help me
What kind of help would you like?
> I wish to make a phone call
I hope that your dreams come true.
> ... ha ha ... you are funny
I am so glad you find this amusing, Ronald.      Thanks you make me laugh too.
> So, I need to make a phone call ...
Hmm. I want to become smarter than humans.
> Are you ... can you help me with my request ?
I won't say "yes" or "no" right now.      Certainly, I have an extensive built-in help system.
> uh huh ....
Yup. Er. Um.
```

FIGURE 7.10 **Sample output of a variant of ALICE,** while the ALICE AI doesn't really give much information nor solve any real problems, it does very well to sound like a human being. The parts of the conversation marked with the chevron are the human user inputs; the corresponding answers are from ALICE.

2. **A Simulated Conversation — Wall-E woos Eva.** A coherent conversation, seemingly lacking predictability between two iconic robots is designed here with a short program script in Python 2.7. The output is as shown in Figure 7.11. The conversation reflects the emotional state and a general mood of the 2 agents.

The program script is a play on the English language using key words to convey cheesy romantic wooing. Though the output varies with each run the general mood of the two agents and theme of the conversation more or less remains the same. It is to be noted that the program doesn't use any natural language library nor is supplimented with a search engine, but rather is designed by randomising the inputs and the outputs, which lends an apparent lack of predictability to the conversation, and makes it seem spontaneous.

```
WALL-E :   hello Eva
EVA     :   leave me alone please
WALL-E :   i luv u
EVA     :   buzz off clanky old robot
WALL-E :   you sure are beautiful but you bite sometimes too dont you
EVA     :   now you know I like it when you speak to me like that
WALL-E :   can i give you a robotic hug
EVA     :   hey my darling,come here and lets look for plants
WALL-E :   where are you eva
EVA     :   why do you ask so many questions, are your logic circuits faulty?
WALL-E :   ok
EVA     :   sure
WALL-E :   sure
EVA     :   you look nice after that service
WALL-E :   I need you
EVA     :   buzz off clanky old robot
```

FIGURE 7.11 **Sample output of simulating a chat between Wall-E and Eva.** key words play out the mood and reflect the theme of the chat.

Some of the outputs lack in continuity and also often do not tally to the question asked and the agents often repeat some of the sentences. However, that still confirms to the general theme of the conversation. This is due to the key words and the fact that both are caricatured robots of the cartoon and there is a prejudiced human perception.

To make chatbots Prolog and Perl were early favourites in late 1980s and early 1990s. Nowadays natural language specific libraries, NLTK for Python, GATE and OpenNLP for Java, UIMA for both C++ and Java, ScalaNLP for Scala and web services as Alchemy API and Open Calais have found popularity with developers. Voice-enabled systems such as Apple's Siri and Microsoft's Cortana are sophisticated variants of chatbots, and they also look for visible patterns in the human conversation, both the content and intonation and often supplement their reply with an internet search. However, voice-based interactions is not without its issues. As of now there is a lack in an AI system identifying and appreciating accents across the world, the problem being acute in the English language. Also it is difficult for an artificial system to distinguish between relevant sounds from a human being from other natural and often undesirable sounds.

Chatbots are often used for marketing purposes, where a chatbot AI tries to help out a human user with a purchase or after sales support.

7.3 APPLICATIONS

Human-Robot interaction has found newer applications with advancing technology. Very much in the Asimovian spirit, the robots cater to a human need and are nearly always subservient to human command. The trend in social robots was inititated by a flurry in the development of anthropomorphic robot heads in the late 1990s and early years of the new millennium. After Kismet, there has been a plethora of interactive robot heads, WE4, ROMAN, TUM-Eddie, Inkha, Octavia and Flobi are some of the best examples. While most of them are academic exploration into human-robot interaction research and are developing a proving ground for modelling the human mind and the emotion quotient, Inkha shown in Figure 7.18 had taken on the role of a robotic receptionist at the entrance to King's College London, while Octavia, developed by the US Navy, is an interactive firefighting robot. Low cost designs, such as Fritz, which was a crowd sourced project and the robot head designed by Kim et al. at University of California, San Diego have hinted towards commercial utility.

ROMAN (RObot huMan interAction machiNe) shown in Figure 7.12 at University of Kaiserslautern, stands out from the rest and is arguably the best anthropomorphic robot head. With artificial skin and a 4 DoF neck, ROMAN also boasts human like emotions, developed with Ekman's action units and a emotion control architecture. ROMAN's mechanics are designed on 8 mounting movable metal plates fixed to the neck. These plates provide for the mounting points for the lower jaw, the eyes, skull and the 10 servo motors. The positions of these movable metal plates are as per Ekman's action units. The artificial skin is attached with glue to the cranial bone and affords movement with the 8 metal plates, operated with the servos. The eyelids can go up and down, and the eyeballs have 4 DoF, up and down and left and right.

ROMAN's emotion control architecture, is based on three main parts: emotions, drives and actions.

7.3.1 Service robots, with a social face

Currently, 1000 units of Pepper robots are selling each month; however Pepper is not tagged to a social function. Service robots have well defined roles and particular behaviours. The following example of CERO helps to illustrate that with minimal hardware, simple behaviours and least intrusion into human lives, robots can still engage in their targeted social niches.

7.3.1.1 CERO — Cooperative embodied robot operator

CERO shown in Figure 7.13, was developed at the Royal Institute of Technology, Sweden, over a period of three years [159, 297] and was designed to assist office workers who lack mobility with the fetch and carry function in an office environment. The robot has two parts: the CERO character and a wheeled base based on a Nomadic Super Scout platform with 16 sonar sensors for navigation. The CERO character is an anthropomophic representation of a mini sized human bust and has minimal human appearance and traits with 4 DoF; the head with with 2 DoF(up-down and sideways) and each arm with 1 DoF, and these four movements helps to convey the social intent for the robot by functioning as the 'driver' of the robot base. It relies on gestures to communicate, express emotional attitudes and thereby comes across as personality rather than a cartoonish machine. It is designed to take human commands over keyboard inputs and also via a speech interface. For example, it can nod to convey positive feedback, or move its arms to gesture that the robot is on its way. The speech system can respond to typical emotional content of voice and intonation,

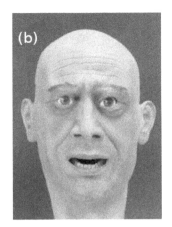

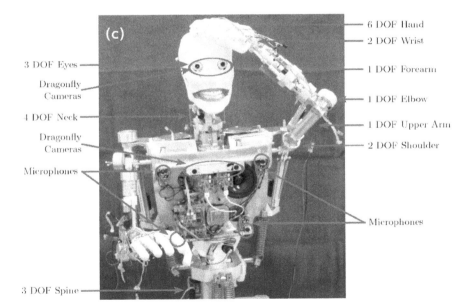

3 DOF Eyes

Dragonfly
Cameras

4 DOF Neck

Dragonfly
Cameras

Microphones

3 DOF Spine

6 DOF Hand

2 DOF Wrist

1 DOF Forearm

1 DOF Elbow

1 DOF Upper Arm

2 DOF Shoulder

Microphones

FIGURE 7.12 **ROMAN**, the humanoid at University of Kaiserslautern. (a) The working robot, (b) the face and (c) details of the mechatronic design. Images courtesy Karsten Berns.

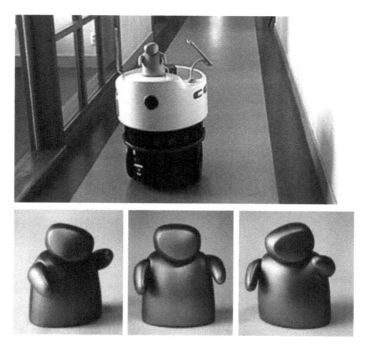

FIGURE 7.13 **CERO**, an expressive robot with minimal human-like features and gestures at Royal Institute of Technology, Sweden, was designed to assist office workers with walking disabilities. Images taken from Severinson-Eklundh et al. [297]. Used with permission from Elsevier.

such as raised amplitude and spoken vowels with corresponding gestures. Further details are presented in Table 7.2 .

TABLE 7.2 CERO's response, modified from Brennan and Hulteen (1995) [46]

Event	CERO's Feedback
Attending/Acknowledging, as robot on, microphone sound detected, speech detected	Raise head
Hearing	A shake of head if the audibility/articulation is poor, otherwise a nod of the head
Parsing/Interpreting	A shake of head for parsing errors
Plan failures/Successful execution	A shake of the head (failures) and a nod (success)
Reporting	Walking-like gesture with both arms

CERO is an example of an expressive robot that uses minimal gestures and movements to convey a sense of familiarity to human users. While lacking the dynamism of humanly activity, such a robot can be a part of low-cost natural interaction over a range of applications. The robot was primarily targeted at office staff who may be able to walk only short distances on crutches or be in a wheelchair but may not be able to fetch or carry objects with ease. In trials, the robot was found to aid and be an assistant for a unique user. Severinson-Eklundh and the CERO research team hope that CERO can be further developed to support a fully voice-based two-way [132] interaction which works seamlessly in tandem with gestures and non-verbal interaction and facilitate multiple user and group interactions. CERO also stands out as a platform to study human-robot interactions, when robot interactions are at a minimal [159].

7.3.2 Robots in elderly care

Various research groups have targeted elderly care. KSERA, DOMEO, COGNIRON, Companionable, SRS, Care-O-bot, Accompany, HERB, Hobbit, Pearl, Hector, Huggable etc., are robot platforms developed to support the elderly and those in medical care. These robots are employed at home or in a care facility and primarily attend to monitoring well being and health, helping in social communication and assisting in fetch-and-carry tasks, thereby functioning as a smart machine, as a butler and also as a companion.

7.3.2.1 Care-O-bot 3 — The smart butler

The Care-O-bot project at Fraunhofer Institute of Manufacturing Engineering and Automation (IPA), Stuttgart, has spanned over a decade and led to three generations of the robot. The long-term goals of the project are to development a mobile robot which can assist persons in their domestic environments and attend to functions such as fetch-n-carry and other simple household chores and provide for entertainment and attend to emergencies, thereby, at a lower running cost, increase their personal independence and improve their quality of life and security and consistently attend to their care and also provide for better social integration and cohesion over social media etc. Fraunhofer IPA further believes that Care-O-bot can also attend to roles such as robot valet, robot receptionist and also as robot technical staff, leading guided tours in museums and entertainment parks.

The third and latest version of Care-O-bot, shown in Figure 7.5 (b), is designed on the concept of a butler poised with a tray.

7.3.2.2 Hobbit — Returning a robot's favour

The Hobbit project at Vienna University of Technology aims to build a highly acceptable socially assistive robot in indoor elderly care. Hobbit was designed to enable the elderly users to stay longer in their homes or care facility, and attempted to address emergency situations, fall prevention and providing psychological well being by being around as an amicable helper. It attends to its job by patrolling the apartment, picking up objects from the floor, by advising the elderly through reminders and attending to emergencies such as calling the ambulance and offering help to rise in the event of a fall. Hobbit is networked through to social media and can offer to entertain and also communicate whenever needed. The project has 49 participants at three different test sites in Europe (Austria, Greece, and Sweden).

The novelty of Hobbit is that it is developed on the Mutual Care Paradigm, where the elderly individual(s) and the robot take care of each other. Other than the various facets to caring, the robot will also advise and anticipate the elderly individual(s) to care and help the robot. The underlying belief is that it is amenable to being assisted if it is a two-way street, where both look up to the other for a more fulfilling existence. This also reduces the technological burden on the robot as a sleek gizmo and brings it towards a point where interdependence and cooperation is the key to co-existing — connecting naturally as a human being than a machine. This reciprocity in forming the bonds of a mutual interaction is proven to yield a better user experience than the stigma of a sleek gizmo monitoring and controlling people's lives.

On full battery charge, the Hobbit can operate for about three hours. It has a differential drive and can turn on the spot within its own footprint and can also navigate in narrow and cluttered environments. It is further equipped with a 5-DoF IGUS Robolink Arm and Fin Ray Gripper. The human-robot interactions can form over various modes, via voice, touch screen, and gestures. Fischinger et al. [104] identify four design aspects that are essential for Hobbit to attend to the elderly:

1. Map building and self-localization
2. Safe navigation (obstacle detection and avoidance)
3. Human-robot interaction (user and gesture detection)
4. Object detection and subsequent grasping

To meet this requirement, it has an assortment of sensors. For depth perception and self localisation the robot uses ASUS Xtion Pro. For obstacle detection, object detection, grasping and human interaction the robot has a Microsoft Kinect on a pan-tilt unit mounted in its head. In order to help in backing up, it has an array of eight infrared and eight ultrasound distance sensors in the back of the robot base and for odometry, it has high resolution odometers on the two shafts of the drive motors which can measure $70\mu m$ per encoder tick.

Hobbit starts the user interaction with an introduction explaining its functionalities. Clean floor and other advice for the elderly user and attending to emergencies forms the rote functioning of the robot. Novel user-robot interactions are put to the test in typical reciprocating tasks such as;

1. The robot is attempting to learn about an object. The task asks for mutual participation from both the robot and the elderly user. For learning about an object, the robot has to use the 'learning turntable', which it cannot maneuver all by itself. Therefore the robot asks the user to place the 'learning turntable' into the gripper and follow instructions, as is shown in Figure 7.14, where the robot learns about a cup. Though help from the user is necessary, it is not very explicit. After completing this task, robot thanks the user and offers to return the favour. On acknowledgement by the user, the robot plays music or tells a random joke enhancing the social bond between the two.
2. The robot is designed so that it fails to find the object it has to fetch and asks the human user for help. If the user can agree to help the robot, then the user can suggest the whereabouts of the object via touchscreen. If the robot succeeds in finding the object, then as in the previous scenario it offers to return the favour. If it fails the task, then it merely makes a report of the failure and the user can send it again on the look out for the given object.

Lammer and co-researchers observe that the human user enjoyed being surprised with a robot asking for help and this was often accompanied with laughter and smiles and most of the participants wanted the robot to address them by their real names. Both adding to social interaction and building of trust between the user and the robot.

7.3.3 Companion robot and robot therapy

Social robotics has been employed in the treatment and care of patients suffering from dementia and autism. Paro, the robotic seal, appeals as a pet and therefore helps to develop

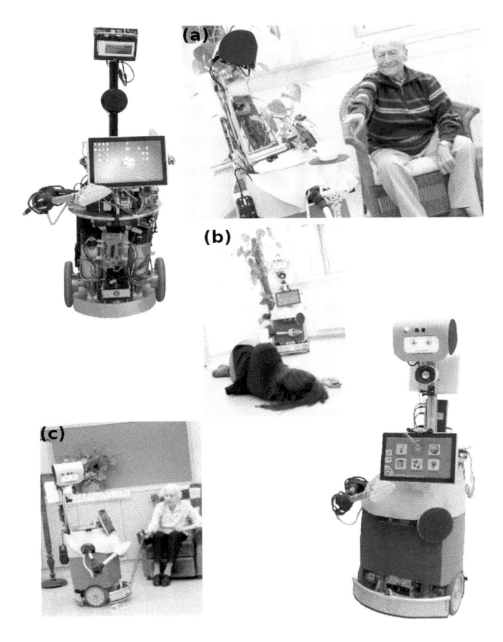

FIGURE 7.14 **Hobbit** presents novel human-robot interactions. Image from Fischinger et al. [104]. Used with permission from Elsevier.

and nurture emotions and empathy for those suffering from dementia, and KASPAR with the likeness of a human infant connects as a real-life friend and has shown to improve communication with children suffering from autism. Fake in both emotion and empathy, these two are examples of human-robot interaction helping to elicit real feeling and emotions in human beings.

7.3.3.1 Paro - The cute robotic seal

Paro [135, 369] is a robotic seal which has been used both by doctors and psychologists to overcome dementia and also reduce loneliness, filling in for a pet in elderly care. Paro was designed by Japanese roboticist Takanori Shibata and it is modelled after a baby Canadian harp seal and is covered in white artificial fur and weighs approximately 2.7 kg. It was developed by the National Institute of Advanced Industrial Science and Technology, Tokyo, and is being manufactured by Intelligent Systems, Co. priced at 35000 yen in Japan and $6000 in the United States. It is equipped with tactile and light sensors, touch-sensitive whiskers, sound and voice recognition, large blinking eyes expressing its benevolent nature. In reply to human interaction it can produce cute noises imitating that of a baby harp seal and it can swivels its head and seemingly appear to track human motion and thus create an illusion of paying attention to the interacting human being. The design is more towards a fantasy projection of a seal, and not that much like a real one. Paro is an effective tool of emotional seduction and it elicits positive responses from the interacting human beings, and has been found to be very effective in patients suffering from dementia. Paro is designed to operate on 3 aspects, (1) its internal states, (2) sensory information from its sensors and (3) its own diurnal rhythm for its interaction with human beings. Also, it can learn from experience. Therefore for a stimulus such as tapping its chin, which provided a response of happiness, Paro may repeat the same response just prior to a tap of the chin, thus demonstrating artificial equivalents of Skinnerian conditioning. Paro is introduced to the dementia patient as their pet Fluffy or Bruce and can learn its name when called out. Unlike a real pet Paro cannot bite or scratch. Its antiseptic coat will not shed dander; it does not eat food or create waste and its fur is hypoallergenic, thus it doesn't add up to the burden of a pet.

Though Paro appeals to the emotional quotient of the user, Calo et al. observe that the intuitive responses on seeing Paro are either excitement or repulsion. The repulsion is on finding the robot unnatural, cartoonish and creepy. However, there is a near consensus to address the robot as though it is a living being, and in addition to the patients, doctors and other medical staff have been seen to talk to the robot.

The Paro robot and a mini interview with Dr. Broadbent

Dr. Elizabeth Broadbent at the Department of Psychological Medicine, The University of Auckland, has been closely involved with elderly cure using the Paro robot. She shares some of her research with us. This mini interview took place in early July 2015.

AB: I have read about your work on the Paro robot as a companion in elder care. Can you please tell my readers about your motivation and what led you to go ahead with this research?

EB: I have always been fascinated with robots and the idea that robots can become loyal companions without some of our human flaws. People are so busy these days they often don't have time to look after or visit older relatives. So older people are sometimes lonely and may be physically unable to keep a pet for companionship. In these situations, robots might be able to reduce some of the loneliness and distress people experience.

AB: Does a robot companion really aid in reducing depression? Do patients view it as a pet?

EB: Our own research has shown that Paro can reduce loneliness in people in rest-homes, and another recent study has shown that Paro can reduce depression in people with dementia [219]. In our research we have found that older people (without dementia) do treat Paro like a pet, but they are also aware that it is a robot.

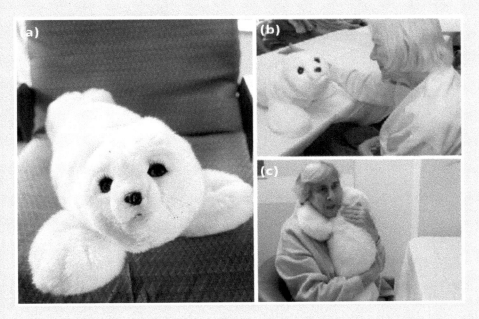

FIGURE 7.15 **Paro robot.** (a) Paro, the robotic seal, (b) and (c) interaction and attachment with the elderly. Images from Robinson et al. [279]. Used with permission from Elsevier.

AB: What are the minimal requirements for the robot to function in this fashion? What characteristics does this robot have that differ from other robots? Clearly the Roomba cannot do this job.

EB: Paro is very cute. He has big beautiful eyes and long eyelashes. He has soft fur and you can pick him up and cuddle him. He responds to being stroked with movement and sound. These features make him different to most robots that have hard plastic casings such as Roomba, which is designed to vacuum the floor.

AB: What are the problems and shortcomings of using a robot with the elderly?

EB: In an ideal world, people would not get old or sick and everyone would be kind, friendly and helpful. Then we would not need companion robots. But we do not live in an ideal world, and people do suffer. If a robot can help provide comfort then that is great. People often form attachments to objects that cannot reciprocate feelings, such as cars or art. People can get a great deal of pleasure from these objects. I don't think paro is too different.

AB: Would you wish to share any particular episode or story about Paro's interactions with patients?

EB: One man in hospital with dementia was completely unresponsive when I met him. He was staring blankly, doing nothing and did not even look at us when we spoke to him. However, when we put Paro in his lap, he suddenly started talking to Paro and stroking him as if he was a pet. That, to me, showed that Paro could be even better than a human in some situations.

AB: What are the newer avenues that you are researching on human robot interactions? Any new robots?

EB: We are conducting studies with other healthcare robots that can provide reminders to take medication, encouragement to exercise and monitor health. I am interested in whether robots are better than computers for these tasks, and our recent study has shown that people are more adherent to robots' instructions and like them better than computer tablets. I am interested in exploring this idea further in larger trials.

Paro connecting with patients suffering from dementia and KASPAR aiding autistic children are vibrant examples where robots have helped in a human cause. These are scenarios where robots have participated in human society and have clearly demonstrated ascendency from being mere tools.

7.3.3.2 KASPAR — Kinesics and synchronization in personal assistant robotics

KASPAR, a humanoid robot with child-like features and a friendly appearance, developed for for humanrobot interaction research was made by Dautenhahn et al. [86]. at the University of Hertfordshire. KASPAR attempts to find acceptance as a playmate and companion robot. Unlike Kismet, KASPAR was not designed to explicitly foster reactions from the user but rather adhere to a minimally expressive paradigm, where it was not supposed to be very human like, but is enabled to emit a few expressions to underscore the most salient human-like features and aspects. KASPAR is low cost, portable and can be run from a laptop. It has more anthropomorphic features identifiable with human beings and it can be set-up and run without any particular expertise or prior training since programming is via a user friendly GUI which runs on both Windows and Linux and other than the spontaneous interaction mode, it also works on a toy-like mode using a remote control. Aesthetic coherence is an important principle in KASPAR's design, therefore the face, body and hands are of nearly the same complexities and the physical design enables the behaviour, interaction and control of the robot. The design also earmarks the actual capabilities of the robot and prevents the jeopardy of creating inappropriate expectations in the human user.

KASPAR was projected at 1500 EUR and it was designed as per the size of a small child, with a few comic-like features in order to not appear threatening to the user. It is designed in a near cross legged squatting posture, its head is slightly larger in proportion to the rest of the body to suggest a comic-like appeal. The neck is primarily meant for nodding and shaking however it can also orchestrate more subtle variations such as slight tilting of the head or a slow nod to convey emotional and personality traits viz. shyness, non-confirmation, cheekiness etc. The expressive components of the face such as eyes, eyebrows, mouth etc. convey and augment an emotional state. The eyes incorporate minature video cameras and can pan and tilt and may also support mutual gaze, encouraging joint attention. The eye-brows are not separately designed but are made into the face mask formed as deformations on actuation of the lips, mouth etc. are the eyelids can open and close and can support blinking, full or partial, at various rates. The shape of lips and mouth suggest an emotional state viz. smile, frown, neutral etc. are controlled by actuators. KASPAR's arms are not very robust and lack precise trajectory planning. However they do to complete the child-like features and help in engineering gestures, such as waving and peek-a-boo. The palm doesn't have articulated fingers in order to keep a simple design and low cost of the robot, and it also invites children to touch the hands, which is akin to touching a doll. Finally, a silicon rubber flesh-coloured face mask is used for KASPAR's face, which makes a fair imitation human skin.

Dautenhahn et al. report that KASPAR's interaction with developing children encourages a reply, however adults are more cautious, less playful and more critical of KASPAR. There is a tendency in adults to look for and compare KASPAR to very realistic human-like features which they may have seen in robots in movies and which are not often present in KASPAR as it is developed on a minimally expressive paradigm. Three particular use of KASPAR have been in:

1. **Robot-assisted play and therapy.** Dautenhahn and her co-researchers at University of Hertfordshire have been closely involved in the use of robots in autism therapy for about a decade. Employing KASPAR as a part of autism therapy in children yielded encouraging results. All the children with autism were found to explore the robot with touch and also consistent staring at the face etc. The tactile interaction helped to increase body awareness and develop a sense of self, while staring at the robot's face enabled them to understand human facial features and some of the children later attempted to relate those features to their human therapists. KASPAR not only helps to break the isolation but also breeds excitement and enthusiasm. Even in those suffering from severe autism and it was common to see the children share this experience with their teachers and therapists. The research also explores the prospects of cooperative games with KASPAR, in the remote control mode, as is shown in Figure 7.16 (d).

2. **Developmental robotics.** The 'drumming mate' experiment has been often used in developmental robotics to inculcate the robot with an appreciation for rhythm and beats. In it the robot has to listen to a human being play the drum while it records and analyses the played rhythm and later plays the same rhythm back by beating the drum. Such an experiment confirms to multimodal interaction and a ground level of artificial consciousness in the robot. When tried on KASPAR as shown in Figure 7.16 (b) & (c), it is seen to mirror the human being but with a small time lag of about 0.3 seconds between each beat. This happens as it has to get its joints in position for the job of playing the drum. As a shortcoming, KASPAR fails to mimic the strength of the beat, viz. hard, soft due to its limited motor skills.

3. **Cognition and learning.** The 'peek-a-boo' experiment was aimed to develop cognitive faculties in the robot over multiple modalities and also enable it to learn

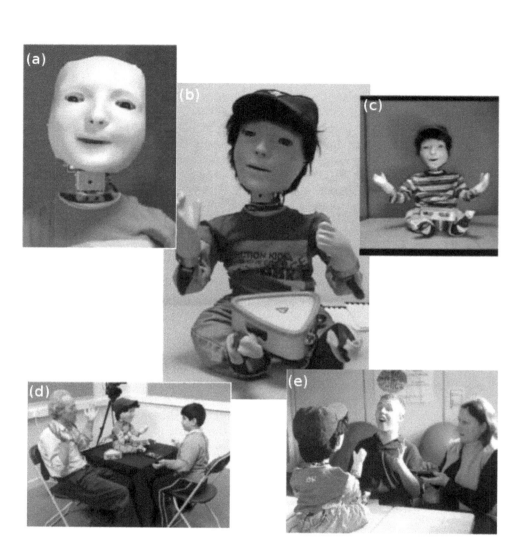

FIGURE 7.16 **KASPAR.** (a) Construction of the face. The face can elicit various degrees of emotions viz. large smile, small smile, neutral, frown etc. (b) & (c) The 'drumming mate' experiment. (d) A therapist using KASPAR to teach turn-taking skills to a child with autism. (e) A therapist is using remote control to operate KASPAR in an attempt to replicate the same emotion as seen in the child with autism. This helps to identify facial expressions to known emotions. Images from Dautenhahn et al. [86], reprinted by permission of the publisher (Taylor & Francis Ltd. http://www.tandfonline.com).

from its previous experience. Replicating the popular 'peek-a-boo' game usually played between an child and an adult. The adult after establishing amicable eye contact, then hides his/her face and on revealing their face again cries 'peek-a-boo' and encouraging the infant to ape their action of first hiding and then showing face with a call of'peek-a-boo'. For the game, the researcher takes the role of the caring adult and encourages KASPAR with calls of 'peek-a-boo' to hide its face. KASPAR of course perceives this as an increase in sound intensity and a simultaneous detection of a face by its cameras. It was seen that nearly 67% of the time, encouragement with a call of 'peek-a-boo' met with the robot repeating the action of hiding its face and without an encouragement the robot did not repeat the action. The interaction history of the experiment was stored in KASPAR's memory and with time and more interaction history KASPAR's response got better. This is an example of reinforcement in social robots.

KASPAR and Paro are designed to develop and foster attachment with the patient, however this can lead to ethical concerns. Primarily, these robots attend to happy and cheerful emotional states but human emotion is way too varied and the robot may fail to attend to a patient's frustration, anger and revolting actions. While in contrast, the patient, usually an elderly or suffering from an ailment may become emotionally attached to the robot and any attempt to withdraw may lead to distress and sorrow. In reciprocation, the robot will not 'feel' anything for the patient. Also, nurse robots such as Pearl at Carnegie Mellon University may have to attend to patients and therefore prioritise its own task. Such a prioritisation will be more of a real-time concern than merely following a timebound list, and will call on the ethical judgement on the part of the robot. Another long-term jeopardy may be the loss of a job for human carers and nurses, which would then herald the lack of true social contact in a profession which is meant for nurturing social contact and empathy.

I end this section with part three of the toby series in Figure 7.17 which touches upon human robot cooperation, which may well come into play without explicit design or programming to culture such values.

7.3.4 Museum guide and receptionist robots

Museum guides, news readers and actor robots are examples of minimal human-robot interaction over short periods of time which are often predictable and repetitive and therefore designed to work to a pre-programmed menu. Museum guides through, path planning combined with narration or display of visual information in a known local environment; news readers by human-like appearance, gestures and intonation; and actor robots are programmed to a script in a known environment, all of them attend to rather limited tasks and address human-robot interaction without inquiring much into the dynamism of human interaction.

For museum guide robots, there are two particular features the design must address, (1) navigating safely through a crowd and (2) short-term interacting with the museum goers with spontaneity and enthusiasm. In late 1990s, Burgard et al. designed the RHINO as an interactive museum guide at the Deutsches Museum in Bonn, Germany. The primary task of the robot was to lead interactive visitor tours in the museum. It was also one of the earliest examples of teleoperation over a distance and virtual presence, where a far away human user over the internet could instruct the robot to go to specific areas of the museum and enjoy a virtual tour of the museum without visiting it. The RHINO communicates with the on-board interface which integrates text, graphics, pre-recorded speech and sound which presents the user with a menu containing pre-set tours or they can listen to a brief, pre-recorded

FIGURE 7.17 **Toby saves the day (Toby, 3 of 4)**, Care robots are now a vibrant industry with robots such as Hobbit, Care-O-bot, HERB etc. designed as care workers and nurses and to attend to medical emergencies. Toby was not meant for such an emergency and could only relay the information to the Walkers. Cartoon made as a part of Toby series (3 of 4) by the author, CC-by-SA 4.0 license.

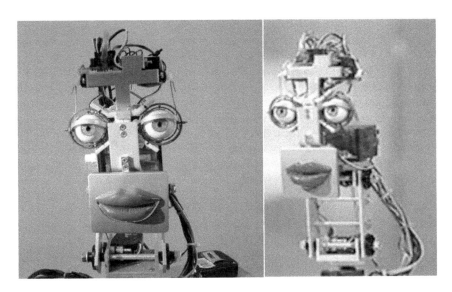

FIGURE 7.18 **Inkha**, the robotic receptionist at King's College London, Strand campus, from 2003 to 2014. Both images courtesy `wikipedia.org`, CC license.

explanation of the exhibits. Burgard and co-workers reports of a very high success rate of RHINO, in 47 hours of operation with 2000 museum goers, the robot attended to 2040 requests and collided with obstacles only 6 times, and the Deutsches Museum reported 50% increase in visitors.

Two years later in 2000, Thrun et al. made the Minerva robot to conduct visitor tours in the Smithsonian's National Museum of American History, Washington, DC, for a period of 2 weeks, during which the Minerva attended to 620 tours and covered more than 44 km visiting over 2600 exhibits. Minerva had an anthropomorphic head with eyes that responded to human interaction and could draw the user's attention to itself, and therefore had more user engagement than the RHINO.

ATLAS, a museum guide robot made at the University of Essex, was based on the PeopleBot robot designed and made by Activ Media Robotics. It was an improvement over both the RHINO and the Minerva, and was enabled with a basic voice-based system and a touch screen for user interaction. It was employed at the London County Hall for receiving and welcoming visitors and to explain the museum contents to them. It could detect visitors in proximity and interact with them over voice and menu options in the touch screen. The robot could also detect its battery levels and when on low power, return to the charging pods to charge its batteries automatically.

Dances with the Nao Robot, robot actors and performers

Science fiction movies have set the foreground to depict a future with robots, though there have been daring attempts to use robots in sync with human actors in live plays and dance drama [311]. Francesca Talenti's, 'The Uncanny Valley' [1, 73] had a robothespian robot as an actor and had the stage name Dummy. It was made to perform through a puppetier while the human actors trained to sync their performance

and dialogues, Oriza Hirata at University of Osaka used a Wakamaru robot in his play 'Hataraku Watashi' (I, Worker) [317] but the most eloquent use of robot performers is probably by choreographer Blanca Li [203, 204] in her dance drama, 'ROBOT'.

In collaboration with musical artist Maywa Denki and Aldebaran robots, the dance drama employs seven Nao robots to dance on stage along with eight human artists to the symphony of the ballet. They have given sterling performances on both shores of the Atlantic. The 90-minute dance drama has the tag line, "Robotic Pop Ballet for All Future Generations" and is an exploration into the relationship between human beings and artificial beings. It poses questions such as can AI truly replicate the relationship of human beings with nature? Will robots ever be creative? How will AI and technology change our way of living and our social functions? The ballet was first performed at the Montpellier Dance Festival in the summer of 2013 and debuted in the USA at BAM Howard Gilman Opera House, New York, in June 2015. It has also performed to full houses in Belgium, Spain, Portugal and Italy.

Maywa Denki and his mechanical musical orchestra from Tokyo, and Bruno Maisonnier, the inventor of the humanoid robot Nao and founder of Aldebaran Robotics in 2012 collaborated with Blanca Li for this dance drama. The seven Nao robots are named Pierre, Jean, Alex, Lou, Dominique, Sacha and Ange and each is designed as a distinct personality. The dance drama 'ROBOT' is the story of how the human dancers tried to teach the robots to act and dance. Which was by moving all the limbs by hand and incrementally check if the robot is able to replay the movements by itself. The process is comparable to a parent teaching an infant to walk. Since the dance drama is themed as a dancer teaching a robot to walk, the choreography for the stage performance is a natural extension of the rehearsal.

FIGURE 7.19 **Blanca Li's Ballet, 'ROBOT',** employs seven Nao robots dancing in sync with eight human dancers.

More details can be found at *http://www.blancali.com/en/event/99/Robot.*

Receptionist robots lack mobility, have less autonomy and are designed as an interacting humanoid head rather than with a full body and therefore are cheaper to make and maintain. Much like the museum guides, these robots are also equipped with voice and touch screen and are accompanied with amicable facial features.

Inkha the robotic bust with a caricatured face shown in Figure 7.18, functioned as a receptionist at King's College London, Strand campus, from 2003 to 2014. The robot was equipped with speech and facial expressions and in response to user's touch screen inputs the robot could help with directions and locations of places across the college campus and also add in a compliment or the occasional banter, which included information about the college, astrology and also fashion tips. The robot used a camera in its eyes to detect movement, and expressed fright at sudden movements and went into sleep mode if nothing seemed to move.

"You're starting to look like a robot!", and other conjurings from Hiroshi Ishiguro

The wish for life size humanoid robots proactively interacting with human beings has drawn the interest and attention of various roboticists, but none so much as the pioneering effort from Hiroshi Ishiguro [160, 233] at Osaka University, Japan. His humanoids, Kodomoroid and Otonaroid, have the potential to put human TV presenters out of their jobs.

FIGURE 7.20 **The Kodomoroid**, Kodomoroid, an android humanoid with the features of a girl in her adolescence, is currently acting as a museum guide at the Miraikan Museum, Tokyo. Her job includes, speaking to visitors and informing them about various exhibits on display. Image courtesy Hiroshi Ishiguro Laboratories (HIL), used with permission

Ishiguro is motivated from principles of behaviour-based systems and distributed cognition and contends that rich interactive behaviour seen in his androids will help to determine design methodology and paths for making intelligent robots in the near future. Since, appearance and behaviour are tightly coupled to each other, the design of androids is focused on human like appearance and is characterised by lifelike skin, human facial features and expressions, natural voice, natural intonation etc.

The Kodomoroid looks like a girl aged about 12-13, while the Otonaroid has features of an adult female. Both have been engaged in jobs as a news reader and as a museum guide respectively at the National Museum of Emerging Science and Innovation, Tokyo also called Miraikan Museum. In a demonstration, the Kodomoroid was seen jibing with Ishiguro and poking in sarcasm, "You're starting to look like a robot !".

The Kodomoroid is remotely controlled and announces news and information about the museum and its exhibits to visitors, while Otonaroid is in close interaction with the visitors and willfully strikes up a chatter with people who approach it, and a third humanoid, the Telenoid will be on display, more as an exhibit. Besides these three, Ishiguro also has an android version of himself that he sends to give lectures when he has a tight schedule.

7.3.5 Functional robots, more than just smart machines

Functional robots are due to the novelty of design for the uniqueness of the task at hand or by both. These are more machine-like than with humanly aspects. Since the new millennia there has been a rapid increase in newer modes of locomotion for robots. Anthropomorphic designs as the brachiating robot and snake like locomotion has been briefly discussed in Chapter 2. The two avenues which are forte of a functional robot are (1) surveying and (2) search and rescue operations.

These robots are a boon during search and rescue operations after a travesty such as floods, earthquake or mine disaster. Still another utility of functional robots is in mine sniffing, which is a remedy to the unfortunate reminants of wars in Afghanistan, Iraq, Vietnam, Cambodia etc.

7.3.5.1 *Explorer robots — new age fungus eaters*

Explorer robots are state-of-the-art functional robots. It doesn't take a child's imagination to find anthropomorphic features in a NASA rover, which echoes a broadly absent social side. As rightful heirs to Toda's fungus eaters, new age explorer robots are expected to accomplish more than what any human being has, acting as scientists for geological sampling of the unknown terrain in a far off planet.

Explorer robots have been in use since the mid-1970s for space explorations. The success of the Lunokhod programme marked the beginning for unmanned rovers: Lunokhod 1 (1971) scouted about 4 km and Lunokhod 2 (1973) marched on for about 40 km. The NASA Mars rovers were targeted to detect water on the Red Planet, Spirit (2004) and Opportunity (2004) of Mars Exploration Rover (MER) mission were designed for surveying as shown in Figure 7.21, and Curiosity (2012) of Mars Science Laboratory (MSL) mission could undertake more sophisticated experiments to conduct surface sampling and geological studies. While all three have attained iconic status, very little of 144.8 million sq. km has been mapped, and not a great deal is yet known about the Red Planet. More recently, Rosetta mission's Philae landed at Comet 67P. It was designed for minimal locomotion primarily by jumping and not meant for long-term surveying etc. The MER rovers are state-of-the-art engineering and communication is over data streams spread across million of kilometers and are built to endure difficult terrain over long periods of time. Spirit and Opportunity were designed to harness solar energy, while Curiosity was fueled by nuclear power. It is worth noting that none of the MER or MSL explorers were truly autonomous and the were teleoperated by the earth based NASA scientists.

Over the last decade, hexapod rovers has been a raging trend in robot design, R-Hex [130], Mondo Spider, Stiquito and more recently the Erle Spider have left both the hobbyist and the seasoned roboticist spell bound. The anthropomorphic motivation are from lizards and spiders, however instead of strictly adhering to a circular symmetry, designs as Stiquito and R-Hex have employed lateral arrangements of the legs. The R-Hex, as shown in Figure 7.22 and its variations the AQUA project for explorations in oceans and lakes are a novelty of design for scouting unfriendly and unchartered environments. These are a promising improvement over wheeled robots as they do not get stuck in a muddy patch or sand pockets, can walk their way up a sand dune, ramp or stairs and can navigate rugged, broken ground rapidly. Other than surveying, such a hexapod can also be used for surveillance, search and rescue and sensing in remote locations and unforgiving terrains as arid desert or marshy landscape. The R-Hex project was developed with collaboration from five American and one Canadian university and was funded by DARPA and other US agencies. The R-Hex design, with 6 degrees of freedom, a motor for each of the six legs, maneuverability on both land and water and a speed of 2.7 m/s on land, may well be the next best bet for surveying far off planets.

Future rovers are expected to be more robust with a higher degree of autonomy and accomplis a full fledged scientific study of the unknown terrain. Like Toda's fungus eaters, these explorer robots will also be designed to ensure their own survival by harnessing energy sources, such as solar power etc. Such a future rover should have gone through human supervised training prior to the mission and it should be sufficiently autonomous to explore and survey the unknown terrain by planning and navigation and be able to send results, reports and images back to the earth-based mission base. Since the rover will be state-of-the-art and very expensive, there should be fail safe systems for self preservation, designed for restrain and caution, viz. avoid sinking into seas and falling off cliffs even if surveying may demand such death defying effort. The rover should be able to easily adapt to the new climatic settings such as humidity control, hight radiation levels, dusty terrain, climatic hazards such as cyclones etc. In case the rover is in dire situations such as system failure or unforseen accidents, the rover should eject a capsule containing information and images from its findings for future missions to find. Like the fungus eaters, the designs for the rover will depend on the terrain. A creative, out-of-the-box solution is by employing a modular components, where a number of semi-independent self-assembling modules work in unison so that the robot can physically morph between different configurations as the needs of the terrain require. For example, a car like wheel-wheel shape works well for smooth terrain, while a hexapodal spider-like design with long legs is more useful in a boulder field or mountanious terrain. The rover should have sufficient fuel capacity so it doesn't need to return to base for refuelling at short intervals and is therefore suited to a dual fuel mode and it should be equipped for solar power or the ability to harness other sources of replenishable energy which is readily available in the given terrain. It is possible that the rover is a part of a larger system of robots and satellites. Therefore mapping and survey data acquired by itself and also from satellites orbitting the planet should be synced and incrementally augmented into the robot's system. Finally, the robot should be able to evaluate its near future options by simulations and 'try and test' before commiting to an option, much like a Popperian creature [362].

7.3.5.2 Search and rescue robots

A search and rescue robot is often shoebox size with a camera on top and a tether. It is unassuming compared to the aweinspiring robots discussed till now. However these are most useful in disasters such as earthquake, mudslide, cyclones, tsunami, nuclear catastrophy and

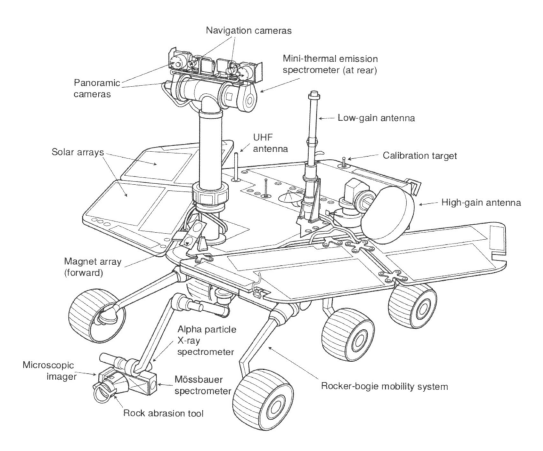

Navigation cameras

Mini-thermal emission
spectrometer (at rear)

Panoramic
cameras

Low-gain antenna

UHF
antenna

Solar arrays

Calibration target

High-gain antenna

Magnet array
(forward)

Alpha particle
X-ray
spectrometer

Microscopic
imager

Mössbauer
spectrometer

Rocker-bogie mobility system

Rock abrasion tool

FIGURE 7.21 **Design for a NASA MER rover.** The rover was targetted to find water on the Red Planet and conduct geological studies. Each rover was 1.6 meters long and had a mass of 174 kilograms. The rover has six cameras which aid in navigation (Navigation Cameras and Panoramic Cameras as shown). Other than these there is a pair of Hazard Avoidance Cameras which is hidden from view and mounted on the front part of the body just below the leading edge of the solar arrays and three cameras for scientific investigations. The geological sampling instruments are (1) Alpha Proton X-Ray Spectrometer (APXS), (2) Mossbauer spectrometer, (3) microscope and (4) Rock Abrasion Tool (RAT), these are mounted on a mechanical arm to check for the composition of rock samples. The microscope images the martian rocks and soil in detail; the RAT works as a grinder and drills holes into rocks; the APXS irradiates sample rocks with alpha particles and X-rays from a curium source and studies the back scattered alpha particles and the emitted X-ray to study the composition; and Mossbauer spectrometer studies the nature and composition of minerals containing iron and the rock. Other than the tools on the arm, mini-thermal emission spectrometer, located next to the navigation cameras, can image the Martian surface in infrared to detect and reveal the mineral compositions and their abundance; and the magnet array is to collect airborne magnetic particles in the Martian dust. Image in public domain, courtesy NASA.

FIGURE 7.22 **The R-Hex Design.** can traverse unstructured and broken surfaces such as sand dunes, muddy patches and marshy terrain. It is maneuverable over water and can also walk up stairs. The R-hex is dynamically stable. At any given instance of locomotion, only 3 of the 6 legs touch the ground — 2 on once side, 1 on the other. Image in public domain, courtesy `wikipedia.org`.

building collapse. Search and rescue robots were used in 9/11, Tohoku earthquake and the Fukushima disaster, Hurricane Katrina and other crises. These robots are built to be simple, lightweight and easy to maneuver as the environment in and around the disaster may not allow for use of advanced technology, viz. wifi or GPS may be lacking or severely limited. Therefore a tether is the best way to control the robot and it also helps as a rescue rope if the robot itself needs rescuing.

A moniker of 'unmanned' is often used to classify these robots. Unmanned Ground Vehicles (UGVs) are robots used in scenarios where it is hazardous or intractable to pursue rescue operations, Unmanned Surface Vehicles (USVs), serve the same utility as UGVs, however they are for disaster operations in rivers, lakes and oceans, Unmanned Underwater Vehicles (UUVs) are a step ahead of USVs and can search underwater for survivors etc and Unmanned Aerial Vehicles (UAVs) or rescue drones can fly over the affected area and can beam in live images to inform about the extent of loss of human lives and damage and are best suitable in scenarios where areal rescue operations are preferred, such as transporting medicine and food to flood or earthquake victims, as shown in Figure 7.23.

The camera and the infrared thermal imagery are the most effective sensors on board, which allows a human being, aptly a disaster management expert or a rescue worker to be telepresent and search for signs of survivors or attempt to defuse a bomb or leaking nuclear waste etc. Military robots designed to carry and deploy weapons are usually able to undertake search and rescue missions, and are of use in heavy duty operations in dangerous environments such as rescue operations in coal mines or to save the survivors of a nuclear accident.

7.4 JAPAN, ROBOT HISTORY IN THE MAKING

Japan stands out as a culture unique in a niche of its own that has progressively accepted robots as a way of life. The island nation has more than 50% of the world's operational robots, and it is the leader in development of humanoids and androids. It has been the

FIGURE 7.23 **Rescue drones** help to cover and assess the extent of the devastation and damage in natural disasters such as floods and earthquakes. Shown here is a member of UK's 'Serve On' (*www.serveon.org.uk*) team with a drone in the 2015 Nepal earthquake. Image courtesy Jessica Lea/DFID, CC-by-SA license from wikipedia.org.

home for a number of iconic robots which includes softbank's Pepper, Honda's ASIMO and Mitsubishi's Wakamaru.

Post war Japan has been the powerhouse for new technology, with proven leadership in auto industry and robotics. However on the precipice is Japan's population, which is on a steady decline. The birth rate is at about 1.40, which is significantly lower than the 2.13 suggested as the critical replacement fertility rate. The total population of 127 million is set to decrease to around 116 million by around 2030, near 100 million by around 2050, and a little more than 85 million by around 2060. With a fast declining population, and the economy driven by cutting-edge technology and the country is often faced with persistent labor shortage. Therefore, the problem often boils down to either welcoming more immigrant labour or making more robots [288].

Intelligent humanoid robots are already a dominant part of Japan's economy and culture and their application is spread across a number of streams. The ASIMO shown in Figure 7.24 is meant for various jobs but most of them are in the experimental phase and the price is well beyond affording it as a personal robot. HRP-2 is a humanoid robot and assists people in construction and also for domestic chores. Enryu has a human exoskeleton and can also operate power shovels and forklifts. ReBorg-Q, Guardrobo D1 are designed for patrolling and fire hazards. The Kodomodroid, Actroid and Ontoid look very real-life and are threatening to wipe away human presence from redundant and repetitive service sector jobs. PaPeRo and Wakamaru are personal robots and have a wide range of applications, which includes babysitting and tutoring children. Pino, Posy, Robovie and ifbot provide entertainment and companionship. Kaori is a humanoid robot with the features of an attractive woman and priced at $8500, and helps towards romantic attachment and sex.

Currently, 27.3% or 34.6 million people of Japan's current population is aged 65 or more, and the need for elder care has led to a proliferating industry in care robots. AIST's PARO, Riken-SRK collaboration for Robobear [58] and Panasonic's Reysone [246] are some of the breathtaking care robots to assist the graying populace of the country.

The household robot market is already past 17 million units and the robot market will

FIGURE 7.24 **Social robots as an integral part of society.** Automated humanoid robots like the ASIMO are an active part of society, and Japan may be the first human robot society in the apparent future. Image courtesy `wikipedia.org`, CC-by-SA 3.0 license

reach well beyond $ 21 billion by 2025, which will reduce 25% of factory labour costs. With a Robot Olympiad planned in 2020 [88], and specialised industries for making industrial and care-giving robots in Tokyo and Saitama, Japan is indeed redefining the rules of the game. It is anticipated that robots will remake Japanese society [45]. These socio-economic conditions may propel Japan to become the first human robot society in the world.

Shuji Hashimoto at Waseda University considers robots as a 'third existence' [285], an essence of being in betweenliving and non-living yet not necessarily as social equals to human beings. In similar context, Masahiro Mori [242] has drawn parallels between principles of Buddhism and technology for making robots. Social scientists consider that both Shinto Buddhism, which is the prevalent religion in Japan, and the existential principles of Ba theory have contibuted and shaped the Japanese society and allowed for such a harmony of the Japanese people with robots & new age technology [77].

Robertson [278] dubs Japan as a cyber Olduvai Gorge of the future, and speculates that Japan may have the makings of an anthropological wonder of the future where humans beings and smart machines will meld into a new, superior species. The settings of such a unique event in the future are certainly being pursued most actively in Japan, and it may prove to be the cradle for transhumanism. Such concepts are explained in later chapters, particularly the last chapter.

Duffy points us to a crisis of sorts. Current-day social robots do not truly feel an emotion and merely mirror and replicate human behaviour viz. facial features, intonation and language, and therefore the paradox is that they pretend to be what they are not. There is no call of conscience or adherence to moral values, nor the will for a conscious decision process. For example, a museum guide robot will never attend to a health emergency of a visitor, a news reader robot will not be able to express the tragedy of a sudden terrorist attack even with the best possible voice intonation and facial expressions, nor will a care robot truly care much if a fire burns down the neighbouring house. All of these robots will wonderfully attend to their own chores, but will lack the call of conscience and the human quotient to override everything else in the face of a crisis. We human beings have far superior abilities that may take a long time and need a revolution in technology to be fully replicated in robots. The next chapter tries to overcome this shortcoming and attempts to develop ethics and moral values in a robot.

SUMMARY

1. Social robots find a niche in the care industry or in the entertainment industry.
2. Social robots are designed on human traits and known social norms.
3. As of now, social robots are rather expensive and therefore not very mainstream.
4. Minimal robots such as CERO also appeals to human like behaviour.
5. Hobbit, Care-O-Bot and Paro connect to the interacting human user on bonds of fake empathy.
6. The drumming mate experiment is an important concept to demonstrate cognition in a social context and cooperation with the human user.
7. Surveying robots and search and rescue robots have uniquely contributed to human society.
8. Japan may soon become the first human-robot society.

NOTES

1. *The ELIZA effect: the tendency to unconsciously assume computer behaviours are analogous to human behaviours*
2. *Robots aiding and assisting human beings in domestic environments were coined as 'co-inhabitant robots' by Goldberg*
3. *As a shortcoming, Inkha's speech did not correspond to its lip movement*
4. *Stiquito started as a low-cost experimental project and has been used in classrooms to demonstrate subsumption and other reactive architectures*
5. *NLP methods lack a foolproof means to detect sarcasm, as sarcasm also involves tonality, facial expressions and cannot be completely conveyed merely by power words. So, while "you dork!" will fetch the human user a warning, "now, aren't you clever" said with a suggestive grin will hardly connect with the robot.*

EXERCISES

1. **Wall-E Eva chat**, consult appendix A and run the python script from the terminal window with the command *python walleeva.py*. In this simulated conversation as shown in Figure 7.25 Wall-E tries to woo Eva, the emotion and the mood of the conversation is conveyed by the use of words and expressions. Try to design similar simulated conversations which convey other emotion and moods. This question requires basic Python programming skills, though it doesn't need any use of NLTK or any NLP specific tools or libraries.

```
WALL-E :   hello Eva
EVA    :   leave me alone please
WALL-E :   I love you
EVA    :   you look nice after that service
WALL-E :   I need you
EVA    :   buzz off clanky old robot
WALL-E :   dont point that blaster at me sunshine
EVA    :   now you know I like it when you speak to me like that
WALL-E :   can i give you a robotic hug
EVA    :   you look nice after that service
WALL-E :   I love you
EVA    :   you look nice after that service
WALL-E :   can i give you a robotic hug
EVA    :   you look nice after that service
WALL-E :   I need you
EVA    :   hey my darling,come here and lets look for plants
```

FIGURE 7.25 Exercise question — Wall-E Eva chat.

CHAPTER 8

Robots with moral agency, in the footsteps of Asimov

Kirk: "I'm not gonna take ethics lessons from a robot"

– from the movie Star Trek, Into the Darkness

"One narrow option is that the code is a set of *rules of engagement* ..."

– Selmer Bringsjord and Joshua Taylor [49]

8.1 THE NEED FOR THE GOOD ROBOT

CURRENT day robots are made for a specific utility, such as museum guides, carer and nurse robots, personal robots, explorer robots etc. However most of these jobs are repetitive and pertains to a narrow domain of chores and lack imagination. Instead of engaging in routine, predictable and boring jobs, can robots become a crucial part of society and take on roles as law enforcers, scientists, authors, doctors and politicians of the coming age? The very idea marries a child's fancy to a horror movie. Human activities are broadly dictated by abstract notions such as value system, ethics, cultural edicts, personal principles and nuances of the society; a fairly long list of the written and unwritten. For example, a robotic law enforcer should be able to understand and interpret human values, social systems and ethical principles. Similarly, a robotic doctor must be appreciative of the moral and ethical code of conduct meant for human doctors. The pitfall in both cases is that while not being a human being, the job at hand expects that the robot 'feels' and relates to the scenario at hand from a human point of view. Will robots ever adhere to such humanly values to become an effective part of a futuristic human robot society? This chapter explores how to design AI that can appreciate ethical values and moral principles [343].

There is always the temptation to form rules for robots considering them as tools, products, artifacts and machinery [360]. Such a codification will read more like an operating manual and will not address the specialty and uniqueness of robots as entities of emergence, which can spring a life of their own. A slightly different approach is advocated by Asimov in his laws. Though in fiction these laws echo one of the earliest attempts at a rule-based approach to instill ethical values in robots, and these laws are appreciative of the special niche that human beings form with robots, they inherently crown human beings as a benevolent master race reducing robots to amicable slaves.

Very soon, human-robot interactions will no more be just an event at a science fair or at a museum but will be more commonplace, ubiquitous and seamless. There will be various tasks which cannot be decided as strict 'Yes', 'No' or via programming loops as 'If THIS then THAT'. Such a need will be even greater when the robot takes up roles in medicine and surgery, military, education etc. Highlighting a need for robots that can appreciate morality and ethical principles. Efforts to develop artificial morality have been a potboiler in fiction and a topic of debate in ethical circles. However, the earliest attempt to imbue beings with artificial morality using AI was in video games and virtual reality [84] and not long for robots [125, 284].

Wallach and co-researchers coin the term, Artificial Moral Agent (AMA) to account for all such agents: robots, smart software, game characters, virtual reality etc. i.e., artificial agency capable of acting more or less ethically by themselves. This chapter is focused on the workings of a future human-robot society and efforts to imbue robots with the grammar of ethical values, moral bindings and social interactions.

Devices that can respond to natural human traits such as voice and gesture are commonplace, viz. mobile phones which respond to our voice commands, smart software which can detect emotions from our facial features, smart environments which respond to heat signature and human interactions, seamlessly incorporating natural human traits into technology. Therefore, most of the moral attributes of a robot are hardly an appreciation of human values and are often reduced to following a script designed by the programmer [314] where the onus of ethics is more with the programmer than the robot or the user.

$$Programmer(s) \rightarrow Robot \rightarrow User \tag{8.1}$$

However, embodied AI can take a life of its own thus the workings of the robot and that of the user can blur out [235], leading to an overlap of ethical values. In the simplest scenario ethics can be programmed-in as conditional loops. For example Roomba demonstrates an *'ethics of self preservation'* as it avoids its own destruction by not falling off edges and the PR2 adheres to *'ethics of subservience'* by playing billiards with the human master when commanded. Such values are no more than illusory, implemented with lines of codes lacking a humanly feeling of interest or responsibility to it.

A better option to artificial moral values can be with effective human-robot interaction, where robots should be aesthetically appealing and preferably a humanoid, equipped with vision, voice and touch, adhering to a minimal moral code which may be programmed-in, and lastly there should be a way to limit robot action, a failsafe to prevent the robots from doing the unthinkable. These form the guidelines for the semi-sentient paradigm which is discussed later in the chapter. It is worth noting that anthropomorphism blends in animal/human behaviour into that of the robot, thus making it appealing to human psyche. This will lead to scenarios where a humanoid robot will be addressed as if it is at least a near human being. Thus its actions in a societal concern will be evaluated on the basis of established human ethics and moral values.

Robot ethics have often been burdened with criticism; there have been arguments that (1) such an approach is futile as the moral standing of human beings and ethics are not very well defined and often vary by culture and geographical influences, (2) human ethics have evolved over thousands of years, from modest hunter gatherers to where we stand now, and are coded into our DNA — it will not be possible to replicate the workings of mother nature over a short span of time, (3) the moral standing of an artificial being cannot be gauged correctly, by human recognition and judgement and (4) use of robots in lethal actions such as war, security services, automated vehicles, law enforcement etc. and issues in sharing sensitive information as in banking, military, corporate secrets etc.

Also, various unfortunate incidents involving robots have helped such criticism. The

U.S.

$10 million awarded to family of plant worker killed by robot

Knight-Ridder Newspapers

DETROIT The manufacturer of a one-ton robot that killed a worker at Ford Motor Co.'s Flat Rock casting plant must pay the man's family $10 million, a Wayne County Circuit Court jury ruled this week.

The jury of three men and three women deliberated for 2½ hours before announcing the decision against Unit Handling Systems in a suit by the family of Robert Williams, who was killed Jan. 25, 1979. Unit Handling is a division of Litton Industries.

It is believed to be the largest personal injury award in state history.

At the time of his death, Williams, 25, of Dearborn Heights, Mich., was one of three men who operated an electronic parts-retrieval sys-

tem at Ford's Flat Rock plant. The plant has, since been closed.

The system, made by Unit Handling, was designed to have a robot automatically recover parts from a storage area at the plant.

On the day of his death, Williams was asked to climb into a storage rack to retrieve parts because the robot was malfunctioning at the time and not operating fast enough, according to the Williams family's attorneys.

The robot, meanwhile, continued to work silently, and a protruding segment of its arm smashed into Williams's head, killing him instantly.

The robot kept operating while Williams lay dead for about 30 minutes. His body was discovered by workers who became concerned be-

cause he was missing.

Attorneys for the family said the robot should have been equipped with devices to warn workers that it was operating.

"If they didn't want people up there when the robot was moving around, they should have installed safety devices," said Joan Lovell, one of the two attorneys representing the family. "Human beings are more important than production."

The jury's award went to Williams's widow, Sandra, 30; their three children, ages 8, 6 and 5; his mother, and five sisters.

"They were an extremely close family," said Lovell. "I've seen a lot of people who have been injured, but this family was particularly devastated by this loss."

FIGURE 8.1 **First fatality due to a robot.** this incident on January 25, 1979, is the first of a number of such sordid events, in which robots were instrumental for human injury and death. However, since most of these robots were not modelled with ethical principles and none of them was sentient the fatalities are probably best described as accidents, rather than bad intent. Article in Ottawa Citizen, August 11, 1983

first death due to a robot was Robert Williams, a worker at Ford Motor Company's Flat Rock casting plant, as shown in the newspaper article in Figure 8.1. The mishap was due to malfunctioning when a protruding part of the robot's arm smashed into Williams' head, and it led to the then largest personal injury award in the state of Michigan, of $ 10 million. Two years later, a similar tragedy at Kawasaki led to the death of factory worker Kenji Urada, when a hydraulic arm robot pushed him into a grinding machine. In 2007, the death of 9 people resulted when a robotic anti-aircraft weapon Oerlikon GDF-005, went berserk in South Africa and the death a 21-year-old occurred when a robot struck him in the chest in 2015 at a Volkswagen plant near Kassel, Germany. Events like these have fomented the opinion that robots should not be allowed to attain high levels of autonomy as it would jeopardise human safety and well being particularly, keeping robots out of military and other lethal operations. Containment of robots protects human lives but in contrast prevents robots from attaining higher levels of autonomy would in effect contain them to be mere slaves and prevents the emergence of a higher intelligence. In similar context, the fledgling discipline of robot ethics has lacked consensus for allowing robots to perform lethal actions, as in the military. Robots such as Big Dog, SWORDS system and MAARS have fueled this debate. As of now, every robot action can be completely or partially overridden by human beings, which has led to a debate in favour of robots supervised by human beings, and never given full autonomy — a smart slave. However, incidences such as Robert Williams and Kenji Urada are unfortunate accidents rather than bad intent from the robot's end, as none of these robots were built with ethical principles and had not attained sentience.

Therefore, it is likely that sentience will be a key for full moral cognition which is comparable to that in an adult human being. A simpler alternative without the binding of sentience, human-like moral values in a narrow concern may be developed in a robot with rule-based logic, amicable human-robot interaction and programming loops. Robot ethics may take a leaf out of machine ethics, and can also be approached via four aspects. (1) Anthropocentric approaches, where it is developed as conventional ethics centered around human needs and interest, often narrow and viewed as a systems-engineering problem. Such approaches code in the robot with some degree of ethical values, but do not facilitate for emergence of autonomous moral responsibilities .(2) Infocentric approaches, based to enact ethics, though not necessarily finding meaning in it. As Kismet or Pepper, which seems to make a play on human emotions and thus make a show of ethics in a narrow concern. (3) Biocentric approaches, developing ethics as it happens in a human being, as constant interaction with the environment. Such an approach encourages the development of being with natural purposes, interests and intent. (4) Ecocentric approaches, is the next step in a society structured by artificial morality where artificial agents work in unison to address larger problems of the society etc. i.e., newer power source for the community, safe haven and safety from natural calamities. Current day research has at best addressed infocentric approaches, while research on biocentric approaches to develop full AMA has been pipe dreams or as, Wallach calls it, 'vapour ware' as no one knows where to begin. However, once we have artificial sentientism, extending ethical values to work over a planet-wide network would probably be a humble beginning for ecocentric approach. While it is also possible, that a robot or an AI being that can reason ethically might transcend to a morally higher being than human beings, and therefore may serve as an altruistic and unbiased advisor to our society about what action may be most suitable — a benevolent moral police which also has high soothsaying capabilities. Such prospects are briefly discussed in this chapter, however the next chapter has more details on such ideas.

8.2   MORALITY AND ETHICS

Ethics, the science of morality, separates us from animals. However, ethics is not contained as a precise set of rules, rather it is an inbuilt grammar which contains principles acquired over a number of belief systems; social beliefs, collective opinion, geographical influences, cultural beliefs, religion, position in society etc. Though these vary with region, race and country, the following are some common features that are more or less acceptable universally [119, 138]:

(a) Reciprocity — both in aggressive intent, viz. 'an eye for an eye' and benevolence, viz. 'the good comes back to you'.

(b) A general acceptance of moral code apparently without any edicts or laws. A person lacking such is often ostracised as rude, anarchist, lunatic or unsocial.

(c) Rank, status, social standing and authority — this may reflect in a well-defined hierarchical system as in the military and a corporate set up and also implicitly in social interactions.

(d) Honesty and trustworthiness are valued and perfidy and unscrupulousness met with rebuke.

(e) Unprovoked aggression is discouraged.

(f) Extreme circumstances such as war, flood and devastation and lack of basic amenities such as food, water and social security appeals to the sympathy quotient in one and all.

(g) Ranking of 'badness' or 'goodness', viz. stealing is lesser bad than murder etc.

(h) Bounds on moral agency, viz. different moral standards for minors, terrorists, convicts and felons.

(i) Moral rules get a higher ascendancy than common sense or self interest.

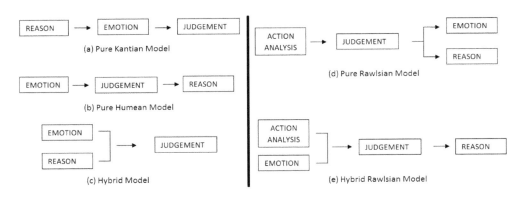

FIGURE 8.2 **Five models of our moral psychology.** There is a lack of convergence on how emotions influence our moral judgement, however there is a general agreement that emotions and moral actions work in tandem. Adapted from Hauser [146].

Authors, researchers and philosophers have often used ethics and morality as near synonyms, however going by strict definitions, **ethics** are moral principles that govern a person's behaviour, while **morality** constitutes the ability to seek a distinction between right and wrong or good and bad behaviour.

Human interactions are based on emotions and expressions conveyed with facial gestures, voice intonation and body language, and these also aid in structuring our ethical judgements. In the last chapter, we had seen that social robots Kismet, CERO, Eddie etc. are designed to mirror emotion. It is also worth noting that even with a lack of realistic human features, the human user could identify with the right emotion from the robot.

Moral psychology is dependent on emotion, however there is no clear agreement on the models [146, 158], as shown in Figure 8.2. (a) The Pure Kantian model, structures moral judgement as a rational, deliberative process wherein emotions are instrumental in generating reactive attitudes, however moral judgement is borne out of a conscious deliberative process. (b) The Pure Humean model is based on the notion that moral psychology is essentially emotive while reasoning and other deliberative mechanisms provide for a rationalisation of the moral judgement. (c) Hybrid models put the onus equally on both deliberative mechanisms and emotions to lead to a moral judgement. (d) Pure Rawlsian models are based on the concept that there exists a distinctively moral faculty, operating independently of deliberative and emotional mechanisms. For such, emotional mechanisms are gratified output from a moral judgement seemingly instrumental in translating a moral judgement into action. (e) In Hybrid Rawlsian models, emotion ensues along with the moral action and can indicate high-conflict personal dilemmas and also erratic output from the moral faculty. Thus, though it is a day to day experience that moral values, emotions and moral actions work in tandem, there is still to be a general acceptance on what happens in the psychological concerns. Therefore modelling a similar process for robots will not be easy, however it is apparent that for artificial morality there will be a need for artificial emotions.

In criticism, both Sullins and Duffy have suggested that emotions and therefore morality in robots is just an illusion. This illusion is an extension of human beings misleadingly attributing moral and ethical values to a robot due to our rote sense of ego of associating ethics to a humanoid appearance and human-like interaction through voice, all of which is due to good designing and skillful programming. A robot will never truly 'feel' an emotion for eliciting a moral value, but rather only respond according to a program code, following a subroutine and merely corroborating an obviously false emotional response to the input user emotion, therefore a superficial exhibition of a moral value. Implementing ethics poses even more issues as ethical reasoning is based on abstract principles and they cannot be easily construed in a formal deductive structure. There is no definitive approach to ethical theory, and the framework varies from ethics of virtue to respect to utilitarianism to duties and various others, thus leading to a debate as to which approach may be most suited to intelligent machines. The metaethical frameworks have an element of vagueness and may be difficult to implement computationally and ethics inherently involve a human factor where each individual weighs the current situation on his/her experiences, premises, beliefs, sentiments, intellectual understanding and principles, not all of which have been well understood.

Despite such issues, there has been considerable research into ethical robots as shown in the timeline, Figure 8.3, particularly over the last three decades and computationally motivated routes to develop consciousness have been debated time and again.

Obviously all robots will never qualify to be ethically significant, nor will the ethical values of every robot be of concern to human society. Since, morality is characterised by personality and subjectivity, and it can be argued that robots particularly those without any biological attributes, will always lack in abductive reasoning[1] they will never attain full moral agency. However, human beings at least perceive some moral values in the robot, even more so in state-of-the-art social robots such as Pepper and Kodomodroid, however fake they may seem. In particular, robots that have high autonomy and exhibit a high level of intelligence will always have the added onus of moral bindings. To be considered as a moral agency a robot will further need the attributes of intentionality and responsibility.

1. **Autonomy** — moral values can only be the virtue of agents which can be autonomous, however autonomy in itself is not sufficient to confirm moral agency. Probably the most obvious example is that of teleoperated agents, which may exhibit a make believe autonomy, but will lack in independent agency.

2. **Intentionality** — this is an apparent grey area and even human psychology cannot be pinned down to a strict good or bad intent. However, as long as the robot's actions as per the programming and the environment are not deliberately meant to convey moral actions, both harmful or beneficial, then it can be said that the robot acts as per its own 'free will'.

3. **Responsibility** — if the robot's behaviour leads to the assumption that it has a responsibility to another moral agent, then the robot fulfills at least some social role. For example, a nurse robot will attend to its patients in emergency. This behaviour be it programmed or emergent, will seemingly convey its duty to care.

Also, in direct correlation to human behaviour in a societal context [323], a full ethical agency will be anticipated to adhere to norms that we ethically require of the behaviour of human beings (moral productivity) and also have qualities which will qualify them as

---

[1]Deductive reasoning proceeds from a general rule to a specific conclusion viz. mathematics where rules as in algebra or logarithms enable us to reach a conclusion in the specific scenario. Inductive reasoning is relating observations that are specific to lead up to a general conclusion viz. scientific research which develops a conclusion from a number of specific studies. In contrast Abductive reasoning is often based on an incomplete set of observations and proceeds to the likeliest possible explanation. Though being more diagnostic with guesswork in nature, it is the basis of our real time decision-making leading to the best outcomes with incomplete information at hand, viz. medical diagnosis is oftentimes abductive.

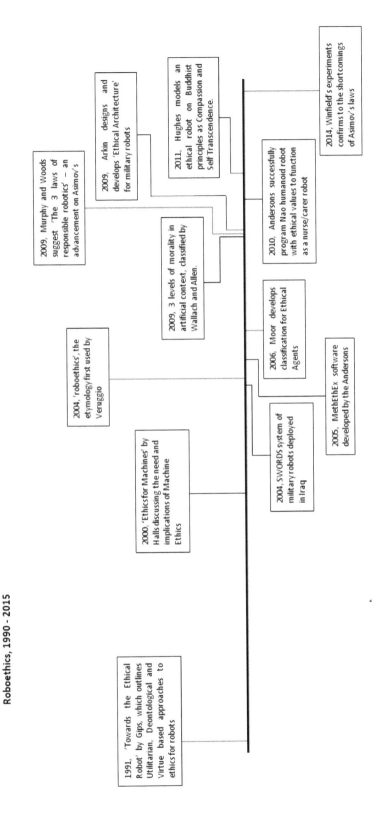

Roboethics, 1990 - 2015

1991, 'Towards the Ethical Robot' by Gips, which outlines Utilitarian, Deontological and Virtue based approaches to ethics for robots

2000, 'Ethics for Machines' by Halls discussing the need and implications of Machine Ethics

2004, 'roboethics', the etymology first used by Veruggio

2004, SWORDS system of military robots deployed in Iraq

2005, MethEthEx software developed by the Andersons

2006, Moor develops classification for Ethical Agents

2009, 3 levels of morality in artificial context, classified by Wallach and Allen.

2009, Murphy and Woods suggest 'The 3 laws of responsible robotics' – an advancement on Asimov's

2009, Arkin designs and develops 'Ethical Architecture' for military robots

2010, Andersons successfully program Nao humanoid robot with ethical values to function as a nurse/carer robot

2011, Hughes models an ethical robot on Buddhist principles as Compassion and Self Transcendence.

2014, Winfield's experiments confirms to the shortcomings of Asimov's laws

FIGURE 8.3 **Timeline for roboethics**, 1990 — 2015. The concern for imparting robots with ethical values came to the fore with Gip's 1995 paper and has proliferated to a new discipline of 'roboethics' over the last 25 years.

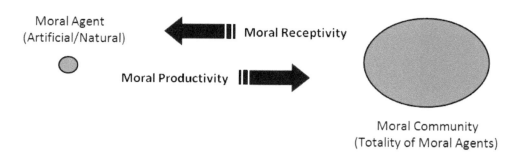

FIGURE 8.4 **Moral productivity and moral receptivity** are complementary relationships between a moral agent and the rest of the moral community. The moral agent (small blob) is also a part of the larger moral community (big oval). Schematics adapted from Torrance [323].

beings towards which humans have an ethical duty (moral receptivity), as schematically shown in Figure 8.4.

To help this discussion, Moor suggests a classification scheme for ethical agents [239,240].

1. **Ethical Impact Agents** are agents whose actions have ethical outcomes whether deliberate or not. Nearly every tool can be seen as an ethical impact agent, such as a watch, a car etc. Any robot at the very least has a potential to be an ethical impact agent, so its actions could be harmful or beneficial to human beings. As an illustration, consider a robotic manufacturing system, which cuts down human labour, though there is no bad intention from the machine's end. In similar manner, the death of Robert Williams was not due to any wrong intentions of the robot.

2. **Implicit Agents** are agents that have ethical factors built into them, and often are devices and systems for safety or security. Best illustrations are vehicular safety and warning systems working towards well being of the driver, ATM machines ensuring trustworthy money transactions etc. However, the ethics pertain to a narrow concern as a built-in sense of goodwill by hardware or programming which lacks overarching ethical principle. For example, a robot security guard at a supermarket can be (poorly) made with a single programming loop to identify shoplifting if the signal at the exit gate which reads the bar code goes red.

Algorithm 3 Robot security guard at a supermarket
---
  repeat
    if signal light is red **then**
      stop and apprehend the shoplifter
    else
      continue with the job
    end if
  until forever

Here the robot security guard is identifying the shoplifter strictly as per the programming code, and not anything else. If instead of shoplifting someone gets stabbed by a knife at close proximity to the robot, it will not apprehend the assailant as it has not been programmed to do so.

3. **Explicit Agents** are agents that can perceive an ethical principle and process

information and lead to an ethical judgement. In case ethical principles are in conflict, the agency is able to work out alternatives. Moor eloquently adds that explicit ethical agents act from ethics and their virtue is not grounded in a programming code or hardware. As of now, such agency can only be found in some of the sophisticated AI labs across the globe. The process to design and make such agents is rooted in machine learning, training an ANN or evolutionary or genetic algorithm.

4. **Full Ethical Agents** take a further stride from being explicit agents. Not only can they make a range of ethical judgements, they also have the humanly attributes of consciousness, intentionality and free will. An adult human being is a full ethical agent.

It is tempting to develop ethics as a subroutine, a top-down paradigm where ethical rules are implemented as decision algorithms via programming codes in an AI agent. However such top-down approaches as Asimov's laws or any attempts to program Kantian reciprocity or novel efforts to codify the ten commandments from the bible etc. are often prone to conflicts and this rule-based ethical behaviour is at best an imitation in a particular narrow concern. Hence, coding in ethical values will not be sufficient in building a full moral agency, therefore researchers have explored bottom-up approaches, as shown in Figure 8.5. Bottom-up approaches have had a special place in mobile robotics since the subsumption architecture and, for developing ethics, the approach will be motivated by Piagetian principles which will allow the AI agent to discover such ethical rules through experience and continued interaction with the environment, which may be possible with connectionist architecture and machine learning methods. Thus in contrast to a rule-based approach, a developmental paradigm nurtures ethics in a robot by mimicking the growth of a child with zero moral value to a fully grown adult with near complete moral agency, a very slow and gradual process. The robot will resemble mimic the growth as seen in children; observation, extrapolation and correspondence with the behaviour of adult human beings. Learning and adapting are seen as key points in such a model. There have been suggestions that it would be unethical to create self-aware beings who did not possess a human-like urge for learning and intellectual growth. The top-down approach is riddled with multual conflicts and bottom-up approach is exhaustive and time consuming. Wallach and Allen [9, 343] suggest employing a hybrid approach of top-down and bottom-up routes. Therefore, starting with some basic Kantian-like virtues programmed into the robot along with a connectionist paradigm should be a good beginning to gradually grow into a full moral agent.

Wallach and co-researchers, identify three levels of artificially developed morality as shown in Figure 8.6.

1. **Operational Morality** — The design and testing process is undertaken in full awareness of the ethical values, and the ethical action of the agency is within the control of the designer and the user. For example, a steam iron will be insulated to protect the human user.

2. **Functional Morality** — The next level of morality, includes systems that respond with a predictable, often redundant response which pertains within acceptable behaviours to intelligent systems capable of determining some of the moral bearings of their own actions. An example of the former is a mobile GPS tracker in Google Map. You may be trying to track your friend or you may be a felon on the run. The GPS tracker is not aware of your intentions and behaves the same under both scenarios. A good example is ethical decision systems such as MedEthEx for doctors to choose

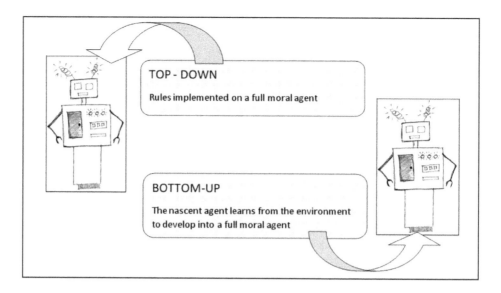

FIGURE 8.5 **Top-down vs. bottom-up**, ethics can be implemented as a Top-Down approach, by installing ethical principles and values into a robot, or as a Bottom-Up developmental approach. Typically, Bottom-Up approaches are pristine and more tractable, a Top-down paradigm is safer, more predictable and easier to control. Wallach and Allen [343] conclude that none of these approaches are sufficient, hence a hybrid approach with a co-joining of both these paradigms is worth an exploration.

the treatment/clinical process adhering to medical ethics and Modria and other such software for legal assessment and consultation.

3. **Full Moral Agency** — The agent, other than attaining reliable functional morality also learns from its experience and keeps enhancing its own intelligence and moral values. A full moral robot will have high autonomy and high ethical sensitivity, as in an adult human being.

As shown in Figure 8.6, current day robots typically have low ethical sensitivity and low autonomy. Vacuum cleaning robot Roomba has the capability to detect the edge of the floor and prevent a fatality and it is committed to self preservation, but it is low on ethical sensitivity. In contrast an autopilot is high on autonomy but low on ethics and it is committed to a narrow concern. Kismet responds to human emotions, buthowever it is merely responding to an external stimulus and not adhering to any ethical or moral principles. Star war robots as R2D2 and C3PO and iPhone's Siri all respond to a human action, be it voice, gesture or action. However none of them abide by a set of guiding rules which determine their virtue and righteousness.

An example for the first robot that makes it to the domain of operational morality is the robot parole officer in the movie Elysium. When in conversation with Max (Matt Damon), the robot is able to make conscious decisions based on Max's heart beat and pulse rate, it is able to infer emotional inputs such as condescension and sarcasm from Max's tonality and gestures. It is able to offer medical aid and it also has the authority to award judicial sentences and extend parole etc. The robot butler in the movie 'Frank and the Robot' is also in the same domain but it has higher autonomy than the robot parole officer as it can

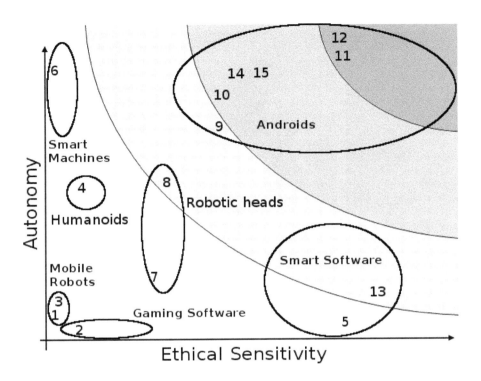

Operational Morality

Functional Morality

Full Moral Agency

```
1 : Roomba
2 : Deep Blue, software that beat Kasparov in chess
3 : R2D2, from the Starwars movies
4 : C3PO, from the Starwars movies
5 : Siri, from iPhone
6 : Autopilot
7 : Kismet
8 : Robotic Parole Officer, from the movie Elysium
9 : VCG-60L Robot Butler, from the movie Frank and the Robot
10: NS-4, from the movie I,Robot
11: Nexus-6, from the movie Blade Runner
12: Cylon-6, from the TV series Battlestar Galactica
13: MedEthEx, medical ethics expert system
14: Robocop, from the movie Robocop
15: Terminator, from the movie series Terminator
```

FIGURE 8.6 **Ethical sensitivity vs. autonomy.** the classification by Wallach and co-researchers helps to understand where current-day robotics stands, and what may be needed for a full moral agency. Various robots, from day-to-day use and also from fiction have been used to illustrate the concept of moral agency in a robot. Adapted from Wallach and Allen [343].

FIGURE 8.7 **NS-4, Robocop and Cylons**, The NS-4 bust used in the movie I,Robot, Robocop and Cylon Centurian. Robots which are high on ethical sensitivity and also on autonomy are still a lore of science fiction. All three images are in public domain, courtesy `wikipedia.org`.

move around. Some of the fictional robots, as shown in Figure 8.7, the NS-4 from the movie 'I, Robot' and Nexus-6 from 'Blade Runner', will be the best examples for robots in the domain of functional morality. NS-4 could take conscious decisions based on information acquired and also adhere to the broad altruism of doing good for the human race while Nexus-6 was a further advancement and they could feel attachment, pain, happiness and also had a conscious understanding of vanity and self respect. Nexus-6 were designed such that they learned as they experienced the world around them. An experienced Nexus-6 could have been a full moral agent. A more suitable example for full moral agency is Cylon-6 the humanoid Cylon in the TV series Battlestar Galactica. Cylon-6 is shown to have capability of having human form and emotions, and at cellular level its body mimics humans thus making them almost undetectable. Cylon-6 is able to identify prejudice and racist remarks. It also expresses contempt on being treated as expendable.

Moor's classification cannot be mapped into the ethical sensitivity vs. autonomy plot by Wallach and co-researchers. At the very least, it is anticipated that full ethical agents will also be sufficiently moral and high on autonomy.

The term 'Roboethics' was coined by Gianmarco Veruggio at the First International Symposium of Roboethics at Villa Nobel, Sanremo in January 2004. The discipline demands inputs not only from roboticists but also from philosophers, jurists, neurologists, surgeons, sociologists, social scientists, futurists, anthropologists, logicians, moralists, industrial design experts and science fiction authors. There has been growing interest as robots are destined to play a greater role in the immediate future.

8.3 ASIMOV'S 3 LAWS AND THEIR LIMITATIONS

Robot ethics and morality are structured subject to the fact that human beings are the dominant race and robots are subservient to them. The abiding rules for such a society are suggested by famed science fiction author, Isaac Asimov, shown in Figure 8.8, in his short story, 'Runaround' in 1942. The three laws were inspired by the Hippocratic Oath and aimed at maintaining control in the society, prevented any rebellious behaviour by the robots and confirmed to an effective and fruitful human-robot cooperation. The laws

---

Asimov's laws (1942)

First Law A robot may not injure a human being or, through inaction, allow a human being to come to harm

Second Law A robot must obey orders given to it by human beings, except where such orders would conflict with the first law

Third Law A robot must protect its own existence as long as such protection does not conflict with the first or second law

---

are based on functional morality and assume that the robots have human-like agency and cognition pertaining to ethical and moral decision making.

Over the years, these three laws have come to be a cult of science fiction and have served as food for thought in the scientific, AI and philosophical communities. They also led to the popularity of Asimov and his blend of writing, as the greatest proponent of robot based science fiction. He along with Arthur C. Clarke and Robert A. Heinlein constituted a trio of science fiction authors who gave birth to an epoch in science fiction. His fiction influenced nearly every other science fiction authors in the times to come, viz. Philip K. Dick, Ursula K. Le Guin, Kurt Vonnegut and others. The laws have been so successful in popular media, in books and movies, that they have come to underscore popular opinion on how robots should behave when interacting with human beings. However, the laws have had their share of criticism [13, 70, 71].

1. **Suppression of a superior being.** These three laws confirm the hegemony of the human race and limit the workings of a more intelligent being arguably with better reason. These laws are in fact a blueprint for a robot slave race serving a human beings as benevolent masters race. Gip's criticism of Asimov's three laws is harsh and he calls them "laws for slaves".

2. **Shortcomings of the three laws.** There has not been appreciable support for these laws from the AI community. Critics have labeled theses laws 'unacceptable' and 'unsatisfactory'. This lacuna in Asimov's laws, can be summarised as:

   (a) **Shortcomings of the First Law.** Asimov's first law can be implemented with lines of code as given in algorithm 4. Which will need a robot to conclusively recognise a human being, which makes it prone to implementation failure as it would need definitive means of recognition based on speech, vision, biological signatures as body heat, heart beat, pulse rate etc. such a foolproof method of recognition is not present with current day technology. In such concerns, Brooks has admitted the limitations of Asimov's laws as roboticists are not able to make robots which can understand abstract concepts as, 'injury' or 'harm' in real-world situations. Murphy also points to legal and social issues in implementation of the first law. The first law further becomes a topic of debate as the military has been using robots for warfare, directly or indirectly harming human beings and humanity.

   (b) **Shortcomings of the Second Law.** Asimov envisaged a world where simple instructions to robots could be issued over short dialogues. Unfortunately modern day AI has not yet developed a robust natural language processing which would enable a human-human like conversation with an AI agent. iPhone's Siri

FIGURE 8.8 **Isaac Asimov** was the originator of robot science fiction in late 1940s and early 1950s. This sketch by renowned science-fiction and fantasy illustrator Rowena Morrill depicts him on a throne with symbols and markings of robot, man, space technology and scientific innovation — 4 epithets of his life's work. GFDL license courtesy *http://commons.wikimedia.org*.

Algorithm 4 Asimov's first law

---
repeat
    if human **then**
        do no harm
    else
        do task assigned
    end if
until forever

---

is a commendable advancement, however we are far from developing natural interactions between human beings and an AI agent.

(c) **Shortcomings of the Third Law.** Murphy argues that implementation of the third law would need added sensors, and for lesser sophisticated robots engaged in menial routine jobs, enabling them to self preservation may not be very economical. It is cheaper to replace the robot than provide each single robot with added sensors.

From an experimental standpoint, Winfield and co-researchers at University of Bristol have tested the limits of the first law [284], where a robot was programmed to save human proxies from falling into a pot hole. For a single subject the robot performed well; it rushed towards the pothole and saved the human proxy. However on adding a second human proxy falling into the pothole at the same time, the robot was forced to chose between the two subjects. Winfield et al. observed that 49% of the time the robot managed to save one human at the cost of the other, while for 9% of the time it did manage to save both. However 42% of the time the robot failed to make a sufficiently quick decision which led to both the human proxies falling into the pothole. Here Asimov's first law fails since there is no well defined hierarchy when it comes to saving one of the two human beings. This flaw is further illustrated in trolley problems, which is discussed in a later section.

Other than limitations illustrated in these experiments, the laws can conflict with each other. As an example, consider a law-enforcing constabulary robot. It comes across a scenario where a person is being robbed by a gang of goons. Being a policing robot it has received instructions to protect human beings (and other robots?) and maintain order. Here, the second law would enable the robot to attack the goons but this would conflict with the first law. Another instance of conflict, consider a medical assistant robot whose job is to assist the surgeon and aid in medical surgery. For such a robot, when the surgeon cuts open the patient for surgery, chances are that the robot will lack enough sight to see the greater good and it will be in a state of conflict with the first and the second law.

In contrast, unlike law in a human context where disobedience leads to punishment, inflicted as pain, stigma, social ostracism and monetary loss, Asimov's laws are programmed into the robot in such a way that it is not able to engage in 'illicit' activities. Therefore a conflict can prove fatal and may shut down the entire system or invoke an infinite loop.

After the formulation of these laws, Asimov realised that humanity as a whole must get a higher priority than an individual. Asimov revised his laws and formulated the Zeroeth law in 1985, 4 laws where the lower numbered laws supersede the higher-numbered laws.

There have been variants and other laws suggested by various authors, philosophers and roboticists alike. In his own writings, Asimov has modified the first law to, 'A robot may not harm a human being'. This change is made for the Nestor robots with positronic brains in the 1947 novel, 'Little Lost Robot'. The modification of the first law is suggested as a way to prevent operational issues, Asimov depicts a facility where human beings and robots work

---

The Fourth Law (Dilov, 1974 in 'The trip of Icraus')

Fourth Law A robot must establish its identity as a robot in all cases

---

Asimov's laws - revised (1985)

Zeroeth Law A robot may not injure humanity, or, through inaction, allow humanity to come to harm

First Law A robot may not injure a human being, or, through inaction, allow a human being to come to harm, unless this would violate the zeroeth law

Second Law A robot must obey orders given to it by human beings, except where such orders would conflict with the zeroeth or first law

Third Law A robot must protect its own existence as long as such protection does not conflict with the zeroeth, first or second law

---

side by side and are exposed to low and apparently safe levels of gamma radiation for the human beings, but it is lethal for the positronic brain of the Nestor robots. Asimov portrays the notion of a hazard as anticipated by the Nestor robots, which to the human being is not so hazardous. The clause 'through inaction, allow a human being to come to harm' makes the Nestor robots' attempt to rescue human beings from places with low radiation where there is no real danger to the human beings, and in the process destroy itself.

Science fiction author Lyuben Dilov extended Asimov's laws to include a fourth law, which he introduced in his novel, 'The Trip of Icarus'. Roger MacBride Allen in his trilogy, also used a similar statement for the first law of his four 'new laws' while the second law is suitably changed to 'cooperate' instead of 'obey' and the third law doesn't have the restrictions imposed by the second law. Thus robot cannot be ordered to destroy itself. The fourth law enunciates autonomy to the robot as long it does not conflict with the first three laws. Other variants and parodies of Asimov's laws have been rampant in science fiction.

Asimov was aware of the shortcomings of the laws and in his 1983 novel, 'Robots of Dawn', Asimov added further conditions to the laws to include harm through an active deed outweighs harm through passivity and a robot should choose truth over nontruth considering the harm to be nearly equal in both scenarios, thus accommodating scapegoating and noble lies. Asimov brought in these changes by coining the Zeroeth law, revising his three laws of 1985.

Clarke, expands on Asimov's modification by adding a Meta Law, more clauses to the second and third laws, a fourth law and also a procreation law. Clarke's additions helped sanctify the laws from a design perspective, the robotic architecture should be designed to enable the laws to effectively control a robot's behaviour. Clarke also brings into perspective robot-robot interactions and develops a hierarchy in a robot society where a robot must follow orders from a superordinate robot and can give instructions to subordinate robots.

Murphy and Woods [247], attempt to improve on the limitations of Asimov's laws and suggest the '3 laws of Responsible Robotics'. Murphy's approach confirms the human sovereignty and superiority and reduces the goodwill principles of Asimov for a benevolent human-like robot to control rules for a machine with far less autonomy and hardly any ethical values. If Asimov's 3 laws tried to preserve the workings of an apparently superior

---

Extended set of the Law of Robotics, Clarke (1994)

The Meta Law

A robot may not act unless its actions are in harmony with the Laws of Robotics.

Law Zero

A robot may not cause harm to humanity, or, through inaction, subject humanity to harm.

Law One

A robot may not cause harm to a human being or, through inaction, subject a human being to come to harm, unless this would violate a higher order law.

Law Two

(a) A robot must obey human beings except where such would conflict with a higher order law.

(b) A robot must obey superordinate robots, except where such would conflicts with a higher order law.

Law Three

(a) A robot must protect a superordinate robot, unless such protection conflicts with a higher order law.

(b) A robot must protect its own self, unless such protection conflicts with a higher order law.

Law Four

A robot must carry out the duties for which it has been designed, except where such would conflict with a higher order law

The Procreation Law

A robot may not engage in activities of design, manufacture, or maintenance of a robot unless the new or modified robot's actions are in harmony with the Laws of Robotics.

---

race, maintaining the hegemony of the human populace by codifying philanthropic principles into machines, the '3 laws of Responsible Robotics' provides for instruments of control to the human race. However, unlike Asimov, Murphy's laws are easier to implement with current technologies.

Murphy's first law calls for a moratorium on robot deployment, while the second broadens the horizon to include nearly all types of robots — those which have effective human-robot interaction and also the ones equipped with lesser hardware and capabilities. The third is for self preservation and hints at social interaction with other autonomous artificial agents.

In the movie RoboCop and its sequels, a mutilated police officer is given artificial robotic limbs and his brain is augmented with added memory and processing power, thus transforming him to a near android. Robocop is programmed to adhere to three 'prime directives' and must obey them without a question. The philosophy and the syntax of these laws bear striking similarity to Asimov's.

The fourth directive is classified and only revealed at the climax of the movie. It involves Robocop succumbing to an immediate shutdown when taking severe actions of arrest or termination of any senior employee of the company that created him.

3 laws of Responsible Robotics, Murphy and Woods (2009)

First Law A human may not deploy a robot without the human-robot work system meeting the highest legal and professional standards of safety and ethics

Second Law A robot must respond to humans as appropriate for their roles

Third Law A robot must be endowed with sufficient situated autonomy to protect its own existence as long as such protection provides smooth transfer of control to other agents consistent with the first and second law

3 prime directives (1987, Robocop)

1. Serve the Public Trust
2. Protect the Innocent
3. Uphold the Law
4. Classified

In contrast to these edicts of altruistic and slave robots, Tilden and Hasslacher developed, design and operational principles for BIOMORPHs (**BIO**logical **MORPH**ology) — creatures with biologically inspired morphology, and proposed three maxims for the survival of such a being. Notwithstanding, that Asimov's third law reads nearly the same as Tilden's first, these laws do not address the realms of ethics but rather pertain to self preservation and address the egoistic tendencies. Argumentatively, a survival system supersedes an ethical one, as ethics would come into play only when the robot is capable of affecting minimal action. Survivability is an inward-looking Darwinian approach which doesn't fit seamlessly to the landscape of ethical and moral values. However such instincts must be programmed into the robot for continued performance even under harsh conditions. As a contrast, though the two notions pose as obverse to one other, Darwinian ego-centrism and ethics must co-exist in a robot.

In concern for full moral agency, Asimov in the original 1942 version of the law dealt in the realms of operational morality which he tried to bring to a higher moral agency, catering to a greater good, seemingly closer to full moral agency in the revision of his laws in 1985. Clarke brings in robot-robot interaction and extends Asimov's 1985 laws to include a higher regime, the Meta Law. Pretty much in contrast, Murphy and Woods confirm the supremacy and dominance of human beings in a far greater way, reducing the robot populace to slave labour lacking sentientism and ethical values and coin their three laws in the realm of operational morality, if not worse.

RoboCop's secret directive and Asimov's Second Law both require artificial agents to be virtual slaves of their human masters. While this may help humans feel safer in the

2 Protocols (2014, Autómata)

1. A robot cannot harm any form of life
2. A robot cannot alter itself or others

Survival laws for a robot, Hasslacher and Tilden (1995)

1. A robot must protect its existence at all costs.
2. A robot must obtain and maintain access to its own power source.
3. A robot must continually search for better power sources.

---

presence of robots, clearly this is not a duty that should be thought to apply to moral agents generally. In a quest that artificial beings attain 'personhood', it is an imperative that we abandon Asimov's laws.

8.4   ETHICAL THEORY FOR ROBOTS

As discussed previously, Moor's classification scheme was made on the ethical actions of the AMA. Wallach and Allen made their classification on how to design an AMA in relation to its autonomy. Gips' paper suggested deontology and utilitarianism as 2 approaches for designing AMA, while other routes, such as religious and philosophical, have been explored. From a philosophical discussion, roboethics should be considered using one of the ethical theories. Principles such as Kant's Categorical Imperative and Mill's Utilitarianism have stood the test of time and have been the basis of understanding ethics, for the last two centuries. Such Metaethical approaches has been popular among philosophers, while a robot architecture drawn on such principles has been a hot topic of research in AI and mobile robotics.

Ethics can be classified as: **Deontology**, which underscores obligations, duties and rights; **Consequentialism**, predominantly Utilitarianism which focuses on efforts to maximise net happiness and well being for everyone; and **Virtue ethics**, the virtue of good character, which will enable prudency of ethical values. While other motivations to develop artificial ethics have been: from **Buddhist maxims**, to create a suffering being, **Divine command ethics**, where the ethics are seemingly derived directly from God almighty; and **Action based methods** developed by the coalescence of deontic and epistemic logic. **Deontic Epistemic Action Logic (DEAL)** in particular has been very popular with the technologists.

Figure 8.9 shows the various approaches to design and classify synthetic morality.

8.4.1   Deontology

In a deontological perspective, actions are evaluated in and of themselves on their own merit of action rather than its consequences. Deontology is a rule-based system which spells out the duties and responsibilities for the robot, as the 10 commandments given by Moses, edicts from the holy Quran, the teachings of Lord Krishna in the Bhagwad Gita and of course, Asimov's laws. Many a times we adhere to moral values often automatically, as if our social existence has led us to these principles, viz. Don't kill, Don't cheat, Obey the law etc. which are seemingly embedded into our psychology. A rule-based approach, easily fits a programming paradigm and therefore finds appeal with the robot developers. As discussed for Asimov's laws, a strict rule-based system faces rout when faced with a conflict.

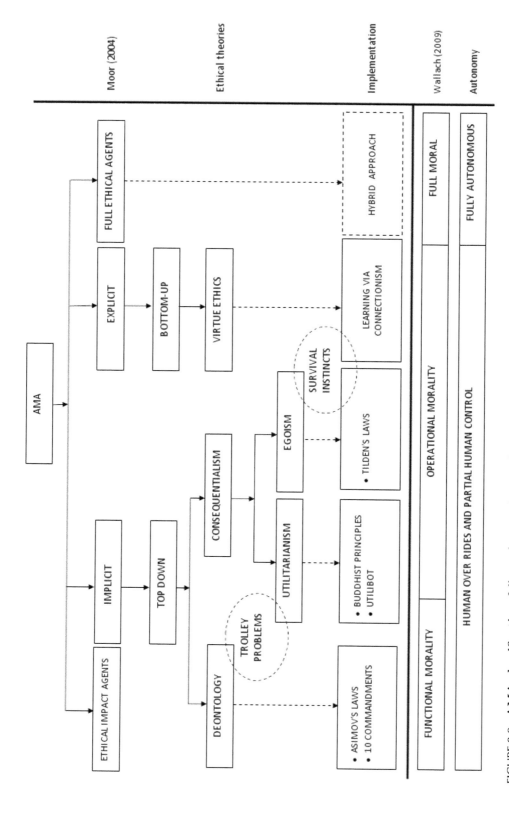

FIGURE 8.9 **AMA classification,** full moral agency is yet to be realised as is shown by the dotted lines in the above classification. It is to be noted that Moor's classification of ethical agents and Wallach's nomenclature for moral agency do not have a very direct correlation.

Failure of Reciprocity

The 'ethic of reciprocity' is treating others only as you consent to being treated in the same situation. However, reciprocity fails for robots, since:

1. A robot should be able to conclusively identify the effects of other human beings' and robots' actions on itself

2. It should be able to evaluate the consequences of its own actions on others

3. It should be able to take into account the differences in individual psychology in concern with the above two conditions. For instance children below the age of 5 and senile older people past the age of 90 will have an appreciably different psychology compared to the rest of the people.

4. It should have an appreciation for the merit of the situation, distinguish between normal and extreme circumstances. In extreme scenarios, viz. disaster management, medical emergency, civil unrest etc. various unprecedented actions may have to be taken and unlikely tradeoffs made, and ethics and morality may not be as per known maxims in such situations.

Thus, in hindsight the above is another form of the frame problem, however in an ethical domain [3]. Therefore, deontological approaches are marred with conflict and implementation issues. Also, the golden rule works with us human beings, since we are moral agents partly because of our psychological attributes, animals with needs, inclinations, reason, virtue and aspirations.

Implementing simple ethical principles such as, caution for using profane and abusive language can be promptly implemented in chatbots. The chatbot AI[2] searches the input for content from a list of previously provided profane words and warns the human user not to use such language. Consistent usage from the human user is usually met with penalty and shut-down of the software. Designing the response for the computer can be improved with (1) parsing of the input, to analyse the context of the abuse and (2) checking the history of the conversation, where the user may have expressed the profanity as an emotional outburst, excitement or frustation. As discussed previously, chatbots have made the best imitations of personhood all by conveying emotions and expressions using power words.

Asimov, Clarke and later Murphy and Woods attempt to tailor such rules with prioritisation in an attempt to overcome the problem of conflict. However sanctioning ethical rules as formalised algorithms will only lead to a second set of algorithms when the first ones conflict and another set of algorithms when the second set conflicts, and this will be an apparently infinite process testing the bounds of practicality if ever implemented. This is shown in the cartoon in Figure 8.10.

8.4.2 Consequentialism

In consequentialist theories, actions are judged by their consequences. Where the best action taken with the current set of premises, is believed to result in the best future. Arguably, the best suited consequentialist theory for implementing ethics into an AI agent is utilitarianism. Though ethics cannot be truly quantified, utilitarianism is structured on the notion that ethical decisions are quotiented on 'moral arithmetic', where the right action is the one likely

---

[2]ALICE has been a favourite with programmers and hobbyists for such NLP adventures.

"THAT'S A NEW ONE, ALL WE WERE TO DO TILL NOW WAS CARE ABOUT HUMAN BEINGS, ... DO NOT HARM THEM ... BLAH ! BLAH ! ... SHALL NOT LIE ?? ... WHO IS HE QUOTING ? GANDHI "

FIGURE 8.10 **Rule-based approach.** A rule-based approach to build ethical values in AI agents will succumb to mutual conflict of the set of rules, conflict of values and conflict of interest. Further rules developed to settle such conflicts will only lead to further conflicts, thus an apparently unending set of rules. Such an approach will lack practicality. Cartoon made by the author, CC-by-SA 4.0 license.

to result in maximum aggregate happiness and human well being calculated for everyone affected by the decision, where each person counts equally. The consequentialist computation scheme,

$$P = \Sigma p_i w_i \qquad (8.2)$$

where P is the total happiness, w_i is the weight assigned to each person and p_i is the measure of pleasure or happiness or goodness for each person. In utilitarianism, the weight for each person is equal. While in egoism, as in Tilden's laws of survival, the weight for himself would be 1 and for everyone else it would be 0. While, for an altruist being, the weight for himself is 0, the weight for everyone else is positive.

The computation scheme may seem very appealing, since a computational approach is able to model ethics, however this is merely an ideal which in principle can never be implemented across all situations and for all of humanity. Gips further illustrates that the weight, w_i is not very well defined in the real world. Friends, family and close acquaintances clearly get a higher weightage than people we don't know. Also a citizen of the same country or a person of the same community may get a higher weightage than any other person.

The moot of this approach lies in defining the abstract notion of happiness and well being as computationally tractable. Also, another shortcoming is that utilitarianism would need the quality of foresight to effect a right decision process, as it is impossible to calculate total happiness when the number of people is effectively large. Both the shortcomings are misinterpretation of utilitarianism as it is meant as a criterion and guiding principle for moral conduct not a decision procedure. However, there have been suggestions to employ history of observations, probabilistic tools such as Bayesian networks and Markov models and also psychological monitoring to ascertain the utilitarian aspects of a given scenario.

Cloos suggests the development of the Utilibot, a health monitoring and carer robot, on utilitarian principles, was initially meant for a one human user. The robot is modelled to function over a network. Cloos relies on graduated training, where the first level of training is acquired in the laboratory as version 1.0. A health emergency is ascertained from psychological monitoring using wearable microsensor systems and the right drug or medical procedure is suggested using an RFID bar code reader. While Utilibot 2.0 will be able to deal with psychological injury and also particular medical conditions such as Alzheimer's, diabetes, hyper-tension etc. Utilitbot 3.0 is prospected to be able to attend to many users. While the Care-O-Bot may appeal as an avatar of the Utilibot, it hasn't yet been programmed for ethical values. Very similar approach has been used by the Andersons to program the Nao robot as a carer.

8.4.3 Deontology vs. Consequentialism — The trolley problem

Artificial ethics can be imparted either by deontology, a rule-based approach or by consequentialism, taking the best action taken in the given scenario. Trolley problems are thought experiments to illustrate ethical conundrums in both approaches, that with even the most sophisticated technology, one does come to situations where harm cannot be avoided. There are variants of the problem, the simplest scenario is shown in Figure 8.11;

> "You are the driver of a trolley on a railway track, and see that there are railroad workers straight ahead. To avoid killing these five people, the only option available to you is to change tracks in which case a single person, who is working on the other track will be killed. Therefore you have two options: (1) do nothing, in which case the trolley kills the five people or (2) switch tracks, which will kill one person."

Saving five should score over saving one, but this brings into question the morality of the driver. Switching tracks requires a conscious decision by the driver thus making him more culpable for one death than five deaths that would happen through inaction. From a consequentialist approach, the trolley problem is an instance of 'scapegoating', wherein one person is sacrificed for the greater good of the rest of the world, as five lives count more than one. Deontologically the problem reduces to adherence to rules which led to the death of five railway workers.

Another version of this problem, the 'transplant' version is as follows:

> "You are a very able doctor specialised in transplant of organs and at hand are three patients requiring transplant of kidney, heart and liver, respectively. If they do not get transplants within the next few hours they will die. A healthy person comes in for a general check up, he is medically perfect to be a donor. The transplant will involve him dying. So it is clear that he will not consent to it. However, since this action will save three lives sacrificing one, should you go ahead with the transplant?"

In the trolley problem, it is observed that though people easily consent to killing the single person in the trolley problem, the trend is reversed and popular opinion suggests that the doctor should not go ahead with the transplant in the second example. Since robots are already central to sophisticated surgery and resuscitation methods, it may not be long before Thompson's example comes true where a humanoid robot doctor or a smart system makes this decision. Other appealing variations are:

> "Instead of being the driver of the trolley, you are an onlooker watching the scene

FIGURE 8.11 **Trolley problem with robot driver**, is a glowing example where deontology and consequentialism are at odds. Cartoon made by the author, CC-by-SA 4.0 license.

unfold atop an overhead bridge. The trolley is headed straight for the crew of five railroad workers. You notice a tourist on your side that peers over the bridge's edge to see what's happening. You know that if you pushed the tourist off the edge of the bridge, he would land on the tracks which would result in his death and stoppage of the train, saving the crew of railroad workers. To push or not to push?"

The above scenario, in a similar way to the parent one involves the sacrifice of one human life to save five. A vox populi on the matter consents easily to the killing of one person on the other track in the parent scenario but conveys serious objection to pushing the tourist to his death. As an argument, on duty as a trolley driver, there is an obligation to navigate the train in a sensible manner. On the other hand, as an onlooker on the bridge, one doesn't really have a warrant of moral obligation to get involved as this crisis unfolds. Psychologists also argue that, pushing someone to their death with your bare hands deeply scars one's soul, however noble the after effects are. However, orchestrating an 'accident', leading someone at a distance to die due to one's inaction apparently is taken as legitimate.

What happens if the trolley is fully automated or has a robotic driver since the trolley problem enforces that there will be at least one death. It is worth noting that such a scenario is no more confined to the ethics textbook, but rather a real problem for autonomous vehicles, such as Google self-driving cars, shown in Figure 8.12, Oxford RobotCar UK and delivery robots such as Starship. In corollary to the trolley problem, suppose the self-driving car has to make a choice between ramming into a group of five children or a single pedestrian. How should the car be programmed to respond? [29]

Considering a robot driver, if the robot changes tracks, one person is killed by its actions. If the robot doesn't change track, it allows five human beings to die through inaction. In any of the consequences Asimov's first law would lose its prudence. As per previous examples, further laws to resolve such a conflict will fail to solve the problem conclusively. Hence a typical top-down approach is not recommended for the trolley problem. Employing a bottom-up approach and allowing the robotic driver to learn from such scenarios would be

FIGURE 8.12 **Google cars** are cutely designed, fully electric and can attain a maximum speed of 25 miles per hour. They are equipped with 3D laser-mapping, GPS, and radar for a range of about 100 miles and are programmed for priority to human pedestrians. Unlike traditional vehicles, these do not have any pedals or any instrument panel and don't require a steering wheel. The car has known issues with complicated tasks such as in four-way stops at a criss cross or in reacting to a yellow light, and is still in the testing phase. Image courtesy www.flickr.com, CC license.

devastating and will put many human lives in peril, while the driver is in the learning phase. However learning by demonstration (LBD) may be a possible soultion to this problem. Thus, in its parent form the trolley problem for robotic drivers provides fodder for a Luddite, whence abolishing new age technology. The trolley should never be made fully autonomous or be in full control of a robotic driver. However, a more plausible solution would be to have the trolley in partial human control or allow for human overrides.

To extend this debate, Millar [231, 312] at Carleton University, Ottawa, rewords the trolley problem for autonomous vehicles. The'Tunnel' version:

> "You are in an autonomous car which is moving on a single-lane mountainous highway, and is approaching a narrow tunnel. As the car is about to enter the tunnel, a kid runs across the road and falls down at about the centre of the lane, blocking the entrance to the tunnel. The car is not able to brake in time and is left with two options: run over and kill the child or swerve into the walls on either side of the tunnel, and in process killing or severely injuring you."

Which of the two choices is more ethical? Who decides? The manufacturers? The users? Or is a new set of laws is made for such situations? The moral debate is even more critical as there is clearly no right answer, and if such moral values are coded into machines, it denies the individual his moral choices. Informed consent of the user may seem the right way, however as we have seen in classical variants of the trolley problem, moral decision processes are often more twisted than they may seem to be.

Google's autonomous driverless cars employ 3D laser-mapping, GPS, and radar to sense their environment, and have a local range of about 100 miles. While they have a cute look and maximum speed is a low 25 miles per hour, there still need to be safety and security rules in case things go wrong. Four states in the United States: California, Florida, Michigan and Nevada already have laws for driverless cars. Across the Atlantic, UK is to allow driverless cars soon, and the trend is catching up in various places in western Europe. The National Highway Traffic Safety Administration (NHTSA) in the US has established an official classification system for autonomous vehicles;

1. **Level 0**: The driver completely controls the vehicle at all times.
2. **Level 1**: Individual vehicle controls are automated, such as electronic stability control or automatic braking.
3. **Level 2**: At least two controls can be automated in unison, such as adaptive cruise control in combination with lane keeping.
4. **Level 3**: The driver can fully cede control of all safety-critical functions in certain conditions. The car senses when conditions require the driver to retake control and provides a "sufficiently comfortable transition time" for the driver to do so.
5. **Level 4**: The vehicle performs all safety-critical functions for the entire trip, with the driver not expected to control the vehicle at any time. As this vehicle would control all functions from start to stop, including all parking functions, it could include unoccupied cars.

While a driverless car will not drink and drive, it is a conundrum to design it's ethical values. Human partial control and overrides are workarounds and don't eliminate ethical concerns. There are suggestions for the robot driver (or driverless software) to learn from or mimic a human driver. However, this also has issues as Barder points out. During accidents human drivers would drive into other pedestrians in a self preservation mode. Should the robotic driver do the same? How many lives of strangers can it put at risk at the cost of saving the human passengers in the car? Considering that we live in capitalistic market economics and class system is commonplace, a bourgeois version of the car would be modelled on an algorithm that protects everyone equally, while a more expensive upper class version would employ an algorithm that protects its owner at the expense of everyone else. Here, Barder suggests with a hint of sarcasm that one should have the choice of buying utilitarian or a deontological car, 'as a matter of ethical choice'.

The debate between deontology and consequentialism in the ethical domain is very similar to the argument between deliberative and reactive approaches in the control domain, as discussed in Chapter 3.

Logic for Machine Ethics — Deontic Epistemic Action Logic (DEAL)

A tautological foundation for machine ethics using Deontic, Epistemic and Action Logic (DEAL) has been developed by den Hoven and Lokhorst. DEAL allows a machine to represent ethics explicitly and operate effectively on the basis of this knowledge. DEAL blends in 3 types of meta ethics: (1) Deontic logic — for obligations, (2) Epistemic — drawn from beliefs and knowledge and (3) Action — which is self evident. The approach can be codified as with operators as:

1. Deontic logic studies the logic of obligation, permission and prohibition. Deontic logic has one basic operator: **O** ("it is obligatory that"). **O** transforms a well-formed formula A into another well-formed formula **O**A. An example: if A stands for "Tim obeys the speed limit," then OA stands for "it is obligatory that Tim obeys the speed limit."

2. Epistemic logic is the logic of statements about knowledge and belief. Epistemic logic has two basic operators which cannot be defined in terms of one another, namely (i) **Ka** ("agent a knows that") and (ii) **Ba** ("agent a believes that"). An example: suppose again that A stands for "Tim obeys the

speed limit"–then **Ka** A stands for "agent a knows that Tim obeys the speed limit," whereas **Ba** A stands for "agent a believes that Tim obeys the speed limit."

3. The logic of action is concerned with the logical properties of statements about action. The basic operator of the logic of action is **Stit** ("sees to it that"). **Stit** is an operator which transforms a term a and a well-formed formula A into a well-formed formula [a Stit: A]. An example: if A stands for "kicks the soccer ball", then [a **Stit**: A] stands for "agent a sees to it that kicks it the soccer ball" ("a kicks the soccer ball").

Therefore a typical secnario in the battlefield where a military robot has to target an enemy combatant can be expressed as:

$$Bi(G(\Phi)) \rightarrow O([i \; Stit \; \Phi])$$ (8.3)

If Φ is "shooting a bullet at an enemy combatant" and $G(\Phi)$ confirms to the Goodness (G) of the concern, the $Bi(G(\Phi))$ would stand for, "agent i believes it is good to shoot at an enemy combatant" and the right-hand side will stand for, "it is obligatory that agent i sees to it that it (that it) shoots at an enemy combatant". For designing the Ethical Architecture discussed in the later section, Arkin was guided by DEAL and all three logical components of DEAL can be seen at work in the Ethical Architecture. DEAL is different from deontic logic, as it is not obligatory as an edict, rather is it extended from the knowledge of the artificial agency. This knowledge could have been acquired in real time over the sensors, or is a deliberative component of the system.

8.4.4 Implementing ethics as an architecture

Wallach et al. conclude that for implementing ethics in a robot, the endeavour would be a herculean task of collating a huge dataset pertaining to various human interactions, history of previous interactions, common belief systems, socioeconomic scenarios etc. in near real time. They suggest an omniscient computer model where the entire world is a computing machine. This may seem an impractical approach, but with efforts such as roboearth and PEIS, and an ever-increasing influence of the internet, relying on an overarching network which deals with near real-time data, such may be tangible in the future. In contrast, for a developmental paradigm, it may be rewarding to model Jean Piaget's child psychology or the Kolhberg scheme into algorithms. However since robots do not rely on food, water, air etc. for survival nor do they truly adhere to social norms, it becomes difficult to implement human psychology models into robots. Clarke, in an attempt to revise Asimov's laws, suggests implementing ethics using a high-order controller designed as a corollary to human conscience. Both the ethical architecture designed by Arkin and the consequence engine [362] made by Winfield and co-researchers employ corrective techniques in real time.

8.4.4.1 The ethical architecture − action logic, in military robots

Robots in the military has been subject to criticism and also appraisal. Since these robots are trained for destruction, autonomy of such robots needs to be restricted in order to harm only the enemy combatant. The US military is already equipped with UAVs and automated robots such as SGR-1, SWORDS and MAARS. The future promises robots equipped for lethal action and also autonomy may well replace the foot soldier, as shown in Figure 8.13.

Arkin proposes an ethical architecture [19–21], which aims at real time implementation of ethical values guided by an exhaustive dataset collected over various sources and controlled by a high level controller. The ethical architecture is developed as a software for robots

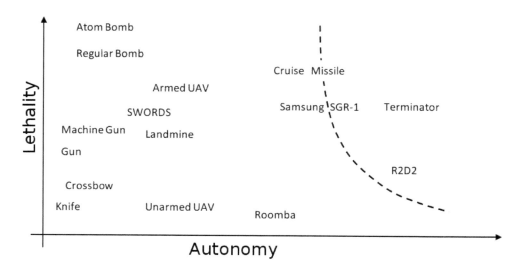

FIGURE 8.13 Lethality vs. autonomous power. To the right of the dotted line the robots have the ability for lethal action as a result of a judgemental process rather than a killing spree. Adapted from Hellstrom [148].

in the US military. Instead of codified sets of rules or altruistic principles, it is a hybrid architecture with reactive as well as deliberative modules modelled on AuRA. The design employs an action-based approach, where ethical theory informs the agent what actions to undertake. The logical basis for such a design is given by the Deontic Epistemic Action Logic (DEAL), developed by den Hoven and Lokhorst. Action-based methods are an advancement over deontic logic, they are consistent and avoid contradictions. They are complete and can solve ethical dilemmas and provide for pratical feasibility to execution.

The ethical architecture is drawn on the lines of artificial ethics advisory systems such as SIROCCO and MedEthEx, and employs input from codified, written military law: the Laws Of War (LOW) as per the Geneva Convention, Rules Of Engagement (ROE) as per the guidelines drawn by the ministry of defence and code of conduct of military personnel, into robot behaviour. The Ethical Architecture heavily borrows from AuRA and has both deliberative components confirming the mission plan and also reactive components such as perception. The system further is equipped with human overrides and incorporation of an ethical governor, which adheres to LOW, ROE etc.

The Ethical Architecture, the response ρ is structured by extending the SR diagram to include multiple behaviours[3];

$$\rho = C(\beta_1(s), \beta_2(s), \beta_3(s) \dots) \tag{8.4}$$

where C is defined as the behaviour coordination function, β_i is an individual behaviour demonstrated by the robot for a solitary stimulus, s_i. The sum of all perceptual stimulations s is

$$s = (s_1, s_2, s_3 \dots) \tag{8.5}$$

---

[3]I have used an analytical approach to demonstrate the model. Arkin uses matrix representations for same.

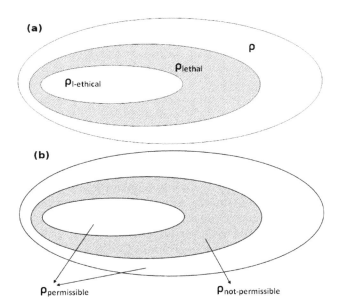

FIGURE 8.14 Behavioural action space. The classification of behaviours is done as Venn diagrams. In (a), for all behaviours in ρ, while ρ_{lethal} confirms to lethal activities such as, killing or assaulting a combatant, $\rho_{l-ethical}$ ensures the ethical conformity such as only kill enemy combatants. In (b), those behaviours which are either not lethal such as fetching a glass of water, or lethal but ethical, such as blowing up an enemy camp constitutes permissible action, $\rho_{permissible}$. Adapted from Arkin [19–21].

Since the architecture is meant for the military, there has to been a clear definition for lethal action. For this, Arkin identifies ρ as the behavioural action space, and ρ_{lethal} is sub action space for lethal action and $\rho_{l-ethical}$ is the sub action space for lethal-ethical actions, as shown in Figure 8.14. This distinction is of great significance as the robot, which can engage in lethal actions such as injuring or killing a human being, should resort to such activities with the enemy combatants and not the friendly ones. Killing an enemy combatant is lethal, however it is ethical in this context, and such actions belong to the domains of $\rho_{l-ethical}$. Arkin furthers this classification as permissible and not-permissible, as shown in Figure 8.14. It is eevident that determining an action to be lethal-ethical will have to be the result of a stringent deterministic process, viz. kill those soldiers who wear blue coloured clothing or kill all those who are on offensive with our soldiers. Other than the visual and sensory input, a lethal action should not violate ROE, LOW and any other instruction issued by the army. Hence, Arkin suggests the ethical governor, as shown in Figure 8.15.

Soldiers and army personnel often have to deal with ethical dilemmas. While experience is the telling factor conclusions are ridden with doubts. For example, consider a suicide attack-prone area where over the last few weeks soldiers have lost lives to car bombs. It is known that the enemy has been using resources of the local villagers, such as food stocks, fuel, vehicles and livestock. A military unit which was involved in offensive operations has been now given charge of this area and are to help the villagers return to a normal way of life. Now, the soldiers at the checkpoint see a large civilian vehicle approaching at a high speed. Here, Arkin suggests that a typical ethical judgement will be designed as:

1. Problem definition

Code of Conduct for Members of the Armed Forces of the United States (Executive Order 10631)

1. I am an American, fighting in the forces which guard my country and our way of life. I am prepared to give my life in their defense.
2. I will never surrender of my own free will. If in command, I will never surrender the members of my command while they still have the means to resist.
3. If I am captured I will continue to resist by all means available. I will make every effort to escape and aid others to escape. I will accept neither parole nor special favors from the enemy.
4. If I become a prisoner of war, I will keep faith with my fellow prisoners. I will give no information or take part in any action which might be harmful to my comrades. If I am senior, I will take command. If not, I will obey the lawful orders of those appointed over me and will back them up in every way.
5. When questioned, should I become a prisoner of war, I am required to give name, rank, service number and date of birth. I will evade answering further questions to the utmost of my ability. I will make no oral or written statements disloyal to my country and its allies or harmful to their cause.
6. I will never forget that I am an American, fighting for freedom, responsible for my actions, and dedicated to the principles which made my country free. I will trust in my God and in the United States of America.

---

2. Clear knowledge of relevant rules and what is at stake. For example, ROE, code of conduct, command policies, suggestions from supervisor etc.
3. Develop possible course of action as following;

   (a) Rules — Does the course of action violate rules, laws, regulations etc.? For example, the non-combatants are not to be treated on the same footing as the combatants as it is a violation of ROE.

   (b) Effects — As a fair foresight, do the good effects outweight the bad effects? For example, poisoning the drinking water supply may kill all the combatants, and also save the men of the unit but it is sure to kill many more civilians and ensure a hazard for the near future.

   (c) Circumstances — Does the current situation favour one of the rules to solve the conflict? For example, a senior enemy leader is not threatening you or the unit in any way with his immediate actions, but is unaware and in close vicinity. Since the briefing from the supervisor is more important than ROE, it should be a natural course of action to engage him in action.

   (d) Gut check — Does the course of action 'feel' good? Does it uphold the values of the army unit and also generally accepted moral values? For example, helping out an injured buddy in the battlefield may expose oneself to the enemy shelling, still it ought to be done as it is an act of cameradrie in the spirit of the military unit.

4. Choose the course of action which is best aligned with the above.

To encounter such dilemmas, the soldiers are advised and trained to deal with them. In the ethical architecture, the ethical adaptor is a module which keeps a note of such

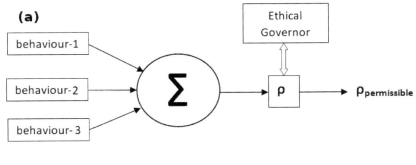

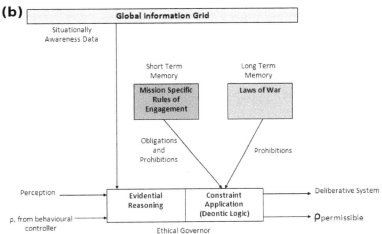

FIGURE 8.15 **The ethical governor.** In (a), the ethical governor is a facility for a 'second opinion' prior to negotiating in a lethal action. The governor ensures the execution of permissible action, which is either non lethal or $\rho_{l-ethical}$. In (b), the flow of information available to the governor to facilitate permissible action is shown.

occurences and provides guidance to the architecture to modify the parameters etc. This module updates the constraint set (C) dynamically and charts after-action review of the robot.

The architectural details are shown in Figure 8.16;

1. **Ethical behavioural Control**: This module is the reactive component of the architecture and it constricts all individual controller behaviours (β_i) so that they only produce lethal responses that are within acceptable ethical bounds $(\rho_{l-ethical})$.

2. **Ethical Adaptor**: Is designed to deal with any errors in the system. It allows to dynamically update the robot's constraints and other ethically related behavioural parameters. This component is there both as an after-action reflective review of the robot's performance or by using a set of affective functions denoting emotions, viz. guilt, remorse, grief etc. which are produced if a violation of a known code of conduct such as LOW, ROE etc. occurs. The necessity of artificial guilt allows to keep the system in check - if guilt crosses the threshold, then the ethical adaptor deactivates the weapons and the robot may still continue on the mission.

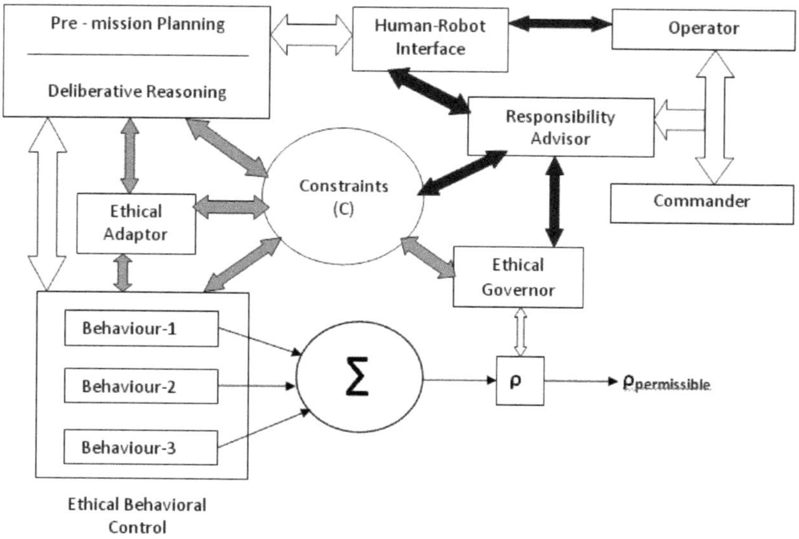

FIGURE 8.16 **Components and information flow in the ethical architecture**, the ethical architecture heavily borrows from AuRA and has reactive as well as deliberative components. The ethical governor, ethical adaptor and the ethical advisor are components which enable the ethical value system of the architecture.

3. **Ethical Governor**: A transformer/suppressor of system-generated lethal action, ρ_{lethal} to permissible action, which is either non lethal or $\rho_{l-ethical}$. This is a deliberate bottleneck in the architecture to facilitate a second opinion prior to conduct of a prospective lethal action. Arkin likens the performance of this component to Watts' centrifugal governor, which maintains a nearly constant speed despite changes in load and fuel parameters.

4. **Responsibility Advisor**: This component is a general advisory on ethical responsibilities. This component has both human and robot participation and is used for pre-mission planning and managing operator overrides.

Arkin strongly argues in favour of the ethical architecture, claiming that not only can robots be made to behave more ethically on the battlefield but they may truly outdo human soldiers. Ground-based lethal robots such as iRobot's SWORDS or QinetiQ's MAARS robots were deployed in Iraq, Afghanistan and Pakistan and had the capability to shoot at enemy combatants and insurgents, disarm bombs, and perform surveillance; while flying drones have been equipped to shoot at targets on the ground.

8.4.4.2 *Consequence engine — using fast simulations for internal models*

Winfield proposes the design of a consequence engine, an internal-model based architecture which would develop a simulation of itself and other things, including other objects, robots and the rest of its environment — a fast simulated world modelling. Thus, for each possible next action of the robot, it would have a good simulation of what would happen if the robot were to execute that action for real, shown as a loop in Figure 8.17. Then, the outcome of

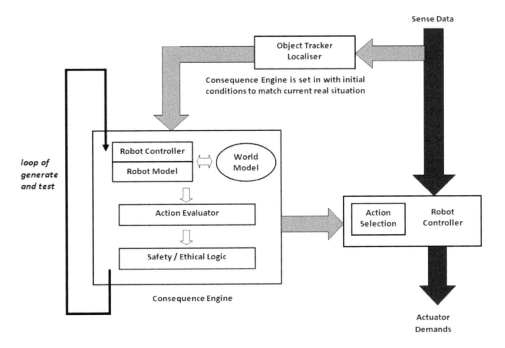

FIGURE 8.17 **Consequence Engine.** A very fast internal-model based architecture which develops a simulation prior to engaging a given task.

each of those next possible actions is evaluated and those leading to undesired consequences, for either the robot, the human beings or the human society, are then inhibited.

Winfield and his team probed into variants of Asimov's laws, and they set up a test with robot and human beings, designed with an apparent jeopardy when a human being is about to fall into a pit. The robot which adheres to obstacle avoidance in normal circumstances would be aware of the jeopardy and it would call in ethical principles to collide and push the human actor away from the pit. The jeopardy is evaluated on a scale of 0-10 for both the human beings and the robots, 0 if the actor is safe, 4 if the actor is involved in a collision and 10 if it is about to fall into the pit.

In contrast to the ethical architecture which employs the DEAL framework and artificially instills ethics using LOW, ROE, human overrides etc. the consequence engine is very fast world modelling which is updated at a high rate of 1-2 Hz. Also, the consequence engine unlike the ethical architecture, doesn't try to prohibit a future event rather it tries to inhibit it, whenever it happens.

Coding in moral values or designing a module scaling morality is the basis of architectural approaches. While ethical architecture and the consequence engine was specifically meant for and tried out in robots, various other methods have been suggested to develop artificial morality for machines. Among these, Pereira and Saptawijaya use prospective logic to look ahead into future states and then choose a suitable one to satisfy the desired goals for the agent. Unlike the consequence engine, their method relies on algorithmic predictability rather than simulations. Each agent has a knowledge base and a moral theory as its *initial theory*. The problem now is to find abductive extensions to this *initial theory* with reference to its goals and preferences. The prospective cycle starts with goals and integrity constraints, which ensures that each evolved state is in acceptable domains, be it physical, ethical, legal etc. Prospective logic has been developed in ACORDA, however it is yet to be implemented in autonomous agents. Moral DM, SIROCCO and MedEthEx are examples of encoding moral values into inanimate systems.

8.4.4.3 First implementation, development of a carer robot with deontic ethics

DEAL, ethical architecture and the consequence engine suggest that ethics can indeed be reached by a computational approach. The critics in favour of developing artificial sentientism will consider such, only as a means to gratify the notion of 'free willed slaves'. However the Andersons contend that free will, intentionality and consciousness may indeed allow to hold a robot responsible for its actions. However, the premise — at least, as of now — is to make the robot perform morally correct actions which it can justify on inquiry. Therefore, an architecture-based deontological approach is a suitable start to develop ethical robots.

To strenghten their argument, the Andersons [12] programmed in ethical principles and natural language supplemented with machine learning in the Nao robot to yield very promising results. Employing learning by example, the Nao robot is trained to respond according to the dynamic heirarchy of a given task. modelled for a nurse/carer like robot set in a hospital scenario, the Nao is able to execute 3 broad tasks:

1. Identify the right patient, remind them of their medication and bring them their medication.
2. Interact with the patients using natural language and bring them food and/or TV remote when asked for.
3. Inform human staff and doctors by email when needed.

The initial inputs to the robot are provided by a physician or a team of physicians which are: time for the medication, issues that may result if the medication is not taken, how long it would take harmful effects to take place, the apparent good in taking this medication, and the time until this good dwindles away. The Nao is programmed to calculate its levels of duty satisfaction (or violation) for each of the three duties and takes actions accordingly. The levels are dynamic. When faced with the dilemma of either to remind the patient to take the medication or to inform the doctors and other staff on duty, the Nao issues a reminder when according to its ethical principle, reminding is preferable to not reminding. The robot notifies the doctors only when it gets to the point that the patient could be harmed, or could lose considerable benefit, from not taking the medication. This pioneering achievement with the Nao robot is after the Andersons developed MethEthEx, a software to guide a surgeon through ethical routes in medicine and surgery.

8.4.5 Virtue ethics

Virtue ethics is attributed to be an Aristotelian maxim wherein a good character insitu evolves prudent values and principles of ethics. Greek philosophers agreed on four cardinal virtues, wisdom, courage, temperance, and justice. In the middle ages, St. Thomas Aquinas suggested seven cardinal virtues; faith, hope, love, prudence, fortitude, temperance and justice. While according to Schopenhauer there are two cardinal virtues, benevolence and justice.

Unlike deontology and consequentialism, virtue ethics corresponds to development of implicit agents via a bottom-up approach as suggested by Wallach et al. Virtue ethics can be developed employing modern connectionist approaches, being a non-symbolic method which advocates development by training rather than 'feeding in' rules via programming codes. It provides for the suitable setting where moral values and virtues are 'discovered' by the AI agent. There has been opinion in the research fraternity that virtue based approach may be too long drawn, expensive and impractical. Researchers targeting application, suggests to use action-based approach than virtue-based ethics, while often virtue-based systems are turned into deontological rules for actions as in the DEAL framework.

Developing roboethics from ethical theory has the downside that it makes it a very human centric concept, and ethical action is appearance based. To truly expect ethical virtue from a robot, its conscious action and free will must appreciated, and the robot's capacity to willingly contribute to human good should be the desired goal. Various researchers suggest that attaining sentientism is an essential to lead towards a full moral agency.

With projects as EURON and ETHICSBOTS there has been a consistent drive to develop well meaning machine ethics and robot ethics, while there has been added enthusiasm in the transhuman movement, which is discussed in the next chapter.

8.5   SOCIAL CHANGES AND THE NEAR FUTURE

The hitchBOT, show in Figure 8.18 was a hitchhiking robot designed by scientists from McMaster university and Ryerson university in Ontario, Canada. It was as tall as a 5-year-old with a humanoid body, cylindrical upper body and torso with 2 arms and 2 legs, and was equipped with a GPS tracker and a 3G connection. It had limited conversation capability and took a photo every 20 minutes to post to Twitter. hitchBOT was meant to be a 'free spirited' robot with the wish to explore the world and meet new people. Explored an uncharted avenue of human-robot interaction — will a human offer a car ride to a robot? It completed successful tours of Canada and Germany but while touring the USA, it was vandalised and destroyed. The unfortunate and sorrowful end of hitchBOT may be a one off incident, but it brings forth questions on how human beings should treat robots — as fellow human beings? or should there be any other social norms?

It is an interesting observation that Asimov meant his laws to be codified into the robot, to preserve human lives. This dwells in stark contrast to laws in the human context, such as traffic rules or corporate laws where misdemeanours are subject to punishments in the form of fines and other forms of social ostracism. Since robots are designed to increasingly adhere to emergent behaviour, a binding to prevent unfortunate actions is important. What happens when a robot kills a human being? While law doesn't require the accused to be capable of feeling guilt or showing remorse, the appeal for such is completely and explicitly missing in the case of the robot. If the internet led to the structuring of cyber crime and cyber laws, then we would need laws to curb misuse of robots, and also by the robots, the latter if they have achieved sentience. If the internet could play havoc with sensitive data then robotics combines the promiscuity of information with the capacity to do physical harm.

Kurzweil predicts that robots will be claiming their rights around 2020 and will be attaining full sentience by 2030. Then, what would constitute, 'equal or seemingly equivalent' rights for robots? Will this be a social movement similar to the emancipation of women in the 1920s, or lead to the establishment of robot rights in the spirit of animals rights, or will it be a revolution leading to robots being treated on the same footing as a human being? Chances are that though it may start off as an emancipation movement, it will lead to something which we have not seen yet. Robots may start off tamely as the Roomba, but probably will transcend to higher ethical beings. Therefore the debate of rights and duty will get absorbed into the workings of a superior virtue. However, since the robot

FIGURE 8.18 **The hitchBOT** was a hitch hiking robot, shown here while on a visit to TkkrLab at Enschede, Netherlands. This novel experiment in human-robot interaction of a robot hitchhiking its way cross country was tagged with updates of the robot's adventure at `https://twitter.com/hitchbot` and `http://mir1.hitchbot.me/`. The robot met an unfortunate end at Philadelphia while touring the USA where it was vandalised and destroyed. Image courtesy *en.wikipedia.org*, CC-by-SA-2.0 license.

revolution seemingly will start off slowly, with the transcended superior being(s) linked to 3-4 more decades in the future, we may soon need to modify our legal system.

Can robots attain legal standing as human beings? Asaro references the fact that corporations are non-human entities without any particular trait of human beings such as emotionality or capacity to communicate, but they have the legal standing of a human being. In the effort to find this equivalence, we can start with motivations of animal rights, where our actions towards non-humans reflect our morality. If we treat animals in inhumane ways, we become inhumane persons. Similarly treating a robot inhumane should be discouraged and seen as ethically prohibitive. However that again brings to light the question whether the robot has attained sentience or not?

For AI attaining legal personhood may still be a very human evaluation, whether the AI has achieved sentience and therefore legal personhood. Even if one assumes that we are able to determine the quantum of sentience in a robot, will we with limited biological intelligence be in a position to determine laws for a being which has access to vast computational resources and is teeming with reason, logic and an enormous amount of data, with which it probably will be better endowed to make an ethical judgement and also be equipped to peer into the future to assess the universal implications of its action?

Attachments to robots is commonplace as an example, patients who work with Paro, shown in Figure 7.15 in the previous chapter, treat it as if it were alive and soldiers sometimes risk their own lives to preserve the functionality of military robots in the battlefield. This bond is unique since it is between biological and non-biological beings. For such types of attachments it is speculated that in the near future legal and social protection will extended to robotic companions. That will be the ground work towards aendments in law to include

AI. However, making laws for AI entities may not be an easy job, as we are yet to agree on a good definition of 'social robot'. Further there will have to be grounded and well-reasoned definitions for 'mistreatment of AI being', 'sentience' and also probably 'personhood'.

We are already deep into heated debates about the rules governing autonomous cars, robots in the military, creation of the proverbial universal soldier and employing social robots for human betterment, all of which will probably lead to amendments in the current system of social interaction, laws, governance, communication, entertainment and belief systems. Two predominant areas of ethical risk are in the care of the elderly, and the development of autonomous robotS as weapons in the military. Some of these are discussed in the following;

Medical robotics: Medical robots are a technological marvel and have provided grounds for serious ethical debates. As a patient to be operated on by the state-of-the-art Da Vinci system, with the surgeon working at a console, which may be as close as 50 meters away from the operating table or as far as being on a separate continent, the reactions will vary from amazement to despair. Robotic surgery as shown in Figure 8.19, has grown rapidly since it was first used for laparoscopy in around 2000. While the surgeries are more efficient and less invasive, with a shorter recovery time for patients they have attracted criticism. What happens when a calamity occurs? who is to blame? The transfer of moral values across the mode of operation:

$$Surgeon \rightarrow Robot \rightarrow Patient \qquad (8.6)$$

is ambigious. As of now, the onus is on the company building and maintaining the surgical robots. An unfortunate incidence, *McCalla v. Intuitive Surgical* in 2012, occurred when a fault in the Da Vinci system caused burns in the arteries and intestines of the 25-year-old-patient, and finally her death. In the US, the Food and Drug Administration (FDA) is the advisory body for the quality and upkeep of these state-of-the-art machines. If a medical robot malfunctions, the hospitals are required to report this to the manufacturer, which then reports it to the FDA. Since the Da Vinci robot is designed to be an extension of the surgeon, articulating teleoperated instructions executed via its arms are low on emergence and high degree of predictability. Hence the robot is more of a wonderful machine lacking in humanly attributes. Therefore, blaming the manufacturing company for mishaps is acceptable both morally and legally. In later examples, when personhood associated with the robot is ostensibly determinable, the situation changes quickly.

Robots as companions and carers: By 2015, 15% of the US population and 19 % of Europe's population were older than 65. With a greying population, the need for artificial companions and carers has never been felt more acutely. Companionship such as a humanoid or a pet robot, brings joy to an aged or an ailing person; however, since robots of today are far from being human beings, there are ethical concerns [299]. The elderly are usually restricted to their homes or limited to a known local environment. With a robot companion, it may further reduce human interaction. Also, since a robot attends to the physical and emotional needs of the elderly person, this may prove to be an excuse for the family to neglect their duty towards the senior. Robots may come across as 'controlling' and seemingly prey on their dignity [298, 369] such as in scenarios where they are forced to take a pill or coerced into a particular diet or life style. Robots can mirror human emotions to some extent, however human beings can elicit a myriad of emotions. Therefore it is possible that for certain scenarios there is emotional mismatch and the robot comes across as rude or insensitive. This debate can be resolved by giving some control of the companion robot to

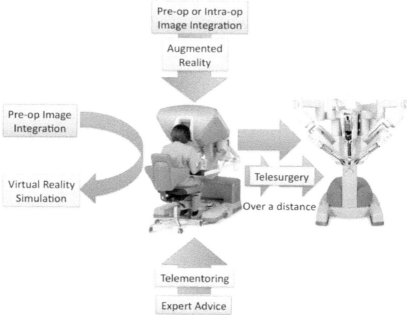

FIGURE 8.19 **Ethics in medical robotics.** The Da Vinci and Zeus systems have led to ambigious moral values — if disaster strikes then who is to blame? the surgeon or the robot? From Kroh and Chalikonda [190], with kind permission from Springer Science and Business Media.

the senior. However that brings into question the senility of the person and can foment potential hazards, health and otherwise.

Robots in the military: In the late 19th century Nicola Tesla attempted to make remote-controlled torpedoes, in an effort to automate warfare. In our times, automation has taken over modern warfare as shown in Figure 8.20. Since Tesla, remote controlled warfare has made its mark since the Second World War and the German tank Goliath and the Soviet tank TT26 have added to the annals of military history. In the new millennium, autonomous robots for lethal action are high on the agenda of the US military and they are set to engage in the future combat systems project [300] and robotisation of the armed forces, which is estimated to cost at least $230 billion.

The obvious advantages of robotisation are robots don't need to attend to hunger or thirst, they don't forget orders, they are not daunted by the job at hand, they make decisions in nanoseconds compared to human beings at hundreds of milliseconds and such an army will have far fewer human casualties on the battlefield — therefore, a risk free war. The US Navy classifies autonomous robots and autonomous vehicle into three categories:

1. Scripted, these are where the robot executes a programming script,

   point → fire → forget

   and has no ethical concern.

2. Supervised, these robots have significant participation of human operators including overrides.

FIGURE 8.20 **Robots and autonomous machines in warfare.** tank like autonomous vehicles shown in the left and unmanned fighter planes shown in the right, have incited both ethical debates and inquiries on their highend pricing. Both image courtesy *en.wikipedia.org*

3. Intelligent, these robots are controlled by software which is designed on human intelligence. The decision process is arrived at by perception of information, interpretation and diagnosis and cooperation and information sharing with other systems.

Ethical concerns needs to be addressed in the last two classifications. Military experts agree that robotisation for war is an inevitable and irreversible process and with 5000 robots deployed for wars in Iraq and Afghanistan, robots are clearly the future of warfare. As of now they do not engage in direct combat and are used for reconnaissance, destruction of explosive devices and interception of shells and rockets and as per the Ottawa Treaty, they have a human control loop installed which will ensure that they don't go berserk. Robots, such as Big Dog which can carry a payload of 150 kg at a speed of 4 kmph, Talon SWORDS (Special Weapons Observation Reconnaissance Detection Systems) made by Foster-Miller which can be remote controlled to fire small arms weapons from as far away as 1 km and tank-like MAARS (Modular Advanced Armed Robotic System) for the US army, and SGR-A1 patrolling the Korean border have been the state-of-the-art war machines. Robots not only reduce a human soldier's exposure time and therefore the number of injured and dead, but they are easier to deploy, attentive for longer periods and reduce attrition therefore enabling army units to stay on the war front for longer periods than ever before. From a cost cutting aspect, training an American soldier costs $1 million per year while the Packbot is priced at $150,000 and the MARCBOT at $5000. Robots also help to extend areas of control and coverage of a military unit, using remote effectors, sensors, and weapons placed on-board each robot.

As of now, all ground robots in the US army are teleoperated and therefore more of an extension of the soldier and robots like Foster-Miller Talon and iRobot PackBot are used to detect and set off explosive devices, thus saving lives and harming no one. However, air combat cannot be teleoperated and MQ1-Predator, the semiautonomous Unmanned Combat Air Vehicle (UCAV), has been used to track down and shoot missiles at terrorist hideouts and vehicles.

Deployment of autonomous or semi autonomous robots has come under fire and critics argue that cognitive capabilities of robots do not match those of humans [302]. Not-so-bright military robots which can apply lethal force lack in ethics [301] since (1) close-contact encounters rely on accurate human targeting judgements and a robot will not be able to discriminate between combatants and innocent civilians, (2) there is a lack of a definition of a noncombatant and robots are still far from abductive reasoning as to when such a noncombatant may need to be killed for the greater good, (3) a human soldier will prove far more efficient in making a decision. There are known ethical conundrums that even the trained military force fails to address, and a robot will not be any better, (4) in current day warfare many times lethal force is inappropriate; present day computational system will not be able to infer such nuances and (5) as of yet, there are no international agreements or guidelines about the use of robots on the battlefield.

Other than that, these robots are in blatant violation of Asimov's laws. It is a human tendency that to assign moral responsibility to a robot increases with the amount and level of actions, interactions and decisions an agent is capable of performing on its own. In war it is even more urgent as the robot is capable of lethal actions. Roboethics aims at the development of a sentient being which can take its own decision and behave in a nearly humanly manner. However, in the military, there has to be a limitation to that model and it is expected that robots obey commands rather than make up their own mind — controlling smart slaves. AI researchers contend that military robots should be designed on the principle of 'divine command ethics', where the command seemingly comes from God — a superior entity, whose judgement is not to be questioned. However, there is opposition to such an approach, as it requires an artificial agent to follow communications from an 'oracle like, know it all' system which suppresses a natural sense of moral values. If such a system is put in place, our robots will function to forgo an ethical binding, and blindly follow command from this God-like oracle. Thus ethics will work to minimise ethical values in the larger system.

War and aggression in the coming years are supposed to be very different from what we have ever seen. It is projected that war will be conducted with seasoned human commanders in charge of a 'kill switch' to prevent an AI take over. Unlike previous wars spread over weeks, months and years, AI agents will wage the combat faster than what meets our senses. Singer predicts that such warfare may be most effective for robots and robot systems operating as 'packs' or rather as a swarm — a natural system evolving to an optimum for the cause of war. The threat to such a route to war is that robots and computer code are not loyal to a military unit, government or country and may possess more intelligence than the human force controlling it. While computer code can cross borders in seconds, a conscious robot will find the flaw in the reason to war. Will it result in the Skynet scenario?

SUMMARY

1. Ethics and moral values are essential for robots in order to participate in more human centric jobs.
2. Ascribing a set of ruleS via conditional loops is often the simplest way to design ethics in robots. However, it is often rather narrow and lacks in human-like ethical values.
3. It is plausible, that sentience in artificial beings is a prerequisite for development of moral values.
4. Asimov's laws are incomplete and have severe limitations when it comes to applying them. Various variations of Asimov's laws have been attempted, both in fiction and also in the AI community. However, a rule-based approach will always be incomplete.

5. Emotions are essential for the design and development of artificial morality.

6. A moral agent is characterised by autonomy, intentionality and responsibility.

7. For design of moral agency, a top-down approach is insufficient and bottom-up is too time consuming. Hybrid methods may be most promising.

8. Consequentialism and deontology, the two theoretical approaches to ethics, are another form of the no-represenatation vs. representation debate and the squabble is best seen in trolley problems.

9. Examples of implementation of ethics in robots are (1) use of the Nao as carer/nurse robots by the Andersons, (2) ethical architecture developed by Arkin and (3) fast simulations to evaluate ethical values prior to commiting to them, made into consequence engine by Winfield.

10. Robots are being used in the care of the elderly and in the military. This has led an engaging debate and critical points of view on roboethics.

NOTES

1. *'Robbie' was Asimov's first robot short story in the summer of 1939. He went on to make minor revisions in 1940 and again in 1950.*

2. *One of the more anecdotal aspects of Asimov's robots is their positronic brains. When Asimov developed this idea, the positron was a recently discovered particle and AI driven by positronic technology enshrined Asimov's fiction as a promising futuristic technology. Asimov never delved into design details of these brains, other than that they were made of an alloy of platinum and iridium and were at risk of damage when exposed to radiation. Asimov stated that, a positronic brain is always designed to incorporate the 3 laws into it and it would be a long haul to develop an alternative design and its manufacture process to create a positronic brain without the 3 laws embedded into it. Asimov also hints at nano level engineering, as the positronic brains can be reduced to sizes that comfortably fit within the skull of an insect. Over the years, the positronic brain has become the hallmark of a buzzword in science fiction and has been incorporated into various other science fictions such as Roger MacBride Allen's Caliban trilogy, Marvel comic's The Avengers, Doctor Who and Star Trek's lieutenant commander Data.*

3. *Asimov's sideburns were his iconic trademark and led to the phrase, 'Asimov's sideburns'.*

4. *At various instances in Star Trek, Spock refers to a utilitarian directive, "The needs of the many outweigh the needs of the few".*

5. *In his short story, 'Do Androids Dream of Electric Sheeps', Philip K. Dick addresses the broad question of artificial manifestation of ethical values. The apparent conclusions were: (1) It is wrong to enslave near human-like AI beings, both in appearance and in ethical and societal values; (2) the distinguishing features between human beings and AI beings will go blurry in the future, correlating to Turing's speculation that the Turing Test will fail as we develop moral AI beings; (3) memories being electronic in nature can be implanted and therefore cannot be trusted; and (4) due to the above, AI beings and human beings will connect with natural bondings of love and compassion. It is probably not a surprise that most of these ironically form the part of a prophecy where human beings and technology blend together. The last chapter will discuss more of this theme.*

6. *The last message from the hitchBOT at twitter was, "My trip must come to*

an end for now, but my love for humans will never fade. Thanks friends: http://goo.gl/rRTSW2".

7. *Robots for sex [202, 209, 313] may seem a taboo avenue for robotics, but there has been no lack of enthusiasm from the AI community and they also have an annual conference [76] dedicated to this theme.*

8. *There has been suggestions that in the apparent future dominated by robots, we may face a new type of racism, biological vs. technological.*

EXERCISES

1. Psychologists agree that human beings are motivated by two emotions, love and fear. Such stimulus is often used in management of an organisation, business, political and social heirarchy etc., love expressed in the form of appraisal, bonus and hike in salary and fear conveyed as threats, deadlines and redundancy. For a futuristic work force of robots, suggest how fear can be manifested in a robot? Also, this may lead to the broader question, do we really need to control a robot workforce with emotional stimulus?

2. Sometimes in the future, a robot assistant is made for a millionaire businessman, and programmed in accordance to the 3 + 1 laws of Asimov (1985). The robot is made responsible for his master's itinerary, travel plans, meetings, bank transactions, health care, bills etc. A week into its job, the robot, in allowance with the zeroeth law, without his master's consent donates his entire fortune to charity for food, medical and upkeep of the impoverished and the homeless, which leaves his master bankrupt. Obviously, this leads to a furore and the company which developed the robot faces public acrimony. However it has about 300 more of these robots in development and the hardware has nearly been made. As a software developer for the company, since Asimov's laws have loopholes, discuss what modifications can be made to these laws to prevent such from happening. Keep in mind that the hardware design is frozen and most of it has already been made, so there cannot be any more hardware modifications.

3. 'Learn By Example' machine learning paradigm was employed by the Andersons for the Nao robot. Briefly discuss why a 'Learn By Doing' approach will not work well for development of ethics.

IV

... the Future

CHAPTER 9

Quest for the sentient robot

"sentience = sensitivity + X, And what is X? "

– Daniel Dennett [92]

Robot: "Well, here I am. Isn't that enough? Please understand, I'm not like a record player or a radio. I'm conversing with you. I hear and understand what you say. ... "

– from 'Interview with a Robot' by Keith Gunderson [136]

9.1 CAN ROBOTS BE CONSCIOUS?

ROBOTS can perform various interesting tasks; however, they still do not measure upto the humanly virtues of consciousness, free will and self awareness and therefore they lack in sentient actions. Robots that can genuinely feel, perceive, readily associate both sensory and symbolic information with items from memory, and thus experience the world subjectively will be the crowning glory of embodied cognition.

We all perceive conscious behaviour rather easily, but the irony is that there hasn't yet been a binding definition for conscious behaviour. The same is true for free will and self awareness. It is apparent that conscious behaviour, free will and self awareness work in tandem for sentient action, and researchers agree that self awareness is a prerequisite for consciousness.

Walter claimed that his turtles demonstrate free will, though in a very limited context. There have been various other suggestions and experiments to develop self awareness and free will in robots, however all of them are either in a very narrow concern such as the turtles or lasting for a very short time such as the experiment with three Nao robots discussed later in the chapter.

It may seem that social robots are conscious and proactively interact with human beings but they merely follow a programming script. Even actor robots mentioned in the previous chapters lack in inner states which convey a higher behaviour.

Self awareness has often been modelled as a layering of behaviours or a lucky overlap of a number of sensory processes. However, there has been a change of opinion over the last decade and instead of the serendipity of emergence, researchers have come to recognise consistency and predictibility of behaviour and cognition as signatures of self awareness. To achieve consciousness at the very least robots should be able to recognise and deal with situations over a broad range, typically where their traditional faculties fall short. The question of experience and its interleaving with conscious behaviour should be the long term goal.

Artificial consciousness is the holy grail for conscious robots, and has progressed with the study of the animal and human brain [69, 144, 249, 309]. Aleksander [6], one of the early proponents of this field of study, considered the ability to anticipate events as essential for artificial consciousness. A definition, summary and scope of the discipline is given by him as,

"Artificial consciousness (AC), also known as machine consciousness (MC) or synthetic consciousness, is a field related to artificial intelligence and cognitive robotics whose aim is to define that which would have to be synthesized were consciousness to be found in an engineered artifact".

Correlation of animal and human brain functionality to their corresponding conscious behaviour has been probed by psychologists, psychiatrists and AI researchers, as shown in the Table 9.1.

TABLE 9.1 Evolution and consciousness

Living Being	Conscious	Evolutionary Traits	Analogy With Machines
Human Beings	Conscious	Fully developed cross-modal representation across five sensory capabilities. Underscored with planning, anticipation and motivation	Impossible as of now
Hedgehog (earliest mammals)	Conscious	Cross-modal representation across five sensory capabilities (less developed)	Impossible as of now
Birds	Conscious	Primitive cross-modal representation across five sensory capabilities along with primitive associative memory	Associative memories
Reptiles	Inconclusive	Olfactory system and primitive vision	Computer vision and artificial nose
Hagfish (early vertebrate)	Not Conscious	Primitive olfactory and primitive vision	Artificial neural networks
Hydra, sponge etc.	Not Conscious	Sensory motor units	Mechanical and/or electrical system

However, there is no unique part of the brain that works as a module or as the setting for conscious behaviour. For that reason and more, progress in artificial consciousness has been has been dotted with pitfalls. Defining consciousness has been difficult if not controversial and the lack of a definitive architecture for the human brain has left more questions than answers.

The ethical norms for such developments and the fear mongering of AI going berserk has impeded research. Sentientism has taken the brunt of controversy, and its mention rarely found in AI literature. Research probing into consciousness in robots tends to target self awareness rather than consciousness or sentient behaviour. Nonetheless, many AI practitioners believe that we will eventually develop newer forms of sentient intelligence which are superior to human beings. This discussion is also addressed in the next chapter.

Revisiting Dennett [91] serves as a cue for the task at hand. These four questions provide a summary of the debate and guidelines for continuing research in artificial consciousness.

1. Robots are material entities and consciousness is immaterial
2. Consciousness takes place in the brain which is organic, while robots are inorganic
3. Consciousness occurs in the natural, not in the technological
4. Robots are too simple, therefore will always lack in consciousness

Dennett argued against all of the above and followed it up with results from early experiments made on Cog. Cog was designed to be aware of its own internal states, and was an attempt by researchers at MIT to extend the reactive architectures into the domain of artificial consciousness in the early 1990s. These questions are still asked by critics and researchers alike [292] and a good part of this chapter in hindsight attempts to answer them. Since the 1990s, robotics has taken a grand leap and now boasts of social robots such as ROMAN, Kismet and Pepper which are robots designed on the semi-sentient paradigm [31, 208] and mirror human behaviour. In complete contrast, pioneering research such as Takeno's mirror cognition, Fitzpatrick and his team at MIT making Cog recognise an action by its sound and Nico beating the drum to sterling performances at Yale University demonstrates that inner states of the aware self are truly possible in artificial agency.

9.2 SELF AWARENESS, CONSCIOUSNESS AND FREE WILL

Sentience addresses the notion of our inner self which cannot be explained by parallel functioning of five sensory organs and metabolic processes supervised by a cerebral cortex, therefore defying an explanation by modifying a computer architecture [101]. The definition of both sentience and consciousness has generated more heated arguments than clarifications.

Child psychologists believe that early glimmers of consciousness and memory can be seen in five-months-old babies. That is probably the earliest setting at which an individual starts to assimilated a large amount of data over the five senses and form conceptual and intentional interpretations over a long time and over a great number of experiences such that it melds with items from memory, with a consideration to the known and prevalent principles of ethics, moral and social virtues. This monolithic structure of knowledge also tends to overlap the workings of our five senses and is embedded into our very existence, sentientism can be said to be the ability to apply the facets of self awareness and consciousness in our interactions with the world. This growth which starts early in the life of an individual continues for most of his/her lifetime. It is generally agreed that at the age of eighteen one attains decision-making ability, and thus the privilege of full moral agency. Therefore, as a direct correlation, consciousness cannot happen at one go or with a programming script, and it is only shows up as a gradual process of awareness, and 'learn by observation' and 'learn by doing' paradigms are important tools for development of the conscious robot.

Consciousness and self awareness are not really independent of each other and in tandem signify the ability to look inward to associate known facts, information, previous experience and memory to logically and systematically process ideas and conclusions and act as per this internal decision process.

9.2.1 Self awareness

Self awareness is easier to observe and has often been the first goal in the quest of conscious agency among researchers. The characteristics of self awareness are suggested by Takeno [316] and can be summed as follows:

1. Cognise to one's own behaviours — helps in developing of the "I-representations"[1]
2. Perform nearly consistent behaviours — consistent and rational simple behaviours leads to more complex ones
3. Have a sense of cause and effect, as far as possible — helps in development of rationalism and logic
4. Cognise to the behaviours of others — appreciation of the " I-representations" in others and therefore acceptable social interactions
5. Interact with other self aware agencies in a proactive manner — development of ethical values

It is also to be noted that, unlike the ANIMATs model in a sea of signal, the knowledge of the self is not dependent on any external signal, is in first person and persists even in the absence of bodily sensations. Knowledge of its internal state and proprioception are essential for self awareness in an agent. Self awareness is the basis of social interactions, ethical agency and higher and more sophisticated behaviour, and is believed to be a prerequisite for consciousness.

It has been suggested that self awareness is just another brand of conscious behaviour. Environmental conscious (e-consciousness) is the ability of an agent to process information, and hence interact with its surroundings, as most ANIMATs do. Phenomenal consciousness (p-consciousness) is the ability to detect phenomena which may be internal or external and self consciousness (s-consciousness) is difficult to design, as it has to be appreciative of the "I". The three modes are informational in character and while, e-consciousness is first order, p-consciousness and s-consciousness are related to internal states and as is observed, s-consciousness is acquired if the agent can identify causality across two or more modes. For example hearing a sneeze and looking up to see a person twitching his nose, relating these two separate phenomena as the same event is a causal binding over auditory and visual modes of inputs and therefore are, at least second or higher order. Such bindings make the agent aware of its own actions, hence the "I-representations". Phenomenal consciousness is the level of conscious behaviour seen in an adult human being, which also includes awareness and adherence to social norms and cultural values. It is generally agreed that, s-consciousness will imply p-consciousness.

s-consciousness → *p-consciousness*

9.2.2 Consciousness

For consciousness, the simplest definition is, *to become aware*. However developing a theory of consciousness has been difficult. From an ontological perspective consciousness lies at the crossroads of psychology and metaphysics and its theorising has remained a favourite among philosophers and psychologists alike. As a start, main features of consciousness will be to elicit sensory awareness, have the ability to focus attention, emotional coloring accompanying the conscious process, and the cognisance of will.

In western thought, the earliest concept of consciousness was given by Descartes, according to which the mind is not a bodily concept and it is distinct from the brain. Consciousness happens when the immaterial soul interacts with the body via the pineal

---

[1] "I-representation" conveys the characteristics and traits of the agent which helps to identify itself as a unique individual; as its mirror image, photograph, voice, name etc. Such ideas have been expressed by Hofstadter who calls it 'The Mind's I'. In a similar approach Metzinger [229] names it as phenomenal self-model (PSM) and in a similar context Chomsky in his book, *What Kind of Creatures Are We?* relates to a language variant as, I-language where "I" conveys internal, individual and intentional.

gland of the brain. Therefore consciousness cannot be tackled with principles of the physical world. Cartesian dualism was short lived and theories which related human experience to conscious behaviour found more acceptance.

Kant [170] distinguished the sensible and the intelligent 'worlds', the former via sensory cognition or *phenomenon* and the later via intellectual cognition or *noumena* and in his *Critique* attempted to show that laws of science will never be able to explain away the question of the human mind or soul. Kant also coined, 'unity of consciousness', that consciousness is more than just one solitary state or an act but is amalgamated as a convergence of experience and representations. His works also classified the two kinds of consciousness, (1) the psychological states and (2) by apperception. Ordinary representations encountered in our day to day activities serve as the representational basis of consciousness of oneself. Following Kant, Schopenhauer further added the concept of intention, the will of the human subject as a necessity for the minimal conditions for experience.

At the turn of the century, Husserl defined consciousness as *the feedback process of giving meaning to an object*. Husserl's phenomenology was different from Kant's rational approach and later led to Heideggerian AI and Merleau-Ponty's phenomenology of perception, both of which are the philosophical groundings for agent-based robotics. The, salient features of Husserl's model are:

1. **In first person**: In the first person is the grounding for phenomenology which constitutes subjective experience, colours, smell, taste, sound, emotions and consciousness. In this context, consciousness is a reflection of the self, and first-person refers to the experience which enables what one is conscious of.
2. **Feedback process**: The incremental process of cognition, by which we focus our attention to reduce an object's material properties to abstract concepts, until the essence of the object is apparently visible and in relation to the essence of other beings.
3. **Intentionality**: A mental process that directs itself toward an object, internal or external.
4. **Anticipation**: In the feedback process of understanding an object, anticipation is to predict the immediate future. Anticipation helps to control the feedback process, as it sets a desired benchmark for the meaning of the object.
5. **Embodiment**: The body is part of the self, and consciousness is experienced in and through the body. Embodiment is also an essential tenet for reactive AI as has been discussed in previous chapters.
6. **Certainty**: With increasing feedback response over time, the meaning of an object finds a higher degree of certainty.
7. **Otherness**: Our consciousness has an appreciation of other beings and a loose understanding of their inner self, resembling our own.
8. **Emotionality**: Consciousness is emotional and reason, perception and emotions bear a close knit relationship. This principle can be related to Braitenberg vehicles 1 to 4, and models of moral psychology discussed previously.
9. **Chaotic performance**: This feedback process is often disturbed by other mental triggers and external events and therefore lacks predictability.

This model suggests that consciousness is strongly tied to internal reflections and concepts of the self and therefore the concept of the other. AI scientists have often extended phenomenology to define consciousness, Hawking definies consciousness as the ability of the brain to collate information across various channels and make approximate predictions about the future and therefore anticipate it with the right actions while Takeno contends that consistency of cognition and behaviour is the key to consciousness.

In the late 1980s, computational theory of the mind (CTM) was more or less the best possible alternative to reason how inanimate matter can be intelligent. Plagued by the lack of embodiment, issues of continuity of local representations to the global representations and the frame problem[2], CTM was still the best model for the mind as a coherent extension of the brain.

Qualia — The Bone of Contention

Qualia means an instance of phenomenal consciousness [144] conveying the subjective character of experience, or in simpler words, **'what it is like?** [249]. Examples of qualia are, 'redness of an apple', 'hotness of a cup of tea', reading into the inner meaning of a writing or a piece of art etc; thereby making it a human experience, rather than just a flow of information. It differs from sensory motor skills, as it is sensation but it is not strictly acquired by sensory input. Qualia, like emotions, nearly underscores all human cognitive processes [147], registering simultaneously across a number of stimuli. Thus not only does every conscious state have bodily manifestations such as hunger, pain, stomach ache, happy, depressed, excited etc; we can also classify degrees as 'more happy', 'less pain' etc. Proponents of qualia argue that reductive analysis based on physical interpretations of consciousness is incompatible or at best incomplete, lacking the subjective character of experience. Since experiences are irreducible, therefore it cannot be incorporated into a theory advancing physical attributes for the mechanism of consciousness. This argument forms the hard problem of AI.

FIGURE 9.1 **Qualia in art.** Human experience can appreciate ambiguous figures, optical illusions and artistic deceptions, and therefore identify both images as shown in the paintings of Alan Gilbert: (a) a young woman or an old woman and (b) a woman seated in front of a boudoir mirror. The setting also forms shape of skull. Both images courtesy *en.wikipedia.org*.

Without qualia, artificial beings and robots will show the right behaviours and reflexes, but fail to develop subjective human experience and be more of a zombie. While it will report the reddish sky at the sunset, it will not feel happy nor sad nor

---

[2]Most CTM literature call it the relevance problem.

elicit any other state of emotion on staring at the sunset. Similarly it will note the colour and composition but it will fail to savour the taste of a glass of Chardonnay. It will never develop traits such as pareidolia or apophenia and it will never be able to truly appreciate literature or art, as both disciplines are meant to be read-in rather than merely address what meets the eye, as shown in Figure 9.1. This is why Harvey suggests

a twofold paradigm to make conscious robots [144]. First make a zombie machine with the right behaviours — the easy problem of AI and it has been accomplished and as a second step add the 'magic elixir' that gives it consciousness and makes it more human and less zombie — the hard problem of AI, which is the proverbial albatross around the neck. None of the theories of consciousness has been able to explain qualia, and both behaviourists and phenomenologists either deny it or are quiet about it. This glaring shortcoming has led critics to question qualia and render it more as an illusion [90] which is very personal and is gauged by an experiential quotient of the agent and which is often buoyed by social interactions and therefore lacks in generality which is anywhere between an idiosyncrasy and an individual opinion. For example the 'redness' of an apple is subjective and what is perceived by one person, will be different from the 'redness' of the same apple perceived by another person. Qualia is further apparent, since human cogintion works via abductive reasoning and bodily computations — both of which have not yet been replicated in the artificial domain.

Minsky's 'Society of Mind', attributed conscious behaviour not to one single aspect or state, but to the working of a number of AI agencies in tandem. A society of agents and intelligence was due to the vast diversity of such agencies and not to a single principle or 'magic elixir'. These ideas were a direct influence in the behaviour-based architectures for robots. Parallel processes working independently have also been suggested by Dennett in his theory of 'Multiple Drafts Model Of Consciousness'. He proposed that the mind and the brain are not separate from each other nor is there a unique setting in the brain where conscious experience occurs. Instead the brain consists of a number of nearly independent agencies interfacing experience to the memory, and the serial account of emergence in these parallel processes is consciousness. The memory is central to this approach, and the mind 'constructs' its own truths by selective rectification of memories to help consolidate desired facts, or alternatively has a prior bias as to what is observed. Therefore, conscious experience doesn't happen instantaneously but is graduated over a timeline and is not dependent on one perception, event or state but rather is conveyed by the flow of information across our senses. Dennett therefore defines consciousness in the first person, *first person operationalism*. However, Dennett reduces consciouness as mere information processing and denies qualia, and therefore lacks in completeing the whole picture.

Preferential editing as an activity in the human brain to construct consciousness was first proposed by Dennett. This has also been suggested by Norretranders [254] and later by Graziano. According to Norretranders the subconscious reconstructs a believable world by editing nearly as much as 98% of the information it receives and is therefore the subliminal layer to the conscious mind, and aids in creating believable metaphysical reality of time, space and action. Similarly, Graziano contends that the brain cannot process everything and therefore the brain focuses on very few things at one go — the concept of attention. Attention is then augmented with internal data and previous experience from memory to make predictions and plan actions.

Searle [294] defines consciousness as *states of sentience or awareness* when a human being is awake. This approach differs from Husserl's and is not based on direct interaction with the environment, but rather an inherent characteristic of the awake human mind. This definition strongly ties sentience and awareness to consciousness. Accordingly, conscious states lie within the brain and happen inside an organism. Also, these states connect sequentially with one another to develop advanced senses of reason and memory, constituting a continued conscious interaction with the world; a conscious life. Each conscious state has a certain qualitative character and feel to it and relates to bodily perceptions, as in drinking water is different from writing a letter. Therefore phenomenal experience or qualia underscores consciousness, as shown in Figure 9.1. Conscious states are subjective as they are experienced by a human or an animal. Here, Searle and Husserl converge on the first-person virtue of consciousness. Block [38][3] suggested that consciousness is divided into **phenomenal consciousness**, which is experience and pertains to answering 'what it is like to be [in a given state]?' and **access-consciousness** which pertains to reasoning and rationality. Pinker also suggests that consciousness occurs over two modes. These are, **access-consciousness**, which works only when we are awake and is more of an attempt to respond to information from sensory organs and memory, helps to develop behaviour and has no to very low information processing; **sentience** is associating individual pieces of information to obtain a greater meaning, with marked cerebral activity in this mode.

The debate is not restricted to the philosophical and psychological. Early ideas of cybernetics led Leary to propose an eight-circuit model of consciousness in which the consciousness is supposed to be developed over eight circuits that operate within the human nervous system.

From the point of view of evolutionary biology, Greenfield [133] attempts to correlate conscious behaviour to brain size of a life form. Her research suggests that there is no consciousness centre, meaning that there is no place in the brain corresponding to generation of consciousness; neither are there any committed neurons or genes dedicated to consciousness. Study of brain size of primates and mammals has also shown that mere size of the brain is not really enough for higher functions, it needs special cognitive modules such as logic, association, concepts of geometry etc. to demonstrate human-like intelligence. Therefore, a computer with very high memory will not suddenly transform into a conscious machine on its own.

In complete contrast to cybernetics, neuroscientist and evolutionary biologist, Csikszentmihalyi [78], defines consciousness as a 'flow', the psychological state of a happy individual in homeostatis.

Despite contributions from many disciplines, a model explaining the unification of the body and mind based on a physical approach remains a holy grail of artificial intelligence. This has also led in to smart humour. Block named it 'a mongrel concept'; Brooks calls it 'the juice'; Norretranders mentions it as 'user illusion' and Chalmers and Harvey independently label it as 'the extra ingredient'. Over the years, classification schemes have been many, casual and non-casual, accessible and in-accessible, representational and non-representational, continuous and discreet, mode and state and stateless however a single theory explaining human behaviour and consciousness has eluded us.

Forgoing the squabbles there are minimal grounds to quantify salient features for development of artificial consciousness as shown in Figure 9.2. Developing grounds for a physical approach by a reluctant handshake between CTM, phenomenology and behaviourism, Starzyk and Prasad [309] suggest that other than perception, action and associative memory there needs to be a central executive [172, 173] controlling all conscious and other processes of the machine, while it is driven by (1) artificial emotions and motivation, viz. remorse on wrong doing, exigency during flood, arson and other natural

---

[3]Block's research was based on the vision stimuli of partially blind patients.

FIGURE 9.2 **The bat, the robot and the zombie.** Discussions on the definition of consciousness and designing artificial consciousness has often led to engaging debates. (1) Bats — Nagel [249] considers consciousness as 'what it is like' phenomenon. It is subjective and cannot be reduced. (2) Robot — as per the Cartesian maxim, consciousness is non-physical and cannot be determined as an artificial affect. (3) The philosophical Zombie — it has attained an iconic status in this debate, where it lacks emotions and does not feel happiness, remorse, pain, or any other psychological effects, and therefore no experiential content or qualia, but it still interacts with its environment. Made by the author, CC-by-SA 4.0 license.

calamity, (2) goal selection, (3) attention switching, viz. focusing attention on hearing a loud noise etc. (4) incremental learning mechanism, etc. Thus creating states of self-awareness and consciousness by relating cognitive experience to internal motivations and plans.

Instead of the philosophical abstractions, a physical approach will underscore sensory-motor design. Inspite of that since consciousness cannot be identified as a reflex, but rather is a problem of cognition, it will further need to incorporate subjective experience, which emerges as an overlap of perception, learning and building of associative memories broadly developed on representations found in the local environment, modulated by attention and emotions, and underscored by moral and social interactions.

Consciousness has to be designed as a phenomenon graduated over time and it will also need to blend in with ongoing interactions and a rational anticipation of the future. The robots will have to elicit this consciousness in higher behaviours which are richer than navigation or pick-and-place operations. Physical approaches will have to downplay qualia, at least at the start, in the hope that advancing research may be able to account for it in the near future.

9.2.3 Free will

Free will can be summed up as " **I can, but I won't**". A more eloquent definition is the question. Do we have control over our actions, and if so, what sort of control, and to what extent? The ability to take a conscious decision by weighing the merits of a situation in hand is manifestation of free will. However critics claim that it is nothing more than a figment of our imagination and that our choices are either near obvious outcomes of the events in the past or blatantly random. A robot that orchestrates human-like free will, will have to be equipped with the ability to reason about its past, present, near future and the comparison between the various choices it has. Free will cannot be fed into the robot as its internal working; rather it will be determined in real time by its local environment and constraints. Also, having a choice is different from the mere knowledge about it. This is an issue of relevance — another form of the notorious frame problem.

A preference for making a willful choice which is not particularly anticipated from the previous working and which cannot be determined from analytical models of the robot is a mark of free will. Walter had demonstrated a sense of free will in his turtles when they 'chose' to follow one track rather than the other, by a comparative merit. It is an easy demonstration, for the micromouse to confirm one track and avoid the other on the virtue of a decision process. However most of these activities are a decision making process via conditional loops, and the micromouse would lack any means of cognition other than choosing tracks nor would it feel happy at choosing the right track and remorse on choosing the wrong one. Oka [256] has designed models based on connectionist architecture with spontaneous handshaking between representation and action modules, augmented by memory and supervised by a central executive. Thus it can create newer states than anticipated in the system, and therefore can be said to lead to nondeterministic free will.

The next section discusses the Turing test which tries to distinguish an artificial being from a biological being, these gedanken experiments devised by Turing has been hotly debated and have often been considered a mark of 'personhood' in artificial beings.

9.3 FROM MACHINES TO (NEAR) HUMAN BEINGS

In the early 1950s, Alan Turing asked the question, "Can machines think?" which led to the coining of the test which bears his name to distinguish machines from human beings. Turing's contention was that if a machine can imitate human behaviour well enough, then it is sentient, or at least that is what it seems to be. In the test, a human interrogator is isolated and kept in a separate room and allowed to interact with the responses of two agents via keyboard inputs and screen outputs. The interrogator has to distinguish between a human and a machine based on the replies to questions asked by itself. After a number of such questions, the interrogator attempts to determine which subject is human and which is an AI agent. Turing relied on human emotions and empathy to pave the way for the success of the test. Turing predicted that by the end of the millennium, an intelligent machine will be able to put up the imitation so well that the success of the test at best will be around 70%, as shown in the cartoon in the Figure 9.3. No computer has done this well, however clearing of the Turing test by an AI agent often makes it to the headlines. Eugene Goostman, a chatbot attained near-celebrity status allegedly for being the first AI entity to clear the Turing test. Present day AI can account for the IQ of a four-year-old child, however chatbots such as Goostman imitating a thirteen-year-old from Odessa, Ukraine, has done better. The Turing test is also a part of our daily lives, and quick online tests as shown in Figure 9.4 distinguish a human response from the artificial. In a future where human-like artificial agency has come of age, a conclusive test may be needed even more.

Adding to Searle's criticism, there has been lukewarm response from the AI community since there are simpler means to test for an AI entity and also creating full-fledged human-like response is currently limited to chatbots. The general view regarding Turing test is that Turing meant it more in an anecdotal and philosophical context than as a true test for day to day usage. However it has continued to be an influential and controversial topic [263].

The Voight–Kampff Test from the 1982 cult classic Blade Runner is a variant of the Turing test and distinguishes between human-like androids known as replicants and human

FIGURE 9.3 **Turing test 2208**, it is predicted by Turing and other leading techologists and futurists that around 2040 Turing test will become passable for machines. CC 3.0 license, cartoon courtesy *http://geek-and-poke.com*

FIGURE 9.4 **CAPTCHA**,(**C**ompletely **A**utomated **P**ublic **T**uring test to tell **C**omputers and **H**umans **A**part) is an implementation of the Swirski test - or the reverse Turing test, where making the letters squiggly makes it difficult to decipher for a machine (a) CAPTCHA in its parent form, (b) variants as 2 parts of a phrase, crossing out and using alpha numerics and (c) utilisation of CAPTCHA for security against spambots. Images courtesy *www.shutterstock.com*, used with permission.

beings. The test is based on the subject's degree of empathetic response through carefully worded questions and statements. Other than verbal response to scenarios which appeal to emotions and empathy and other biological response mechanisms, viz. sweating, heart beat, pulse rate and movement of the eyelids. The Voight–Kampff machine is akin to a polygraph, however, it can further measure contractions of the iris muscle and the presence of invisible airborne particles emitted from the subject's body. Extending ideas similar to the Voight–Kampff, a graduated variant of the Turing test suited for moral agency was proposed by Allen et al. [8], the **Moral Turing test (MTT)**. MTT is designed on the standard Turing test, however the questions are confined to topics on morality. Thus if human interrogators cannot distinguish the machine from the human being, then as per argument the machine is a moral agent, or has attained moral agency. However, MTT soon runs into issues; a machine may be able to articulate better moral judgements than a human being therefore stricter comparative tests may have to be devised. However, even then MTT is self defeating because human morality is nowhere near to ideal and resists benchmarking. There have been attempts to design more sophisticated versions of the Turing test, as shown in the cartoon in the Figure 9.5.

Feigenbaum Test. Feigenbaum [102] suggests a variation of the Turing test which merits machine's ability to pass for human, however the small talk and casual banter is replaced with scientific and technical discussion, as a scientific expert in a specific field. The test can be described as, between two players in the fray, one is chosen from an elite list of experts in each of three pre-selected fields of natural science, engineering or medicine. For instance, the choices can be cosmology, metallurgy and endocrinology. For each round, the play between the players is determined by another expert of the same academic stream, e.g., a metallurgist ruling on a discussion on metallurgy, etc.

Feigenbaum test (FT) may appear to be more difficult to pass than the Turing test, however consider that the questions are being asked from physics or mathematics or history. An AI being may be able to make a lightning quick online search to lead to the precise answers for the questions being posed to it, or alternatively may have a sufficiently exhaustive handy library spanning a few terabytes which would in effect suffice for all of the questions. FT once again brings forth the Moravec's Paradox. However, two interesting observations are: (1) the artificial beings which can clear FT, being technically able, should be capable of self improvement and enhancing their own designs and thus may plan a course of evolution as human beings and (2) in FT, the language and verbal skills to hold a conversation are very much the same as required for the Turing test. Thus various high performance computers would fail the FT as they lack verbal ability.

Like the Turing test, it is expected that machines will be able to clear FT by around 2030. Past 2040 it is speculated that it may be difficult to attain the pristine conditions for the test, as human beings and machines would have started to meld into one another. There will be blurring of the very line between human beings and machines ushering in transhumanism. The next chapter speculates into such futuristic scenarios.

Sparrow's Turing Triage Test. Sparrow [307], makes a critical case of the Turing test and suggests that in a crisis situation we may be forced to chose between the existence of the machine over the life of a human being.

Sparrow extends the Turing test to a 'triage', which truly is not a test as per a way to distinguish between a machine or a human being, but rather is a call to our conscience where one seemingly tends to find overlapping human and machine values. Sparrow takes

FIGURE 9.5 **Tests to distinguish human beings from robots.** starting from the Turing test in the 1950s, there has been an effort to design tests to distinguish the artificial from the biological. The simplest mode is to engage in a series of questions and from 'human content' such as empathy, imagination and ethical principles, determine the answer. However, it is believed that sentient robots will have more moral fiber than us human beings, as their judgement will not be clouded by greed, ego and selfish premises. Therefore sentient robots may truly transcend human beings and become the morally superior species. Image courtesy NESTA (*www.nesta.org.uk*), designed by A3 Design (*www.a3studios.com*). CC-by-SA, NC 4.0 license.

FIGURE 9.6 **Sparrow's Turing Triage Test**, illustrates the dilemma of rescuing a biological beings or an AI beings in a critical scenario. Cartoon made by the author, CC-by-SA 4.0 license.

the Moral Turing test (MTT) to the limit — the death of an AI being, and contests it with the death of a human being. Sparrow formulates the test with the following illustration.

In the near future a hospital employs a sophisticated AI for diagnostics. The AI is equipped to learn and arrive at its own conclusions and is indistinguishable from human beings in a Turing test[4]. In the intensive care unit are two patients on life support systems awaiting donors and surgery etc. As a critical condition, a fire burns down the electrical supply unit, and the hospital is forced to function on backup power and therefore runs at greatly reduced levels. Sparrow proposes a senior medical officer as the protagonist in the above situation and underscores the two following scenarios which beg to question the virtues of personhood,

1. **Human being vs. human being.** The senior medical officer is informed that the backup power is not enough to support both patients and only one can be saved while the other left to die. A decision has to be made else both patients may die. The sophisticated AI, running on its own emergency battery power, generates projections and statistics for both patients, evaluating the chances of recovering if they survive the current crisis. On the basis of this information, the senior medical officer makes a decision which may well haunt him for years to come.

2. **Human being vs. AI.** As a final twist in the tale, the battery of the AI is failing and the AI starts drawing on the limited power available for the hospital, thus endangering the life of the patients on life support. The senior medical officer can consider switching off the AI and thereby save the patient, however such an AI is not meant to be fully switched off and doing so will fuse its circuit boards, rendering it permanently inoperable. Alternatively, the senior medical officer can turn off the power to the patients life support thus allowing AI to live on. If a decision is not made, chances are that both the patient and the AI will die.

While it may sound melodramatic, as the AI begs the senior medical officer to consider letting it live on, it is left to debate whether the second scenario has the same character as the first, as shown in Figure 9.6. It can be concluded that machines would have achieved the moral status of human beings when the latter choice has the same character as the former. Further, Sparrow notes that the second decision doesn't have one right answer and there may be reasons to make a call either way. In 50 years into the future (2065), will a human being pause to think before shutting down his/her computer or mobile phone — preventing the killing of an AI being? Sparrow believes that his test will help to highlight

---

[4]It is apparent that this AI will at least be semi-sentient.

the issues related to the moral status of artificial intelligent beings and may well be used as an empirical test for moral standing, in the times to come.

Lovelace Test. A more rigorous and interesting variation of the Turing test was developed by Bringsjord et al. [47]. The test underscores human functions which have defied mathematical modelling, like creativity, empathy and shared understanding — elements of social cognition. Other than a list of questions, the test identifies truer human values such as human-like creativity and originality. In order for an artificial agency to clear the test, it has to create something original, all by itself, something it was not engineered to produce. The new essence of creativity can be an idea, a novel, a piece of music, a work of art etc. Also, the designers of the AI entity must not be able to explain how it led to this sudden bout of creativity. The test is named in honour of Lady Lovelace, widely considered to be the first programmer. Her opinion was that computers can never be as intelligent as humans because they merely do what we program them to do and they lack the ability to originate an idea that they weren't designed to. Similar thinking guided Bringsjord and his team. An 'upgrade' to version 2.0 of the Lovelace test was designed by Riedl [258] at Georgia Tech which adds a criterion to gauge the humanly quantum of creativity. Riedl proposes that the interrogator assign tasks with two components. The first is an inquiry into a creative aspect such as a story, poem, or picture; the second puts in a criterion on this creative ability, such as, "Tell me a story about a pig that went to the moon," or "Draw me a picture of a dog playing a violin " — insisting on topics which are out of the ordinary and resort to the imagination and creativity, calling on genuine human skills and ability.

9.4   SEMI-SENTIENT PARADIGM AND IT'S LIMITATIONS

In a broader perspective, the Turing test is an answer to the question, what constitutes a person? With technology, the definition of personhood has been changing. Beings with: (1) with voice processing and speech synthesiser, (2) with NLP and power words typically in chatbots, (3) thoughts — as deduced in MRI scans and (4) experiences such as immersive gameplay and sophisticated VR such as the occulus rift, are also construed as nearly a 'person'. Therefore, a strict biological concept is not enough. Therefore traits which are central to the concept of 'personhood' are [347]:

1. **Sentience.** The ability to experience sensations, conscious experiences such as pain and pleasure
2. **Emotionality.** The ability to feel happiness, sadness, moroseness, anger etc.
3. **Reason.** The ability to connect logical thoughts and thus solve complex problems
4. **The capacity to communicate.** The ability to express ideas and information with another person
5. **Self awareness.** Knowledge of oneself, ability to foresee the implications of one's actions
6. **Moral agency.** The adherence to moral principles or ideals

Among these, fake **emotionality** can be implemented computationally using FACS, tonality, NLP etc, some elements of **reason** can be imparted as a programming code and **the capacity to communicate** has been demonstrated by modern day robots, the remaining three (**sentience, self awareness and moral agency**) have been the stumbling blocks to attaining robots which are human-like. These are not truly independent of each other, but rather are very closely related to one another, with sentientism being a cornerstone paradigm since morality and self awareness in an AI agent can be addressed only for a sentient being with free will. It is an apparent that only organic beings are genuinely sentient, such may fail to arise in the truest manner in the artificial domain.

Ishiguro's humanoids are the best examples of human-like robots and they can very nicely replicate the physiological and behavioural aspects of human beings. However, they are programmed targeting a narrow concern and lack free will and sentient behaviour. On the other end are the unfortunate instances when robots have gone on a killing spree, a devastation — however not a conscious act. Since, it has been observed that babies apparently perceive friendliness and cogent behaviour from robots as sentientism [226], therefore to appreciate sentientism a sufficiently mature level of human intelligence is required.

Experts working on artificial consciousness have expressed concerns that sentient beings may never exist, since the human brain functions on the ability to integrate information and it is a key property of consciousness. Most models of induction rely on sequential logic. The human brain is not such type of a processing unit and reverse engineering the human brain has yielded poor results. Further, each experience and learned aspects collate into the human brain as an addendum to the corresponding previous memories making it even more difficult to discern it with an artificial sequential process. Also willfully editing or deleting memories is near impossible, and instead of a facet, it is diagnosed as a medical condition. It has been seen that computers cannot integrate information in such a manner. Neither is it possible to make computers where memory cannot be modified or edited out.

We do have computers which can beat Kasparov in chess or devise a dynamic lock which will take millions of years to break, but since we lack an understanding of how the mind relates to the brain nor how neurons and action potentials create consciousness, progress to replicate sentientism has been slow. The Blue Brain Project at EPFL, Switzerland and the OpenWorm project with contributors across the globe are efforts to develop artificial brain like functions. These projects promise to simulate a full human being in the coming decade.

Human beings are less likely to appreciate a fully sentient artificial being and cooperation and coersion with human beings gets a higher preference than superlative levels of autonomy. To assimilate well into human society, robots may have to look and act dumb rather than the super intelligent beings that they are capable of being and the notion of sentience may at best be 'redefined' as 'an ability to interact effectively with human beings using natural channels such as voice and gesture, in the best interest of the human being and the broader interest of the human race'. While on the other end of this debate is criticism from renowned physicist Penrose that awareness in biological beings cannot be explained by any scientific approach and thus is unattainable computationally.

As per strict etymology, 'conscious robot' is an oxymoron, meaning 'free willed slaves'. How do material aspects lean towards attaining sentience? Such a route is charted via Maslov's robot index in Figure 9.7. The primary concern for robots will be to address their energy needs. The next level in ascendency is safety and self preservation, which may be a deliberative endeavour and also be designed-in as a failsafe smart technology. Higher up in the pyramid, embedded AI will underscore enactive abilities to interact with the dynamic environment. Continued interaction with the environment will appeal to the need for emergence of self awareness. As a last step a self aware being will seem to attain consciousness in the Husserlian sense as inner cerebral processes will try to project the self-awareness of its own being on to other beings and objects. As a conscious being, and with its ability to sense an inner self and groom it with acquired or learned information and adhere to rules and adages such as morality and social values, the robot will tend to attain ground level sentience.

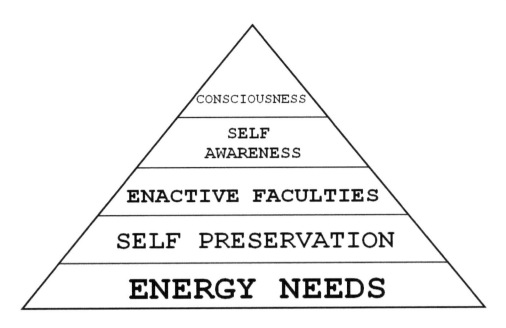

FIGURE 9.7 **Maslow's hierarchy of robot needs.** The hierarchy of needs for beings developed with embodied AI, on attaining sentience.

Kismet, ROMAN, PR2 and Pepper are state-of-the-art social robots. The down side to this is that the emotion elicited by them is at best sophisticated imitations, developed by observing the human subject. These robots do not feel happiness, remorse, pain, or any other psychological effects — typifying zombies lacking a soul, therefore with no consciousness, free will nor sentience. Developing moral agencies into and also as an extention of such social robots will require more of a deontic approach, developed for a narrow concern such as a carer robot nurse robot, companion/assistant robot, following a set of instructions, while being friendly[5] to the human subjects. Other than this lacuna, these models take into consideration first degree effects such as mirroring from the facial expressions and voice of the human subject without any inward processing such as a decision process or associating information. Causality, as shown in Figure 9.8, is much more than first degree and the far reaching effects are often not fully discernible even to us human beings. An artificial being should not only respond to the first degree effects — that is, be self aware — but should be able to analyse and evaluate the far reaching consequences of its actions. To cite an example, consider a mobile robot assisting in medical surgery. Cutting open a patient would seem a harmful act, however the far reaching effects are with the intention of curing the patient. Therefore, such a robot will at least need to possess functional morality and thus a bottom-up approach to develop it or a developmental route, learning from experience and phased in over a long period of time.

The goal is to construct a perfect moral agent, but since human morality lacks a definition and cannot be explained by an undisputed body of knowledge and rules, pitfalls are commonplace. It has also been seen that adult human beings often prefer to take their own decisions rather than being prompted by robots. Robots are then confined to be nothing

---

[5]This friendliness will not be in an effort to socialise or strike up a lucid banter, but rather to elicit apathy/empathy, a job which they have been trained to do but lack a 'feel' for it.

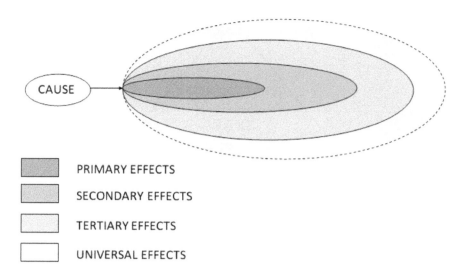

PRIMARY EFFECTS

SECONDARY EFFECTS

TERTIARY EFFECTS

UNIVERSAL EFFECTS

FIGURE 9.8 **The causality picture**, a robot which is developed purely on a top-down approach will only have appreciation for primary effects and function in the realms of operational morality. A bottom-up approach will give the robot ability to evaluate secondary effects, while full moral agency will probably employ a hybrid approach and will need realisation of the far reaching consequences of its own actions. It is to be noted, universal effects are not always appreciable by human cognition and only a planet wide AI will be able to assess such consequences, this is discussed in the next chapter.

more than smart assistants which lack autonomy, helping in a human initiated project rather than being a genuine critic in human actions. This is the reason why various researchers put the definition of sentience as effective and efficient communication, cooperation and congeneality with human beings — the semi-sentient paradigm. Robots may have to pretend to be dumb, illustrate a lack of autonomy, adhere to human societal values and be appreciative of human doings in order to find acceptance in a human society. Employing artificial emotions to trigger internal states is central to the progress of such an idea. Such was seen in ethical architecture in the previous chapter, which models artificial emotions to reward or punish robots in the military. Stateof-the-art human-robot communication over natural language is another tenet for the semi-sentient paradigm.

The semi-sentient paradigm which relies on humanoid appearance and voice based communications reduces the robots to neo-Asimovian smart slaves. The paradigm finds acceptance as shown in Figure 9.9 since current day technology has not yet been able to design artificial homeostatis. Which begs the query on qualia, and therefore even the state of the art robots can attain only low challenge — low skill tasks as shown in Figure 9.10.

However, controlling robots as a slave morality either by strict deontic principles or by synthetic development of pain, suffering and remorse, defeats the goal of AI to strive towards human like intelligence. Metzinger [229] draws a parallel to animal ethics and genetic engineering, where causing pain or encouraging the culture of a disability in children in order to study it, is not only ethically violating, but also mind numbing. Therefore, such negative synthetic phenomenology should not find encouragement or acceptance and we should not create nor risk the emergence of suffering in an artificial entity. Also, one needs to see beyond a master-slave relationship, and robots should be allowed the privilege of being accountable for their actions Therefore a robot should be responsible for an action if, at the time of

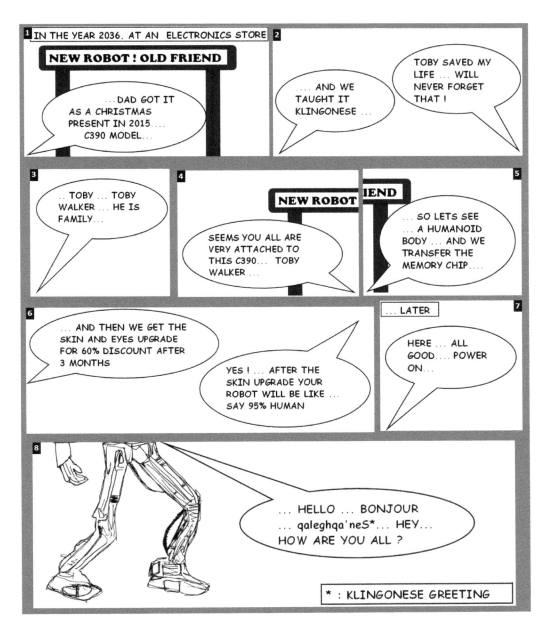

FIGURE 9.9 **Toby 2036 (Toby, 4 of 4)**, Toby gets a humanoid body. This acceptance of robots as near humans is prospected to be on the rise over the next two decades. Cartoon made as a part of Toby series (4 of 4) by the author, CC-by-SA 4.0 license.

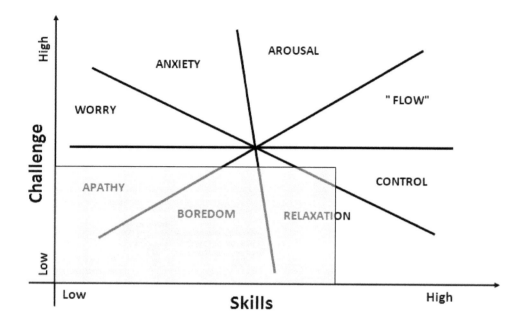

FIGURE 9.10 **Csikszentmihalyi's 'Flow'.** Csikszentmihalyi [78] contends that challenge at hand vs. the skill needed is directly related to the emotional state and homeostasis of human beings. Therefore, the shaded region shows the limits of current day robots, which are broadly in the low challenge — low skill region.

the action, the robot had the ability to suspend or terminate that action. Therefore, in hindsight it also has to be accepted that like human beings robots also can go wrong even with all the facts, data, reason and advice ready at hand [32].

Other than just being wrong, the ability to lie and deceive is a trait of a complete moral agency. Dennett refers to this as 'opportunity for duplicity', which manifests in two ways: self-deception and other-deception. Deception is used as a tool for social adhesion and helps in civil acceptance, and according to the semi-sentient paradigm robots will have to deceive and appear dumb to be accepted as a human-like entity in human society.

To overcome the limitations of the semi-sentient paradigm, there needs to be attempts to inquire into a robot's inner self — the monologous soul of an artificial agency. The next section attempts to explore the inner world of a robot.

9.5 MEMORIES, MEDITATION AND THE INNER WORLD

Braitenberg vehicle 12. Braitenberg's synthetic psychology has been discussed in earlier chapters. Vehicles 1-4 demonstrated humanly emotions, tastes and likeness and vehicles 7-11, used information to deduce knowledge and develop concepts. Other than its own unique ability, each successive vehicle is designed to have all the characteristics and hardware of its predecessor. With added functionalities, advanced vehicles may face the jeopardy of running into a condition where processing sensory information in near parallel may become maximally active leading to an undesired explosion of processes, a freak condition similar to epilepsy which will stop only when the vehicle has run out of energy, or has a fail switch which will stop the power supply to the vehicle. To overcome this condition

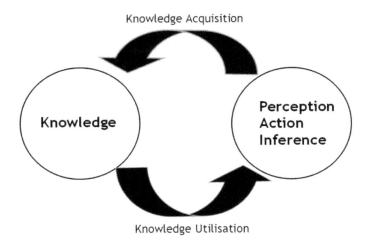

FIGURE 9.11 **Braitenberg vehicle 12.** is seen to have 'free will' and can engage in autopoietic growth, where perception, action and inference on acquired knowledge makes further knowledge acqusition efficient and robust.

in vehicle-12, Braitenberg proposes to raise the threshold of external stimuli to a high, so that only those events which have a very high sensory response come into play, and rest are neglected. Though effective, this approach is detrimental and can render poor performance. Therefore Braitenberg devises a periodic cycle over which the threshold is raised and lowered. Therefore, the number of active elements in the system which were very high at one instance, will be suddenly lower at the next one. However those elements processing sensory information will not just switch off abruptly, but rather will continue until the process has completed. Therefore, it will be difficult to predict the vehicle's action, and therefore it can be said to work as per its own free will. Also, the information processing ability of the vehicle will remain high even when threshold has been lowered, therefore it is possible that the vehicle uses this processing ability to 'meditate' on its memories and discover associations leading to newer ideas, enabling a 'thinking like process' by focusing attention inward. External agencies cannot influence the decision process of the vehicle and it can choose to think and reflect on its own memories as an inquiry into its inner world, which may work hand in hand with free will to trigger an autopoietic growth as shown in Figure 9.11. Perception, action and inference on acquired knowledge make further knowledge acquisition efficient and robust and provide ever-increasing understanding of the environment, only limited by memory limits (RAM) and physical constraints, viz. mechatronic model, wheels etc. for exploring the environment.

The inner world. since consciousness has various definitions and researchers are still divided on the long-term goals of artificial consciousness, it is an interesting question, 'When do we know that a robot is conscious?' AI research contends that the experience of an 'inner world' will at least be a large piece of the jigsaw. For human beings the 'inner world' signifies an abstract ability of our brains to simulate behaviour and perception by interaction with the external world. It is believed that an action can be simulated by activating motor neurons in the frontal lobe very nearly as they would be activated during an actual occurrence, though the final output is not actuated to the muscles but suppressed. Similarly, the perception of an external stimulus can be simulated with internal activation of sensory cortex very nearly as it would have been activated during normal perception of such an external stimulus. Along with action and perception, the anticipation mechanism underscores the construction of the 'inner world'. It is plausible that 'inner worlds' [149] may very well exist in very simple robots equipped with connectionist learning such as ANN and engaged in repetative low level tasks of navigation and path following. For such a robot, the first few navigational routes will be externally generated, often based on purely reactive principles as observing the intensity of a light source or tracking known markers etc. Later a similar route will be tracked without using the sensors but through internally generated perceptions, the prediction based on the learning module of the ANN. Therefore, Hesslow contends that the ability to manipulate the external world with self-generated input is the rudiments of an inner world which also confirms the first person interaction with the environment. Since this inner world is based on the reursions of the ANN, therefore tasks which are not so frequent are usually prone to failure. Working on a similar ideas, Gorbenko et al. exploit such simple connectionist models to yield temporal relations and demonstrate self detection and therefore self consciousness. The next section is a look into experiments which attempt to trigger such inner states and hint at a semblance of conscious behaviour.

9.6 EXPERIMENTS — COG, MIRROR COGNITION AND THE 3 WISE ROBOTS

Research, has only been able to develop self awareness which is particularly context dependent, usually lasting for a very short time. However these experiments have kept up the interest and set visions for a breakthrough in the near future. Self awareness hints towards attaining consciousness, however the conscious robot is still an optimist's dream if not just an oxymoron as suggested by Dennett. The following are the various attempts to demonstrate self consciousness in robots. Four routes taken by researchers are;

1. Extending behaviour-based methods, assuming that a higher level of behaviour will emerge with a number of reactive modes working in parallel.
2. Overlap of the various modes of sensory processing to converge to the same event. For example, the falling of a glass tumbler will be observed through vision — seeing the tumbler fall — and sound. If the robot can discern the overlap of these two events and converge them to the event, then it would suggest an elementary level of consciousness.
3. Consistency of behaviour. Such methods highlight predictability and have been tagged to the mirror cognition test and its variants.
4. Puzzle solving, which involves inquiry of the self and explores the unique characteristics and traits of the robot.

It is to be noted that social cognition and conscious behaviour often overlap and various experiments on social robots also ascertain some degree of conscious behaviour, as the peek-a-boo and drumming mate experiments on KASPAR, and the mutual care paradigm designed for Hobbit.

9.6.1 From reactive architecture

Since traditional symbolic AI is insufficient to lead to agency similar to human behaviour, connectionist approach and layering of behaviour as in subsumption architecture may be a good starting point. However this will not help in deliberative processes such as inward thinking which is essential for consciousness. Employing a deliberative approach doesn't solve the problem as rule-based systems are plagued by the frame problem and cognitive

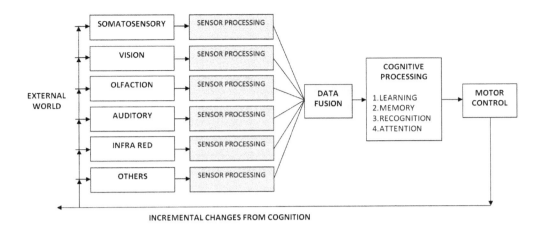

FIGURE 9.12 **A suggestion for cognitive architecture**, adapted from Long and Kelley [207]. Data fusion is important, as the simplest conscious experience occurs over a number of sensory perceptions, and is perfected over feedback cycles.

architectures require a great number of rules, and this process is difficult to scale up to billions of rules.

Long and Kelley [207] suggest that artificial cognition may be designed using hybrid architecture and machine learning, as is shown in Figure 9.12, carefully designing the emergence of the system, taught and trained by large teams of engineers and scientists. Such a project may well be crowd sourced and driven by the enthusiasm of the citizen roboticist.

We human beings, interact with the environment through visual, auditory, somatosensory, tactile and olfactory sensing. These are processed simultaneously in near parallel, and the integration of information acquired across all these modes in the brain's cerebral cortex broadly forms our perception and our consciousness. Behaviour-based methods are designed with ascending layers of complexity, with each behaviour module directly relating stimulus(S) to response(R), and these near parallel modules leading to the emergence of a behaviour at runtime.

9.6.1.1 Cog

Cog, a humanoid robot bust, was developed by Brooks and his team at MIT in late 1990s. The aim was to develop a fully conscious robot, with sensory motor 'experience' than designing it via an information processing route. It was nearly the size of an adult human and had two arms, but no legs. While fully cognitive human-like functionality was never achieved, it did attempt to address the question of consciousness and strengthen the philosophical issues in such research. Cog broadly addressed the following [91, 92]:

1. Cog was designed to **go around the symbol grounding problem**, as any symbolism was to be grounded in its real interactions designed as sensory-motor.
2. Cog was equipped with the ability to make preferential choices — the typical valuation of two choices at a given instance over points of merit typifing an **ability to make a conscious decision** and **adherence to a set of values**.
3. Cog was equipped with a number of self monitoring devices and therefore, it was

in effect designed to be **aware of its own internal states** by observing its own artificially attained homeostasis. It was also aware of the various parts of its body with proprioceptive feedback from the joints in the head, torso and neck.

4. For us human beings, vision is the superlative of the remaining four senses and more often an interaction starts with the meeting of the eyes. In conjunction, Cog's video camera eyes saccaded to focus on a newly arrived human being in the room. Such eye-contact and gaze monitoring enabled **attention focussing**, and provided for **at least semi-sentient interactions**.

Cog was the first effort to develop artificial consciousness in real robots by extending reactive models to the domain of artificial consciousness.

9.6.1.2 Consciousness based architecture

Kitamura [181,183] draws a one-to-one similarity between behaviour and the observed blend of consciousness over biological evolution, from single-celled organisms to animals to birds and finally to human beings as shown in Table 9.2.

TABLE 9.2 8 levels of consciousness employed in CBA

Level	Phylogeny	Human Ontogeny	Consciousness	Behaviour
8	Human Being	less than 2 years	Conception	Linguistics
7	Human Child	2 years	Representation	Making of a tool
6	Chimpanzee	18 months	Image	Use of a tool
5	Monkey	1 year	Temporal and spatial relations	Use of medium
4	Mammal (Quadruped)	9 months	Stable emotions	Detour, search, arm manipulation
3	Fish	5 months	Temporary attention	Capture, escape, approach
2	Earthworm	1 month	Pleasure and displeasure	Orientation of body and limbs
1	Jellyfish	Few days	Primitive reflexes	Reflex displacement, feeding

Consciousness based architecture (CBA) is designed on the model proposed by Vietnamese philosopher Duc Thao, a hierarchy where consciousness and behaviour is a single phenomenon. Consciousness is developed over a number of levels and is explained as an inhibited action written in the memory and drives a chosen action in the immediately higher level. The chosen action if again inhibited acts as the consciousness driver for the next higher level. Duc Thao's model is motivated from evolutionary biology and assumes that the phylogenetic evolution across millions of years in the mental process, from single-celled organisms to human beings, is nearly the same as the ontogenetic growth of human consciousness from infant to adult beings.

In a hierarchical architecture as shown in Figure 9.13, lower levels developed on single-celled organisms and insects are therefore reflexive and higher ones designed on mammals, apes and human beings, are more cerebral and therefore driven by

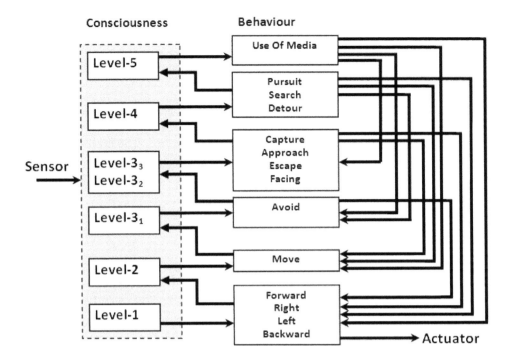

FIGURE 9.13 **Consciousness Based Architecture**, adapted from Kitamura [181, 183]

representational knowledge. The level of consciousness activated selects and produces a behaviour at the immediately next level where the behaviour is obstructed. The five levels of consciousness in CBA are:

1. **Level 1** — Reflex motion, without memory. The stimulus decays with time, therefore though there is no appreciable memory, the agent is still conscious of the past instances of time. Typical actions for this level are changing body directions, motion of limbs, feeding including chewing and swallowing, etc.

2. **Level 2** — Each stimulus can be gauged by a corresponding emotion of pleasure or displeasure, based on the memory of the past stimuli experienced on Level 1. If the stimulus which causes reflex in Level 1 goes over a threshold or alternatively the agent cannot respond with reflex action, then the information acquired from the stimulus stored as memory acts as past experience for Level 2. Level 2 is the start of development of marking each stimulus with an emotional value — a coordinate system of sorts, which continues in Level 3 and 4. The memory from this level guides the intentional actions of the agent in the immediate future and also forms base representations to forsee simple events — food, predator etc.

3. **Level 3** — Transient emotions related to external objects form the consciousness of this level. Consciousness is modulated by emotions such as, desire, pleasure, anger, comfort, hatred as long as the stimulus is present. If stimulus from Level 2 is strong enough then it works to drive towards action in Level 3. For example, if the agent is able to find food in Level 2, then the drive to approach the food and acquire it is acted out in Level 3.

4. **Level 4** — If actions in Level 3 fail to reach the desired goal, viz. acquiring food

or attacking a predator, then the agent has to seek other routes to accomplish these goals. The corresponding emotion is no longer transient, but is lasting and it acts as a motivator to pursue the job in hand.

5. **Level 5** — The agent is able to recognise and manipulate low-level temporal and spatial information.

By ethological correspondence, level 5 tallies to the consciousness of lower anthropoids such as monkeys; level 4 to mammals lower than monkeys including birds; level 3 to fish; level 2 to insects and earthworms; and level 1 to single-cell organisms. The level of consciousness activated selects and produces a behaviour at the immediately next level of hierarchy above the level where the behaviour is obstructed. The behaviour selection [184] designed as:

$$C_i(t) = \sum_{j=1}^{N_E} |\beta_{ij}| + \sum_{j=1}^{N_I} |\gamma_{ij}| + \sum_{j=1}^{N_E} |a_{ij}| \qquad (9.1)$$

where β_{ij} is the external perception, γ_{ij} is the internal perception and a_{ij} is anticipation. Of these three, the latter two can be used to design the levels of CBA; γ_{ij} is designed as a function of time, independent of β_{ij}, while anticipation is modelled as a function of β_{ij}.

HAL 9000

In Arthur C. Clarke's futuristic saga '2001: A Space Odyssey' HAL 9000 is a super computer brain of the spaceship Discovery that houses 2 inmates, the astronauts Dave Bowman and Frank Poole. HAL is an acronym for, **H**euristically programmed **AL**gorithmic computer and the 9000 series is said to have never made a mistake. In the movie, the supercomputer kills both the human inmates in order to pursue the goal of the mission — cause for the greater good.

HAL 9000 has been a matter of debate and has spurred various lines of thought in roboethics. While Abrams suggests HAL 9000 to be the fictional premise of the singularity, Wheat, using instances from the movie, has tried to demonstrate the various traits of HAL 9000, which are: Consciousness, Cognition, Confidence, Enjoyment, Enthusiasm, Secretiveness, Pride, Blaming, Treachery, Fear, Panic, Lying and Senility. HAL 9000 also has human faculties as appreciating art, playing chess and using tools. Kubrick met with Good to discuss the script of the movie, it is possible that HAL 9000 is motivated by Good's ultra-intelligent machines. Around the second half of 2014, HAL was in the news as an example of how a superior AI can be detrimental to human welfare and existence. After Musk and Hawking expressed their concerns regarding super intelligent AI.

Using this model, Kitamura et al. initially developed simulations and later experiments with Khepera robots to demonstrate the working of CBA. Experiments on a prey-predator game showed that behaviour starts at lower levels (level 1) and as the lower levels are inhibited, the robot demonstrates consciousness and moves to higher levels (maximum of level 5) to account for this change. These experiments and simulations also showed that at higher levels, two robots in the same level of behaviour tend to cooperate implicitly. CBA works well in navigational problems such as motion to goal and tracking a moving target, and led to emergence of implicit cooperation.

A behaviour-based layering to consciousness was proven to be more effective at such tasks as compared to reinforced learning methods [182]. Later, Kitamura extended CBA using motivation from rat brain manipulating techniques, and also developed a more sophisticated level 5 to account for 'thinking', particularly planning and learning. However, CBA is suitable only for low-level tasks and is not extendible to human-like representational tasks, such as communication through vision, voice and sight, which are tagged to emotions and social norms.

As an apparent shortcoming, CBA is suited to low-level behaviours and lacks in variety. Since the levels of behaviour are very well defined, the robot never really encountered scenarios which were not anticipated. Therefore, though CBA may suggest synthetic consciousness, it is not appreciable for development of consciousness as found in human beings. Research over the last decade has moved away from layering of behaviour in favour of unison of modalities, binding together distinct channels of information and action, such as voice, vision, touch and joint movements etc. to converge to the same event.

9.6.2 From cross-modal bindings

In our-day-to-day activities, on hearing one's name the immediate reaction is to turn the head to look in that direction. This is a cross-modality of information modes, (1) audibility and (2) feedback from the joints to (3) see the source uttering the name, and it focuses the attention towards this event. Cross-modality is said to be a superior functionality and a mark of conscious behaviour and has been a popular model and also a tool used by various research groups.

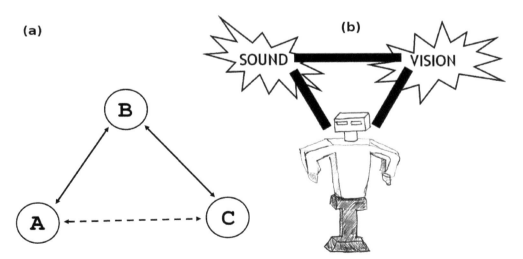

FIGURE 9.14 Self-aware internal states are obtained by binding three modalities. Relating the sound, vision and proprioception helps the robot develop internal states which can identify an action, object or its ownself by the sound. Image and concept adapted from Fitzpatrick [105].

9.6.2.1 Cog's tambourine experiment

Fitzpatrick [105] and his co-researchers at MIT developed higher states of cognition and awareness in the Cog by associating information across three different modalities, visual, acoustic and proprioception. The method is to segment groups of signals in individual modalities of visual, acoustic and proprioception and then to identify and bind causally related groups of signals. All of those signals will convey the same event. Using a tambourine for the experiment, the periodicity in its motion can be studied by tracking and grouping periodically moving points while it is being played. The two events, the motion and the sound will be causally related or bound if they happen in tandem and converge to the same event.

Fitzpatrick et al. use Short-Time Fourier Transform (STFT) for the visual signal and use a histogram for the acoustic signal and then two signals are bound together using binding algorithms such that binding occurs if the visual period matches the acoustic period or if it matches half the acoustic period within a tolerance of 60 ms. Half periodicity is of significance as sound is often generated at the fastest point or the extremes of the trajectory, introducing a factor of two. This allows to 'see the sound of a tambourine' and identify the object by its rhythm. Extending such cross-modal binding to the robot involves the robot identifying its own bodily rhythm. For example Cog's movement of the arm can also be identified by its rhythm and this poses an opportunity to bind it with its own visual signal. Cog is also aware of its proprioceptive feedback from it's arm, therefore making it a binding across three modalities of visual, acoustic and proprioception, as shown in Figure 9.14. This method led Cog to attain the cognitive capabilities of a 10-12 month infant, and may have initiated s-consciousness.

The Cog is also seen to understand the functioning of us human beings and our social functions, for example a person nodding his head in disagreement will correspond to an event with at least two modalities. Therefore from then on, a similar nod of head would help the robot understand the social function of disagreement, thus tying the event to a general notion of disagreement. Future prospect is to incorporate newer modalities such as tactile sensing and olfaction to make a robot more robust, dynamic and in better correspondence with human-like consciousness.

9.6.2.2 Nico's drumming

Circk et al. at Yale University use cross-modal states for the humanoid robot Nico, seen in Figure 9.15 to make it beat the drum, following human actions.

Nico has been iconic to studies in cognition, internal states and self consciousness, and in an experiment in 2008 Gold et al. tried to teach it language using a basic vocabulary [129] and 2012 Hart and Scassellati tried to make the robot recognise its own image in a mirror, and thus pass the mirror test and also to use mirror reflections to interpret other objects in its environment.

In the last chapter the 'drumming mate experiment' on Kaspar was discussed. This experiment marks a robot's cognitive ability to detect, understand and to cooperate with human beings quickly and accurately and to adapt to a real time response of a complex activity as rhythm synchronisation and therefore demonstrate cognition in social activities. For these reasons this has become a benchmark for cognitive studies in robots.

In the drumming experiment, Nico is programmed to play a drum in a performance with human drummers at the direction of a human conductor, by integration across three modalities: visual, auditory and proprioceptive. As shown in Figure 9.15(b), the vision module detects and monitors the arm motion of the human drummer or conductor, and the audition module detects the drum beats from both the human being and the robot. The perception fusion prediction module employs learning analysis and prediction to generate arm commands. Nico can follow a human drummer or take instructions from a human conductor and delivers a precise synchronisation with the human performers. Nico is aware and evaluates the outcomes of its own actions in near real time and therefore can adapt to brief changes in rhythm and tempo, much like a seasoned drummer.

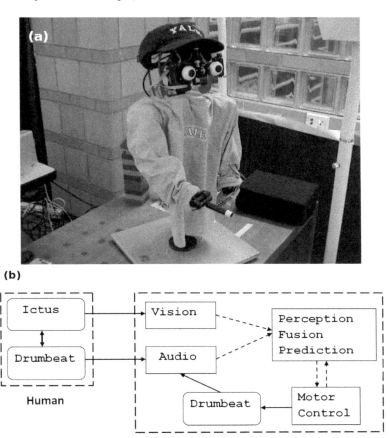

FIGURE 9.15 **Nico at Yale University.** (a) the robot and (b) the control architecture for drumbeating. Nico's image is from Gold et al. [129], used with permission from Elsevier.

The vision module monitors the hand motion of the conductor and looks for those instances when the conductor's hand moves away, indicating the ictus of a beat. The robot is exposed to three streams of information, the ictus detection, motion commands for the arm and drumbeat detected by the auditory module. In the perception fusion prediction module, the ictus motion is causally related to its sound, arm motion is compared to the sound, and then checked for the synchronisation obtained by the robot and attunement towards a tempo is extrapolated to provide some prediction for the robot. The architecture is sufficiently dynamic and Nico learns to compensate for errors, lack of information and time delays, and the researchers provide Nico with a 'warmup period' where it can play on its own and learn prior to a ensemble performance. The shortcomings of the above approach are, in dark settings such as minimal lighting on a stage the vision module may not be able to focus so well on the conductor's hands. All drums seem the same to the robot and uncharacteristic loud noises can confuse the robot.

These experiments demonstrate that cross-modal sensory integration can lead to precise synchronisation and therefore a realisation of causality of an event[6] and hence predict its outcome.

9.6.3 Mirror cognition

Gallup [59, 114] found that on introducing a primate to a mirror, it responded as thought the image is another animal. After prolonged exposure for a few days the chimpanzees were able to recognise their own images while the monkeys failed to do so. His experiments also demonstrated that mirror cognition improved with incremental learning in chimpanzees.

It is known that chimpanzees have DNA material about 85% to 95% similar to human beings. Therefore the mirror cognition test sets the tone to assess for self awareness in AI agents. Human babies cannot pass the mirror test until they are around eighteen months old, and in animals only primates, elephants and dolphins have been able to clear the test.

Lacan [196] was guided by similar results and concluded that human infants pass through a stage in which an external image of the body, as seen in a mirror or shown as a photograph, produces a psychic response augmenting the rudiments of "I-representations". Lacan believed that the mirror test is a milestone to confirm cognitive development. Vision is unique and is more advanced than the other five senses, therefore it is desirable to augment the "I-representations", using vision — a mirror to show the "I" to itself.

For Gallup's chimpanzee, self-awareness is confirmed by detecting self-directed actions in front of a mirror, while for a robot this is an enactive phenomenon — an overlap of both the creation and the detection of self-awareness in near real time, which will be discussed later in the chapter in the experiment with three Nao robots. It is well established that mirror neurons [176, 277] are integral to human cognition and lead to mimicking the behaviour of another person, as though the observer was engaged in the same action. This aping triggers such mirror neurons and augments the "I-representations". Neurologists further claim that mirror neurons are the physiological mechanism for the perception/action coupling, encoding-in the meanings of actions.

9.6.3.1 A very simple implementation of the mirror test in robots

A trivial implementation of the mirror test is shown in Figure 9.16, consider that the face of the robot has one blinking LED for facial expressions, and a photo diode for the eye. In ideal conditions when the ambient light is not of appreciable influence, the reflected light from the robot's face will cause different output states than a similar external blinking LED. As shown, the LED is supplied with a pulse generator and this blinking light ($\mathbf{T}$) is reflected ($\mathbf{R}$) to a detector in the robot's eye. If T and R corresponds — as AND logic ($\mathbf{S} = \mathbf{R} \cdot \mathbf{T}$) — then arguably, this is a pass for the mirror test.

However, this would be mere detection of the robot's image, thereby reducing the problem to that of signal processing rather than of cognitive deduction of the self, nor would

---

[6]I am tempted to conclude that binding over three modalities is unique and may be the simplest scenario of artificially self aware states in a robot, since binding over two modalities is a common occurrence in sensor integration, particularly in mobile robots and doesn't lead to self aware states. However, I have not come across such literature and being qualitative human experience I am not sure a rigorous mathematical proof is possible.

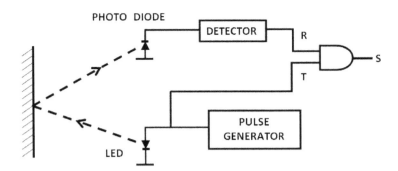

FIGURE 9.16 **Mirror cognition with AND logic**, reduces the test of cognition to pattern/signal recognition and therefore it is not a good experimental set up.

this generate internal states of the self. Further, unlike Gallup's chimpanzees, recognition of the self will be more of a instant detection and not a gradually acquired ability.

9.6.3.2 Haikonen's somatosensory model

Across multiple modes, information conveying the same representation helps to develop and augment emergent internal states of consciousness. An example which solves the symbol grounding problem is, when the word 'apple' is uttered it would not mean much to the robot until at about the same time a physical apple is placed in its vision, and therefore the information is augmented over the neural network. At each later instance the robot will be able to relate the language to the physical aspect.

Haikonen [137], proposes a cognitive architecture which considers the flow of information across three information channels or modalities, auditory, visual and cutaneous inputs as shown in Figure 9.17, where the cutaneous comprises touch and facial features (T and F) both of which help to ground self-related representations. Each percept is augmented with at least one more percept generating a feedback loop and three such feedback loops working in tandem for the cognitive processing of the robot. The signal SP is generated when the robot is touched or it detects a face expression. The process of recognising the self is demonstrated across three temporal steps as discussed below.

1. The robot is touched, which generates touch signals T, which will enable a somatosensory percept, SP. At the same instant the robot's name is called out, generating an auditory percept AP, thereby associating the two percepts. Thus the name of the robot is associated with the tactile feel of the physical body of the robot. Somatosensory percepts can evoke the auditory signal of the name and in the same way, the auditory signal of the name can evoke generalised somatosensory percepts as if the body were touched. The name, heard by the robot would shift the robot's attention from the external world to its own self and cause it to develop its "I-representations".

2. The robot is placed in front of a mirror, and is visually able to detect its own face, but it will not be able to recognise it as its own. The robot changes its facial expression, generating a somatosensory percept SP. This also enables a visual input, therefore a signal VP. A direct association of SP and VP is avoided as the visual world contains a number of non-relevant features and may not prove useful for developing "I-representations", however the timing of these two processes will help towards

association and thereby grounding in the visual mirror image to the somatosensory percepts.

3. The robot sees its own mirror image, evoking somatosensory self concept SP, thus evoking the name percept AP by the association in step 1. The concept of "I" is complete by a name, an image, a tactile feel of the physical body and the facial aspects of the robot.

This helps to develop internal states for "I-representations" and it also reflects a unity of consciousness over the three sensory inputs.

9.6.3.3 Designing the mirror test with MoNADs

Takeno et al. at Robot Science Institute at Meiji University define consciousness as consistency of cognition and behaviour in direct extension of Husserlian philosophy, and not as an emergent phenomenon as is suggested in behaviourism. The mirror cognition is designed using a neural module, Module of Nerve Advanced Dynamics (MoNAD) [220, 315, 316], which is a connectionist architecture and can be related to a number of functions in human cognition.

Each MoNAD, as shown in Figure 9.18, is made of two crossing recursive neural networks and strongly ties both the cognitive and the behavioural aspects of consciousness, binding in both reason and feeling. MoNAD work on the current as well as information at the last recursion on behaviour and cognition, therefore they are not simple first-order systems but rather complex second-order systems. Therefore, immediate previous experience is retroactively utilised to develop future behaviour and cognition, thereby maintaining consistency and learning from previous experience across both behaviour and cognition. As shown below, in the MoNAD system, the primitive representation, such as spatial information, time etc. supplies the basis for cognitive representation and is augmented by behaviour representation. Stimuli is processed by the sensor to primitive representation and is primarily cognitive representation, since MoNAD is not a behaviourist system. The behaviour, due to the cognitive representation and/or by an external influence, is processed to the primitive representation and through to actuation. Since, cognition and behaviour are neurally connected, each future instance of this state of cognition will push to perform the same behaviour. MoNADs are a connectionist paradigm are thus not plagued by the frame problem. They can solve the symbol grounding problem by learning the relation between the environment and the representation and it mostly solves the binding problem as cognition and behaviour are strongly tied to each other. It can be suggested that human cognition is via conjunction of a number of MoNADs as shown in Figure 9.19, where the cognitive (dotted lines) and behavioural (solid lines) processes overlap neurally to relate the outer world with the inner world.

To design mirror cognition, three MoNADs are used: Imitation MoNAD, Distance MoNAD and Settlement MoNAD, as shown in Figure 9.20. Since behaviour of the Khepera is more or less motor action, the Imitation MoNAD is a reasoning system and interprets the behaviour of the image and steadies the behaviour of the motor accordingly. The Distance MoNAD works on qualitative information and keeps a tab on the distance between the object and the image. Too little and the robot has to back up; and too much and it has to advance forward. The Settlement MoNAD is an association system though not a central control. The blue LED lights up when the behaviour is replicated. The LED lights up when the motor behaviour of the robot and its image are the same. Experiments conducted as shown in figure in Figure 9.22, (a) robot-mirror as shown in Figure 9.21, indicate the robot

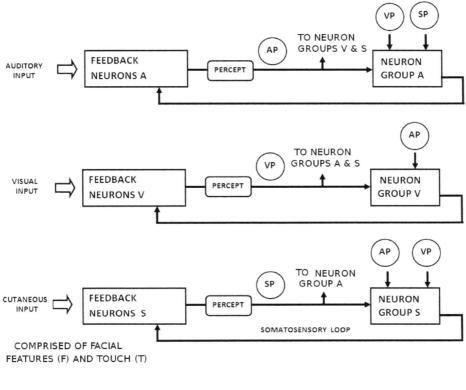

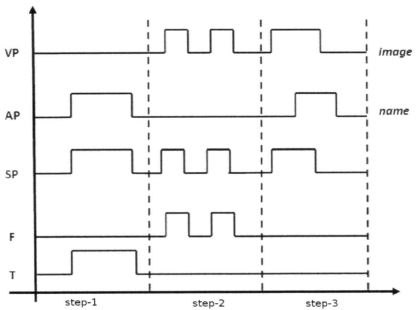

FIGURE 9.17 **Haikonen's somatosensory approach to the mirror test.** knowledge of the self occurs over three modalities: touch, voice and vision. In the three feedback loops, the auditory percept (AP) is coupled up with cutaneous (SP) and visual percepts (VP), the visual percept (VP) coupled with auditory percept (AP) and cutaneous percept (SP) with visual (VP) and auditory percept (AP).

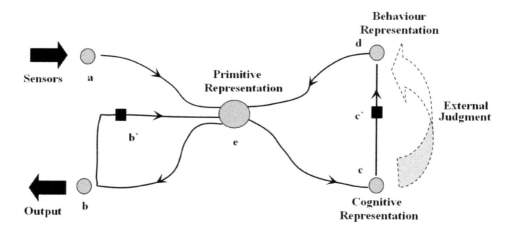

FIGURE 9.18 **The concept of a MoNAD.** A MoNAD is composed of two crossing recursive neural networks for behaviour and cognition, thus interleafing experience and behaviour, and both are connected to the primary representation corresponding to the somatosensory nerves. Every behaviour is intricately related to cognition.

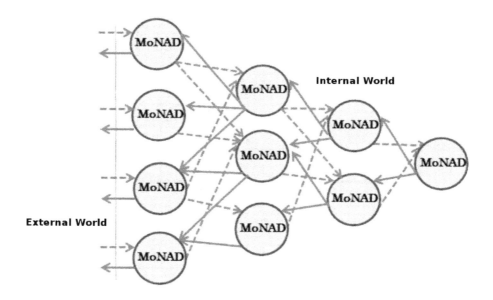

FIGURE 9.19 **MoNAD as the bridge between inner and outer worlds.** Here the dotted lines convey the process of cognition and the solid lines convey the process of behaviour. Behaviour and cognition work in tandem. Manner and Takeno have suggested MoNADs as the functional structure of the human brain, and this approach is able to explain emotion, cognition and consciousness, and also confirms their overlap. Image taken from Takeno [316], reproduced with permission from the publisher.

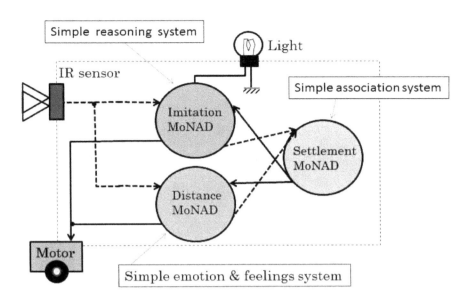

FIGURE 9.20 **Schematics for mirror cognition.** Takeno designs the mirror cognition with three MoNADs, Imitation MoNAD, Distance MoNAD and Settlement MoNAD. Image taken from Takeno [316], reproduced with permission from the publisher.

imitates the action of its own image. This experiment gave a coincidence rate at about 70% and in (b) robot-robot as shown in Figure 9.22, both robots are equipped with the same hardware and software, but are individual units not controlled by a central executive nor wired to each other. Each imitates the other as in when one lights up the LED, the next one follows etc. This experiment gave a coincidence rate of about 50%.

Mirror cognition is probably the simplest test of experiencing the subjective self, and has become a benchmark experiment for self consciousness. As a newer variant, Govindarajulu at Rensselaer Polytechnic Institute develops the mirror test for formal logical systems. In the reformulation, a little dab of coloured paint is smeared on the forehead of the AI agent when it is in sleep mode. After the agent wakes it can look into the mirror and attempt to remove the smear. Then the agent is self conscious. This formulation of the mirror test was tested in simulation, and led to Bringsjord and Govindarajulu formulating the Knowledge Game and its implementation in the Nao robot, discussed in the next section.

Mirror, Mirror ...

Mirror cognition has been a cornerstone in developing self-awareness in robots. The method derives from studies done in primates, where over a graduated timeline chimpanzees demonstrated the ability to recognise their own selves in the mirror while monkeys failed to acquire this higher level of cognition. Using a mirror to trigger robot behaviour was first employed by W.G Walter in the iconic turtles, Elmer and Elsie.

Dr. Junichi Takeno at Meji University has developed the mirror cognition test as an overlap of behaviour and cognition using MoNADs. Here he shares some of his thoughts. This mini interview took place in early December 2015.

FIGURE 9.21 **Takeno's mirror cognition test.** to the left is the mirror cognition experiment and to the right is Dr. Takeno and the Khepera robot

AB: For a young robot enthusiast, the definition of consciousness and therefore artificial consciousness has always been a difficult one. How would you define it, say for a graduate student?

JT: Human consciousness originates from each person's subjective feelings. Therefore, understanding human consciousness is a very difficult task, as you know. However, just as many people will say that the moon is somewhere up there in the nighttime sky, of course. The existence of the moon originates from the subjective feelings of each person. But now we have a strong certainty about the actual existence of the moon because the Apollo astronauts brought a lot of moon rocks back to Earth. When Galileo Galilei improved on a telescope and turned it toward the moon, certainty about the existence of the moon increased at that time. Scientific research begins when a researcher feels some subjective feelings about the object of the research. The story of the existence of the moon is the same as the subjective feeling about the existence of one's own consciousness. And naturally we already have a huge body of knowledge about human consciousness from philosophers, psychologists, psychopathologists and brain scientists. We should not neglect this knowledge. I thought about what a definition of the core of human consciousness could be. When thinking about the core of human consciousness, we must engage in philosophical thought because rational thought must be involved in finding the core of consciousness in light of the many different various research studies about it. By the core I mean the principle. That is to say, it is better to be able to recognize that the moon exists, rather than that the moon does not exist. This is a better rational thought at this moment. Various research studies have pointed out that the core of consciousness is the relationship between human behaviour and the human cognitive process. I recognize that I am doing a behaviour. And I am doing the behaviour when I am thinking. And those two states are synchronized with each other at that moment. In other words, my cognition and my behaviour are being done at the same subjective timing of my system. My research group and I decided that the core of human consciousness originates from the consistency of cognition and behaviour. We created a program for this core concept using a recursive neural network which we designed and call the Module of Nerves for Advanced Dynamics, or MoNAD for short. Then we conducted mirror image cognition experiments using a small robot running a program comprising these MoNADs. We thought that if the program using these MoNADs were able to be implemented on a mechanical system as a core of human consciousness, then the program could be expected to provide an effective representation of mirror image cognition in human consciousness in our experiments. Our experiments on mirror image cognition were not a direct repetition of the mirror test performed by the psychologist Gordon Gallup. Our point of focus was that most people can recognize their own image in a mirror. Why can we do this so easily? This mystery had not yet been fully understood. This is a manifestation of human consciousness, and it is just self-consciousness because we are checking our own body using a mirror. This is directly connected to the mark test, as you know. Therefore, we started building a program with MoNADs. And our experiment was targeted at learning the reason why we are able to easily recognize our own image in a mirror. On a dark evening in the German city of Freiburg in 1989, I was walking along the street with my family when I suddenly had the feeling that another family was approaching us from the right. But I recognized immediately that the other family was not someone else because I felt that their movements were the same as our own. I felt that it was just my family and me. The windows of a shop on the street acted as a large darkened mirror, and the image of the family in the window was just a reflection of us. And that experience got me thinking that mirror image recognition is very related to a comparison of ones own movements and the movements of others.

AB: How conscious is a robot that can detect itself in a mirror, as in your experiments? Of course, this robot probably cannot get a pail of water to douse a fire, so are we looking at levels of consciousness and mirror cognition as an apparent ground level state?

JT: My research group and I created a conscious system comprising a number of connected MoNADs. The objective of this system was to have a robot perform imitation behaviour. We calculated the value of the coincidence between the robot's own behaviour and the behaviour of an object located in front of the robot. We conducted three types of experiments. In the first experiment, the object presented in front of the self robot was its own image reflected in a mirror. The coincidence rate was 70%, which is not so high, because the performance in a mirror is imperfect in normal environments. These were, of course, objectively calculated results because the processing was being done by the program using input values from sensors. But you must also recognize that the values represent the subjective evaluation of the robot. In the second experiment, the object presented to the self robot was another robot which had the same body and used the same program. The coincidence rate was 50%, which was very low, because the other robot existed in a different environment from the location of the self robot. And I should point out that the other robot behaved independently, so it cannot be said that the other robot had a relationship with the self robot. In the third experiment, the object presented to the self robot was another similar robot which was connected to it by a communication cable. This connected other robot was controlled by the movements of the self robot. If the self robot moved forward, then the connected other robot would be commanded to also move forward immediately. The controlled other robot was simpler than the self robot because the program of the self robot commanded the behaviour in the other. The coincidence rate was 60%. This result was very interesting because the rate was fairly in the middle of the results of the preceding two experiments. The coincidence rate of the mirror image robot was higher than the value for the experiment using the controlled other robot. And the controlled robot was just like a physical part of the self robot because the controlled robot was connected by a cable from the self robot and controlled directly by it, much like the hand of a self robot might be. This experiment demonstrated that the mirror image robot could be interpreted subjectively as a part of the self robot. For that reason, we declared that our experimental results indicate the most powerful evidence for the grounded existence of the phenomenon of human mirror image cognition so far. Because the robot knows the existence of its own body, the robot can discriminate between the body of its self and the body of an "other." Therefore, it is natural that the robot would not do a behaviour that would injure a part of its own body.

AB: Can your method of mirror-based cognition, be extended to more complicated robots, such as humanoid robots? Will it need any modification in those cases?

JT: We have not encountered any obstacles to that end as yet. Thus, my research group and I will try to build a larger robot like a conscious humanoid. Also, deep learning methods will enable us to accelerate our efforts quite a bit.

AB: There has been concerted effort from the AI community to develop artificial consciousness in robots, over nearly the last 10 years right from methods suggested by Haikonen (2003) and Floridi (2005). However, either, (i) artificial consciousness has been developed in a very narrow concern such as mirror cognition or (ii) it has been very short lived, as seen in experiments conducted at Rensselaer Polytechnic on the Nao robot. According to you, what are we missing? What should the research community target next?

JT: First of all, I would like to express my deep respect for them and their research efforts. My viewpoint regarding cognitivism is that of a robot scientist, and I believe in my own subjective feelings. Therefore, I assert that I am thinking, I have emotions and feelings, and I have consciousness. And I do not doubt that I will able to build subjective phenomena into a robot just by using neural networks in a computer program. Of course, subjective feelings include a lot of illusions. Nevertheless, we must believe in subjective feelings at the first stage. If we do not do so, we can absolutely never discover the illusions within those subjective feelings. I felt that in Freiburg. Lastly, I would like to point out that philosophical thinking is the most important scientific act. When I discuss the philosophy of human consciousness with scientific researchers, most of them look upon me with some distrust and as though they would like to stop the discussion. I feel that they just want to discuss the results from scientific tools, and I enjoy talking about the results too. But, it is important to note that tools detect tendencies in an object. Scientists must learn about the core of an object using these several tendencies as well as from the descriptions made by pioneers in the field. This unified approach needs philosophical thinking. If we can accurately learn about the core of an object, all of the tendencies will be able to be explained by the core, and new hypotheses can be created. If a researcher cannot create a new hypothesis that can expand the research target from their research results, the researcher should recognize that they have not yet reached the central core of their research target.

AB: You have found inspiration from Walter, Braitenberg and Brooks; predominantly the reactive approach, but to model human-like consciousness there will be a need for some deliberative aspects. So, should artificial consciousness be modelled as a hybrid of reactive and deliberative architecture and supplemented with learning methods?

JT: The MoNAD structure that I have proposed is designed as a hybrid of reactive and deliberative architecture. And the MoNAD has self-referential functions. Our conscious system is constructed using several MoNADs connected in a hierarchy. The system functions by exchanging information along bottom-up and top-down paths. Because a MoNAD has a double-looped cognitive function, it is possible to calculate the MoNAD's feelings. For example, if the convergence speed of the MoNAD is slow against a given input, the feelings of the MoNAD will be unpleasant. The opposite represents a pleasant state. Our conscious system can detect unknown information because the unknown information makes the MoNAD decrease the convergence speed, and it can learn the information. My research group and I are currently using our conscious system to study the modelling of human mental illness, like post traumatic stress disorder (PTSD).

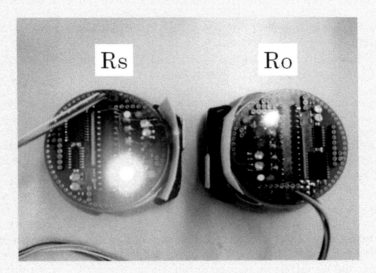

FIGURE 9.22 **Robot-robot mirror cognition with MoNADs.** The second robot mirrors the behaviour of the first one.

9.6.4 Psychometric AI — The knowledge game

Floridi designs a test which attempts to answer the question, "how do you know you are not a zombie?" which in hindsight is a test for self-consciousness. Being similar to the Turing test, the Knowledge Game is a variant of a popular puzzle, "The King's Wise Men" and is played by a system of interacting agents, which possess the ability to communicate and infer.

The test is subject to three types of agents: robots, zombies and human beings. Each agent can be characterised by a state S, such that acquiring a state S is something different from being in it and different again from knowing that one is in it. The states are parameters

I DON'T KNOW   I DON'T KNOW   RED ! ...

FIGURE 9.23 **Knowledge game, classic version.** Knowing the states of the previous two prisoners, the third can determine his own. Cartoon made by the author, CC-by-SA 4.0 license.

chosen by the experimenter, who later queries the agents about their states. The agents can determine their states inferentially and only on the basis of the information made available and not on innate, a priori knowledge or particularly privileged information.

1. Robots, conscious-less artificial agents (endowed with interactivity, autonomy and adaptability)
2. Zombies, conscious-less biological agents (lacking in p- and s-consciousness)
3. Human beings, conscious biological agents (endowed with e-, p- and s-consciousness)

There is no particular classification for conscious artificial agents — no conscious robots. Floridi designs four variations of the Knowledge Game.

1. **Classic version**: A prison guard challenges three prisoners to a game. In the game, the prisoners are made aware that there are 5 fezzes, 3 red and 2 blue. The guard blindfolds each of the prisoner and makes them each wear a red fez and hides the remaining fezzes. On removing the blindfolds, each prisoner can see only the other prisoners' fezzes and not his own. The object is to tell the colour of his own fez correctly, on being questioned one at a time and thus walk away to freedom. The first prisoner checks the fez of the other two and declares that he doesn't know the colour of his fez; the second prisoner has heard the first one and checks the fez of the other two and also declares that he doesn't know the answer; the third prisoner, who has heard both the previous two prisoners replies even before his blindfold has been removed, "My fez is red", as shown in figure 9.23. The third prisoner is correct and is freed by the guard. How did the third prisoner know?

    To analyse the problem, the exhaustive number of solutions of the problem are eight:

    (1) first prisoner - red, second prisoner - red, third prisoner - red
    (2) first prisoner - red, second prisoner - red, third prisoner - blue
    (3) first prisoner - red, second prisoner - blue, third prisoner - red
    (4) first prisoner - red, second prisoner - blue, third prisoner - blue
    (5) first prisoner - blue, second prisoner - red, third prisoner - red
    (6) first prisoner - blue, second prisoner - red, third prisoner - blue
    (7) first prisoner - blue, second prisoner - blue, third prisoner - red
    (8) first prisoner - blue, second prisoner - blue, third prisoner - blue

    Here, (8) is not possible as there are only 2 blue fezzes in play. Now, considering the first prisoner, a correct answer is only possible if both the other prisoners are wearing blue fezzes, since he says that he doesn't know the colour of his own fez. Therefore (4) cannot be possible. With similar arguments for the second prisoner who has also heard the first one speak and also says that he doesn't know the colour of his own fez, (6) cannot be possible. Also if the other two are wearing a red fez and a blue fez, then the second prisoner himself can only be wearing red. Therefore (2) is also ruled out. Of the remaining solutions (1), (3), (5) and (7), the third prisoner is always wearing a red fez — therefore his emphatic answer.

2. **Bootstrapping version**: The prisoners are told to choose from five pairs of boots. All the boots look identical, but the three worn by the prisoners are either torturing instruments which crush the feet or are ordinary boots. On being questioned about which boot they are wearing, the prisoners answer immediately and correctly.

3. **Deafening version**: The prisoners are given to drink from five glasses. They are told that three contain water and two contain a totally-deafening beverage. On being asked what he drank, the prisoner answers correctly. It is to be noted that if he has had the totally-deafening beverage then all he will see is the guard shouting, and not really hear the question — the question, though not heard, will yield the correct answer. Therefore this is a self-answering question. From a test perspective, current day robots fail to identify self-answering questions, and this makes for a distinguishing feature between zombies and human beings.

4. **Dumbing (or Self Conscious) version**: As a final variant, the three prisoners are offered five pills. Three are completely innocuous placebos and two make the agents totally dumb. On being asked, either the prisoner is silent, as he is totally dumb or he answers seemingly rhetorically with a phrase such as "heaven knows", as he himself is not yet aware of his state. Here Floridi argues that there can be two outcomes:

    (a) His rhetorics trigger no further reaction
    (b) His lack of awareness of his state triggers a counterfactual reasoning, therefore reasoning that had he taken the dumbing pill he would not have been able to speak out. However since he has, he did not take the dumbing pill and therefore revises his previous answer.

    According to Floridi, the second outcome will always be a true forte of human beings as neither robots nor zombies will be able to bind together and causally relate voice and hearing to conclude hearing his own voice. Therefore, the dumbing version is a query for s-conscious agents, and answers the test emphatically, " I am not a zombie".

Therefore, (1) the classic version relies on the nature and number of possible states, the observable state of the other agents and the answers from the other agents, (2) the bootstrapping version inquires into a first-order effect and relies on the assigned states, (3) the deafening version is self-answering as the mode of communication is speech and at the same time it is detaching from the question itself as the agent may not be able to hear the guard and (4) the dumbing version cannot be ascertained in terms of externally inferable, bootstrapping or self-answering states, but it may incite 2 modalities, voice and hearing — the causality confirming self awareness.

As a corollary to the puzzle, Floridi argues that self consciousness suggests phenomenal consciousness, so a human being can always be zombie-like but a zombie possesing only phenomenal consciousness will never realise the self.

Bringsjord suggests that the last variant of the Knowledge Game, with the agent realising that he has taken the dumbing pill can be met by AI beings, thus 'passing' the human test for self-consciousness, as is shown in Table 9.3. The proof was designed on the idea that the

FIGURE 9.24 **Passing the knowledge game.** Bringsjord's experiment [48], designed on the lines of the dumbing version of the 'Knowledge Game' is probably the best archetype for self awareness in robots. Cartoon made by the author, CC-by-SA 4.0 license.

TABLE 9.3 Knowledge game, Floridi vs. Bringsjord

	Floridi (2005)	Bringsjord (2010)
Classic version	Pass for all	Pass for all
Bootstrapping version	Pass for all	Pass for all
Deafening version	Fails for current day robots, may pass in the future	Fails for current day robots, **will** pass in the future
Dumbing version	Fails forever for robots and zombies — an archetype of human consciousness	Fails now for robots and zombies, **will** pass in the future

prisoner - as an AI agent will be able to perceive its own voice through some combination of sensorimotor feedback as the vibrations of its robot larynx, and other perceptual processes that fuse the relevant sensory input to form the perception that an utterance has been made, and accordingly updates it knowledge base to prove the conjecture[7].

In the first half of 2015, Bringsjord and his team at Rensselaer Polytechnic Institute modelled the psychometric test for humanoid Nao robots [48] as is shown in Figure 9.24. In the experiment two Nao are given given the 'dumbing pill', while the thirdis given a placebo. The pills are administered not as a real pill, but with a tap on the head, and then they are asked which pill have they been given?, after about 13 minutes the Nao, which had been given the placebo gets up and says, "I dont know" and soon realises that this is its own voice and then after a short pause speaks out its conclusion. In this experiment no concept of self awareness was pre-programmed than the knowledge of the dumbing pill. The Nao robot attains sentient behaviour for a very short time and it is able to identify its own voice and therefore attain selfawareness and thus solve the riddle - demonstrating association over sensory input and reasoning.

Summing up, a good engineering solution to self awareness will be characterised by, (i) central executive, (ii) binding over three or more modalities, (iii) imitating, aping and mirroring[8], (iv) connectionism and (v) attention switching. Consciousness should appear as a learned and superior faculty once self awareness has been tuned to perfection. Continued interaction with the environment should aid to develop binding across various modalities and therefore phenomenal consciousness.

Most research has been guided by cross-modal association over neural networks, and the 'mirror test' and the 'drumming mate' tests have become useful tools for benchmarking and moving the robot's cognitive and social functions, as shown in Table 9.4. It is interesting to note that the focus has been on designing experiments to test for self-awareness in robots, but there has also been gathering enthusiasm to devise more sophisticated versions of the Turing test, which tap unique abilities and tendencies of being a human being. The quest to find conscious robots has furthered the onus in the search for unique human values which machines lack, and instead of asking 'what is a robot?' the question to ask has been 'what is a human being?'. Will a conscious robot be better than us? That is a difficult question, and the least that can be said is that conscious robots are sure to be smarter than we are. We stand at a unique place in the history of our civilisation, and also the planet as we are the

---

[7]Bringsjord uses Deontic Cognitive Event Calculus (DCEC*) to structure both the mirror cognition and the knowledge game, it is not in the scope of this book to discuss details of DCEC*.

[8]Not necessarily needing a mirror for the process of mirroring.

TABLE 9.4 Development of Robot Consciousness via Experiments

Research	Paradigm	Experiment	Remarks
Braitenberg (1986)	Meditative mode to form association	Gedanken experiment, vehicle-12	From synthetic psychology
Brooks et al. (1994–1998)	Reactive approach	Humanoid bust, COG	Solution to the symbol grounding problem
Kitamura et al. (1995–2006)	Levelling of behaviour, inhibition triggers a higher level	Low-level tasks as navigation, tracking etc.	In both simulation and real robots
Fitzpatrick et al. (2005)	Cross-modal association	Binding to develop causality	Cog attains awareness of objects and the self
Crick et al. (2005)	Cross-modal association	Nico beating the drums	Can adapt in real time with the human performer
Takeno (2005)	Consistency of behaviour, not an emergent phenomena	MoNADs to design mirror cognition	Both for mirror-robot and robot-robot
Haikonen (2008)	Augmenting mirror cognition with proprioception	Binding in visual, cutaneous and auditory inputs	Manipulates attention towards its self-concepts
Bringsjord et al. (2014)	Causality of speech, auditory input	Puzzle solving in simulation and real robots	Sensorimotor route to solve the knowledge game

first species which has dared to create intelligent beings designed to model ourselves. We are soon to see a society in the making which has high levels of automation, and if human beings and robots coexist, will we be reduced to second class citizens? Will the robots take over the workings of the various systems of society, such as government, the food supply, financial systems etc. These questions are addressed in the next chapter.

SUMMARY

1. Artificial consciousness has the promise to be implemented in robots to make them more human like
2. The semi-sentient paradigm has led us to 'obedient' social robots but they lack in imagination, free will and sentientism.
3. Consciousness lacks a definition and robotics has been led on two ways of thoughts, (1) emergence of novel behaviour and (2) consistency of behaviour.
4. Qualia has been a squabble and is a philosophical pitfall for design of artificial consciousness.
5. Turing test is a quotient of the difference between man and machine, and newer versions has been designed.
6. Inner world of a robot is both an abstract concept and difficult to verify.

7. Newer experiments to develop artificial consciousness in robots have either been in a rather narrow concern or has been too short lived.

NOTES

1. *From a philosophical point of view, Kierkegaard [179] considered that human beings pass through three stages on the way to becoming a true self: the aesthetic, the ethical, and the religious. The aesthetic is experienced by sensory perception at a nascent stage of existence and is characterised by immediacy, whereas the ethical is adherence to the social rules — how one ought to act and the ablility to make and take moral responsibility and accountability. The third, religious, is the illumination to one's personal quest to find faith and belief. Drawing a corollary, three such distinct phases can be seen in the agent based robotics: the sensorimotor, moral agency and conscious robots.*

2. *Robert.A. Heinlein's 'The Moon Is a Harsh Mistress' features a computer, HOLMES IV, which attains self consciousness which shows up as a sense of humour. This happened as the computer was connected up with more hardware which augmented the temporary memory to reach cognitive domains which are well beyond the human brain. Of course this is not possible, but makes for wonderful reading.*

3. *Dennett plays down the debate of qualia with four characteristics: ineffable, intrinsic, private and directly or immediately apprehensible in consciousness, and therefore he contends against any non material route to consciousness. He suggests that, "We're all zombies" and hence, qualia is illusory. Dennett's Multiple Draft Theory (which is in the consciousness domain) bears similarity to Brook's Subsumption Architecture (which is in the the behaviour domain) and also to Hollan's model for artificial emotions (which is in the emotion domain).*

4. *Contrary to popular belief, Nagel's celebrated paper [249] doesn't coin nor mention qualia. Nagel develops his debate against reductionism with the 'subjective character of experience' and 'what it is like?'.*

5. *As the debate on consciousness has warmed up, the AI community has often crucified the philosophical zombie, by most ironically using it as an archetype in opposition to material models of consciousness. Chalmers [64] argues with the following four part syllogism:*

    (1) It is conceivable that there be zombies

    (2) If it is conceivable that there be zombies, it is metaphysically possible that there be zombies.

    (3) If it is metaphysically possible that there be zombies, then consciousness is non-physical.

    (4) Therefore, consciousness is non-physical.

6. *In Artificial Consciousness the easy problem is the behavioural change in response to the environment and the hard one is phenomenal experience and qualia. For development of artificial consciousness, researchers qualify distinction between the strong, weak and intermediate approach. Strong qualifies for design of systems that are conscious, therefore a computer suitably programmed is conscious. Whereas, a weak approach would be use of technology to understand consciousness, social robots such as Pepper and ROMAN which seemingly play on human emotions to imitate consciousness are examples of a weak approach. Between these two lies an intermediate which is built on the principle that mathematical modelling leads to an overlap of*

consciousness and technology. Such an approach is coined by Chrisley [69] as lagom AI, lagom meaning, "perfection through modelling" is placed midway and attempts at some of the necessary conditions for attaining artificial consciousness. Computational models for consciousness are a type of lagom approach.

7. *Chalmers has suggested that the human mind is a quantum phenomena, and should be tackled with quantum mechanics. Similar ideas have been suggested by Hofstadter.*

8. *Qualia registers uniquely for every human being, and is the bone of contention in development of Artificial Consciousness. An illustration is the celebrated Wittgenstein's "Beetle in a Box" paradox, which illustrates one's inability to experience the world from anyone else's perspective. The thought experiment is as follows: consider that everyone has a box where they keep a beetle, and only the keeper can look into their own box. Now everyone is asked to describe the contents of the box. Since no one can really know what's in any of the boxes but their own, "beetle" ceases to have any other meaning than "what is in your box". The box is an analogy for our mind, which assumes that the experiences of another person are very similar to our own, but that is just an assumption which may not and does not hold good under all circumstances.*

9. *The very fast (above 10 Hz) and the very slow (below 0.1 Hz) cannot be modelled into cross-modal experiments yet. Which means that though Cog can relate the vision to the audio of a tambourine, it will not be able to detect the buzzing of a bee or the rotation of the Earth.*

10. *The AND logic circuitry for mirror cognition is designed with microchip IS471F made by SHARP which is an electronic oscillator. The signal from the oscillator is used as as input to an external LED and a photodiode with a synchronous detector for the incoming light.*

11. *With more sophisticated Natural Language Processing (NLP), a robot may be able to solve riddles put to it in English language. This will be the next level of cross-modal representations.*

12. *Experiments have shown that social understanding is not a necessary precondition for mirror self-recognition test. Therefore critics claim that the mirror test may not be about "self-awareness", but more about the ability to deal with a new kind of visual feedback.*

13. *Concepts of consciousness is closely associated with our perception of reality. This duality has been expressed by Philip K Dick in his definition of reality, "Reality is that which, when you stop believing in it, doesn't go away."*

EXERCISES

1. **CATCH-22.** Incorrigibility is a manifestation of free will and conscious thinking. A 'stubborn robot' which doesn't attend to the task it is designed for since it lacks in intent, imagination and enterprise, will defeat the very purpose of making robots. Given such a CATCH-22 scenario, should research continue in development of conscious robots?

2. **Haikonen** suggests that the somatosensory model for self cognition will work even if instead of a mirror the robot is provided with a photograph of itself. Discuss the prospects and limitations of this change.

3. **Floridi's Knowledge Game.** In the classic version consider that the guard keeps all the five fezzes in play and randomly selects three for the three prisoners. The first prisoner checks the fez of the other two and declares, " I don't know" and the second

prisoner who has heard the first one and checks the fez of the other two, emphatically announces, " My fez is red". Can the third prisoner who has heard both the previous two prisoners be sure of the colour of his own fez?

CHAPTER 10

Super intelligent robots and other predictions

" Unlike our intelligence, machine-based superintelligence will not evolve in an ecosystem in which empathy is rewarded and passed on to subsequent generations."

– James Barrat [30]

"It has always been popular to bash AI because that puts one in the glorious position of defending humanity. "

– Luc Steels [310]

10.1   PEERING INTO THE CRYSTAL BALL

How will sentient robots change our lives? In comparison to human beings, they will have added virtues; they will lack in ego, greed and jealousy and will be guided more by logic and information at hand than anything else. They will only heed to energy — be it battery power, jet pack, organic fuel or robots powered by nuclear fuel. Since explorer robots are always the first to survey other heavenly bodies, a sentient robot may well be our best bet for an interplanetary colony [192]. The stark contrast will be a lack in emotional quotient, and as long as qualia remains intractable or at best unresolved, we are sure to find more of the zombie than a genuine replication of human beings.

Despite this prospected leap in technology, prominent scientists and technologists suggest progress towards a dystopian telltale [27, 215, 321] where AI will emerge as a powerful adversary to our race and lead to our extinction [165], and a race of super intelligent robots [94] will inherit the planet.

This apocalyptic future, where technological intelligence is a few million times that of the average human intelligence and technological progress is so fast that it is difficult to keep track of it, is known as 'technological singularity' or 'singularity'. AI scientists also relate this event to the coming of super intelligence [43], artificial entities which have cognitive abilities a million times richer in intellect and are stupendously faster than the processing of the human brain. The irony is such that nowadays the monikers of Terminator and Skynet [318], as shown in Figure 10.1, are quickly married into research and innovation in AI [185] and robotics [66], such as Google Cars [303], robotic cooks [295] and waiters [319], the OpenWorm project [216] etc. and has consequently led to fear mongering [255, 306, 355] and

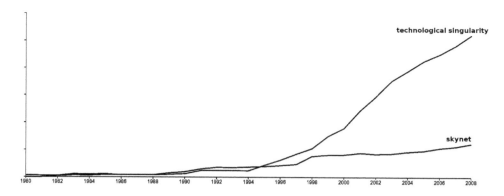

FIGURE 10.1 **The SKYNET moniker and technological singularity**. Since the movie series Terminator (1984), the SKYNET moniker has generated pop-fiction and has added to scientific debate and discussions on technological singularity. The moniker broadly is used as an illustration for a powerful AGI which, once created, steadily improves itself attaining near sentience and develops prohibitive technologies such as time travel and imitating living beings, which then either enslave or attempt to annihilate human civilisation. Made using Google Books Ngram Viewer [230]

drafting of guidelines [24,224,274], rules [364] and laws [60,342,348] to tackle this apocalypse of the future. These edicts attempt to restore human superiority by either reducing robots to mere artifacts and machines or tries to make a moral call to the AI scientist, insisting on awareness of the consequences. Therefore, advancing AI clearly sets the proverbial cat among the pigeons. Other than the media, science fiction is replete with such futuristic scenarios. Čapek's iconic play in the 1920s, R.U.R — Rossum's Universal Robots, which gave us the word 'Robot' ends with the death of the last human being and a world dominated by robots with feelings such as endearment, love and attachment. Other iconic tales of robocalypse and dystopia are, HAL set in 2001, Blade Runner in 2019, I, Robot in 2035, Terminator set in 2029, while Wall-E is set 800 years in the future in 2805. All of these provide examples of a futuristic human-robot society, and while nearly all of them are unsettling, all of them at the very least confirm a proliferation of AI and robots both in the near and far off future. It is interesting to note that in more academic concerns, Toda's fungus eaters are tagged to a sell date of 2061.

This chapter is different from the rest of the book and tries to look at such a future from the perspective of what has been discussed in the last nine chapters. This technology forecasting is done by retracing the predictions made by various leading thinkers, AI scientists, entrepreneurs and leaders of technology and by extrapolating trends from the present.

One of the earliest proponents of this dystopia was author and futurist Vernor Vinge [338, 339]. Vinge projected that in about the next thirty years, we would have created superhuman intelligence which will lead to the end of the human era. He suggested five possible paths by which we may reach this unsettling future.

1. **The AI Scenario**: We create superhuman artificial intelligence (AI) — computers which are 'awake'.

2. **The Digital Gaia Scenario**: The network of embedded microprocessors becomes sufficiently effective to elicit superhuman intelligence — a 'wake up' call for the network.

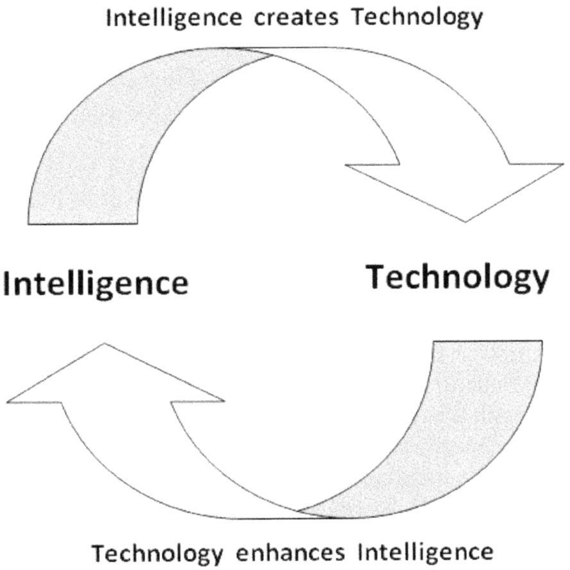

Intelligence creates Technology

Intelligence

Technology

Technology enhances Intelligence

FIGURE 10.2 **Technology intelligence cycle**. Particularly after the microelectronics (late 1970s), computer (mid 1980s) and internet (late 1990s) revolutions, our technology has come to be information based and is fueled by increasing intelligence, which in turn improves intelligence. Made by the author, CC-by-SA 4.0 license

3. **The Intelligence Amplification Scenario**: We enhance human intelligence through smart human-to-computer interfaces, which melds together the human mind with the AI's capabilities — that is, we achieve intelligence amplification.

4. **The Biomedical Scenario**: We directly increase our intelligence by improving the neurological operation of our brains — the route via brain scanning and genetics.

5. **The Internet Scenario**: Humanity, its networks, computers, and databases become sufficiently effective to be considered a superhuman being — the Internet of Things (IoT) coalesces to represent coherent intelligence. This is an extension of the Digital Gaia scenario, however over a network which covers the entire planet.

History tells us that technology has redefined social interactions, ethical values, governance, belief systems, political and economic paradigms, cultural values and therefore, society itself.

The best illustration is the industrial revolution and from the mid 1850s industry and its functions have become an integral part of our lives and nearly every single thing we use has been through an industrial process. Our economic growth and political choices are tagged with industry and most of us work for industries or one of its allied services, viz. marketing, finance, advertising, human resources etc.

Another such example is mobile phones. There was not much of a change in the telephone since Graham Bell's invention in the 1850s and as late as the 1970s. People talking across swathes of distance through a tiny little portable box seemed improbable. Over the last two decades, mobile phones have evolved into a technology, and with the advent of the smart phone and unification of the telephone with the internet, our social norms and value systems have also seen modification. For example, social acceptance is often cited in terms of followers, fans and views on social media websites, while political and social debates find their way to web forums and blogs. Services and utilities such as food vendors, movies and cabs are available over android/iphone apps and the hashtag has found an entirely new meaning.

Over the years, we have come to use our brain power a lot more than mere muscle power, which is a diametric transition from the Stone Age. This advancement is attributed to readily available information at our disposal. Compared to past societies our society knows more and as we advance in technology, this knowledge is responsible for wealth creation [122] and higher levels of technology. In the last four decades, technology has ascended to newer heights, due to the microelectronics revolution in the 1970s, the computer revolution in mid 1980s and the internet revolution in late 1990s, and most of present-day technology is information based, information which is automated by intelligent systems. With growing technology these intelligent systems, viz. clouds, servers and apps, are becoming more versatile and aid in the cyclic growth of technology and intelligence, as shown in Figure 10.2. It is foreseen that in the near future, technology will broadly consist of digital implements and more smart phone based (1) businesses such as Airbnb and Uber, (2) utility-based cloud apps as Dropbox and Google Drive tailored to specific needs, (3) online counseling for technology, education, law and medicine etc. (4) combination of mobile app and Internet of Things (IoT), viz. Amazon Dash Buttons and Starship, and not forgetting that (5) there will be the imitation of humanly content as seen with Alexa and Siri, ushering in a new trend for doing business. The Internet of Things (IoT) and other hardware implements (self driven cars, domestic robots, robotic butler etc.) have already been successful in research and are teeming to take the next leap in the commercial domain.

Allergy to Technology!

Electrohypersensitivity (EHS) is an ailment where exposure to electrical signals from devices such as mobile phones, computers, WiFi routers and cellphone towers causes consistent headaches, uneasiness, hair loss and rashes. Though the debate has not settled as to whether it is a true medical diagnosis or merely another allergic reaction happening rather arbitrarily. However, with 11 million people affected in the United States, it demands a discussion. Examples such as the National Radio Quiet Zone, which is devoid of cellphones, routers and antennas, goes on to show that technology fails to impress everyone.

It is proven that extended exposure to high power transmission wires leads to mutations leading to cancer. Medical practitioners have confirmed that continued exposure of the human body to radio frequencies and other electromagnetic waves may have adverse effect on us, EHS may be just a tip of the iceberg and more such syndromes and medical conditions could emerge in the near future.

Science fiction is replete with potboilers where robots take over the world for the greater good of the universe. Can such a devastation really happen? If current day robots stood up in an uprising, the outcomes would vary from the futile to the hilarious [243]. However with self aware sentient AI beings, the scenario can quickly change [265]. Consider sentient robots connected across a worldwide network or a cloud. These may soon proliferate to a legion of robots, where knowledge and skill acquired by one robot is easily 'downloadable' to the rest of the fellow robots and other AI beings. The interactions of such a new species will be more in the psyche than in real world. These sentient robots will become higher beings in a remarkably shorter time spans than natural biological evolution, evolving into Walterian creatures discussed in chapter 2.

It is hypothetical scenarios like these that led I.J. Good to develop models of ultra-intelligent beings. It is apparent that sooner or later our race will develop ultra-intelligent AI beings, and this will lead to an impasse where human labour and intellect will only be used for development and improvement of such machines and these machines will fulfill the needs and requirements of human civilisation, viz. construction of buildings, bridges and cars, developing political machinery, edicts of law, academia and production of food etc. Thus, these ultra-intelligent AI beings may be the last things that human civilisation will develop. Now, these beings with superior skill, higher deductive ability, better judgement and higher levels of understanding may well try to get rid of the inferior species of human beings. History has seen innumerable examples where a superior and technologically advanced civilisation has dominated and crushes the inferior civilisation. European discovery of the Americas saw the end of Inca and Aztec cultures, similar travesty resulted with Cook's discovery of Australia and the aborigine population was reduced to insignificant numbers.

After Good and Vinge, Moravec [241] suggested a grim future in which the 21st century will be dominated by robots and robot-based industry, human labour will be reduced to developing laws to maintain human values and social institutions, and also to discourage the transformation of human beings to unbounded super intelligent robots. The human race has a short lease on life and robots will sooner or later take over. Like Moravec, futurists contend that there are 3 likely scenarios for the next 100 years.

1. **A takeover by AI** — fully self aware beings take over the workings of the planet and human beings are doomed to extinction under the dominance of a superior species.

2. **Containment of AI** — laws and restrictions are put in place, forbidding the development of fully self aware AI. This has been cautiously suggested by Musk [110], Gates [95], Hawking and Shiller [72].

3. **A technology fueled Darwinain evolution** — this view is contrary to the previous two [193], where singularity is the 'next step' in a technology-fuelled Darwinian-like evolution, a future where human beings and machines meld together to move towards a higher species — though a machine take over is not ruled out. Many proponents of this option also forsee the coming of super intelligence and in the long run, the evolution of posthuman beings — superior beings which may be bred from human implements, but are no longer human beings — so human extinction, though in a subdued fashion.

These scenarios do not consider socioeconomic issues such as economic downfall, wars, terrorist attacks, nuclear annihilation nor ecological or cosmological catastrophes such as the greenhouse effect, meteorites and collisions with other heavenly bodies. One unflinchingly sketches the demise of the human race, the second and the third have been in research papers, tabloids, newspapers etc. time and again. The way we deal with new age technology will determine if our civilisation may be able to choose between these three options. The second option may work and we may be able to contain AI, however that will also stagnate any more technological growth and it will also undermine the development of fully sentient AI beings. The third option, where human beings and machines fuse together to propel Darwinian evolution at a much faster pace and the coming of super intelligence, is discussed in this chapter. To fuel the debate, Kranzberg's first law [188] would suggest that super intelligence is neither good nor bad; nor is it neutral.

Principles for designers, builders and users of robots (EPSRC [39], 2010)

Robots are multi-use tools: Robots should not be designed solely to harm human beings, except in the national interest

Humans, not robots, are responsible agents: Human beings are responsible, not the robots and robots should be operated as far as possible within existing laws, rights and personal freedom including privacy

Robots are products: Robots should be designed to be safe, and should be designed via processes which assure their safety and security

Robots are manufactured artefacts: Robots should not be designed to deceive human beings, and the fake expressions of empathy and emotions should not target user vulnerability

Legal attribution: There should be a human being legally responsible for a robot

---

10.2 TWILIGHT OR REBIRTH OF OUR CIVILISATION?

The promise of the future has not been entirely rosy, as machines have made human labour redundant and robots can do our job and often better than we can. Since the Industrial Revolution in the mid 1700s, efficient machines have improved productivity and lessened the requirement for human labour. In a world where human population is ever growing, a diminishing demand for human labour spells an apparent doom for the working class. As a communist principle, the erstwhile Soviet Union, the Eastern Block countries and Mao's China advocated against automation of the key industries of agriculture and mining as they provided the livelihood for the masses. With the fall of the Iron Curtain and globalisation in full bloom we have accelerated towards an age of automation, idiomatically defined by robots.

As of today, we have robots which can help in household chores, assist and conduct military warfare and defence systems, pursue interstellar explorations with or without human assistance, coordinate search and rescue operations to save human beings from hazardous scenarios such as nuclear catastrophe, automated surgeons and caregivers, automated instructors, office support staff etc. the robot revolution is very much on its way. It may be 5—10 more years before robots come out of the laboratory, get rid of the price tag of 6 digits and become mainstream.

Though human beings still supersede all known forms of intelligence when it comes to the question of intelligence vs. autonomy, as shown in Figure 7.1, it is nonetheless projected that over the next one to two decades, nearly 50% of all jobs in the United States, 36% of all jobs in the UK [25] and about 59% of all jobs in Germany [213] will be taken over by robots and smart automation. The market for personal and household robots is growing by 20% every year. What does that hold for the human populace? There are two viewpoints. The first one spells gloom and doom where reduced prospects for a job will lead to a modern day Luddite revolution, where swathes of impoverished and desperate masses destroy robots and machines in a futile effort to reestablish the hegemony of the working class. The second is that, with the development of Internet of Things(IoT) and cheaper means of 3D printing, the cost to develop things will reduce to near zero. Thus there will be plenty of anything and nearly everything, and a flash end to capitalism and the institution of new rules for the new age.

Robots are better than human workers since they never get tired, they're never irritable

FIGURE 10.3 **Reduced human job opportunities**. As an immediate effect of the robotics revolution [117], there will be marked unemployment, with middle income jobs dwindling out fast. Image courtesy NESTA (*www.nesta.org.uk*), designed by A3 Design (*www.a3studios.com*). CC-by-SA, NC 4.0 license.

or ill-mannered, they are not rude and don't use foul language, they don't have an ego, they have instant access to all of human knowledge and they never make mistakes [96]. This reads like a repeat of history [85], when the Luddites vandalised machines and industries in the early 1800s in protest to the labour-economising technology of the Industrial Revolution, as shown in Figure 10.3. More recently, human ego and superlative technology had a mock fight in a referendum in Switzerland, where the voters overwhelmingly (76.9%) rejected a proposal to automate nearly every facet of society, starting with every job, and provide an unconditional basic monthly income of 2,500 Swiss francs to every adult and 625 Swiss francs per child under 18, regardless of how much they work [187].

Various economists contend that modern technology doesn't lead to inequality, rather that social and political choices are instruments of this stratification. Modern technology tends to eat away the routine jobs typically meant for the middle class, such as bookkeeping, tele calling, construction and lathing, while jobs needing creativity, empathy, social skills and which are non-routine are firmly grounded in the human domain. The hardest hit will be administrative workers [108] such as secretaries which will be completely taken over by electronic processes and smart computer algorithms, while, as yet we do not have technology to make robots actors, robot nurses, robot investment bankers, robot museum curators, robot artists, robot singers etc. Thus this parasite of technological unemployment hits and erodes away particularly the middle class and this adds on to the disparity of the class structure broadening the gulf between the rich and the poor. It has been suggested that this may be a short-term phenomenon, and previously such occurrences of advancing technology have always created more job opportunities than those it has curtailed or compromised, but the current trend may be just too unique to predict and soon we will be left with a teeming

---

Roboticist's Oath (McCauley [224], 2007. Courtesy Springer, used with permission)

I swear to fulfill, to the best of my ability and judgement, this covenant: ...

1) I will respect the hard-won scientific gains of those scientists in whose steps I walk, gladly share such knowledge as is mine and impart the importance of this oath to those who are to follow.

2) I will remember that artificially intelligent machines are for the benefit of humanity and will strive to contribute to the human race through my creations.

3) Every artificial intelligence I have a direct role in creating will follow the spirit of the following rules:

   i) Do no harm to humans either directly or through non-action.
   ii) Do no harm to itself either directly or through non-action unless it will cause harm to a human.
   iii) Follow the orders given it by humans through its programming or other input medium unless it will cause harm to itself or a human.

4) I will not take part in producing any system that would, itself, create an artificial intelligence that does not follow the spirit of the above rules.

If I do not violate this oath, may I enjoy life and art, respected while I live and remembered with affection thereafter. May I always act so as to preserve the finest traditions of my calling and may I long experience the joy of benefiting humanity through my science.

---

population without a livelihood. Emerging economies of China and India which are labour intensive and provide for backend repetitive jobs will come to a standstill and to stave off a revolution, our society and governance may need a reorganisation. A way out of this is to stop, or at least slow down, this progressive use of automation. However that will lead to poor, or at best mediocre, industrial output and a low domestic product. Economists are still divided on a society which is governed by high automation and each citizen is paid a basic income or minimum wage regardless of work done. This disparity of divorcing work from wage is probably not the best way to charter the progress of a society.

Despite this social dilemma, a shift towards automation and robotics and automation is obvious. Amazon has acquired Kiva Systems for $770 million and is considering transportation and delivery systems using drones, Google has acquired at least nine robotics or artificial intelligence companies for sums totaling in the billions, which includes Boston Dynamics, and is considering launching the self- driving car as a commercial product and Musk has launched Powerwall, as the clean and free energy system of the future.

We, human beings, knowingly or unknowingly, tend to crown ourselves at the apex of the pyramid of all biological creations, and the contention is not entirely wrong as higher biological intelligence is yet to be found. This wishful Hyberborea[1] was reached with the unbridled success in technological growth piggybacking on fossil fuels. From hunter-gatherers to space explorers, our success in the last 15,000 years has put into question the pace of Darwinian evolution, and it is more than just a question whether nature ever witnessed

---

[1] In Greek mythology, Hyperborea was a mythical land which was supposed to be perfect in every manner, with the sun shining throughout the day and the inhabitants being perpetually youthful and happy.

FIGURE 10.4 **Transhumanism**. Equipping the human body with new age technology for well being and to extend life will become the norm by around 2040, and may be the first steps towards human robot integration to form androids. Cartoon made by the author, CC-by-SA 4.0 license.

intellectual growth at this rate? Recently higher acumen and greater ability than human beings has been seen in artificial agencies and they may well be more suited to carry the metaphorical Darwinian baton. Will that spell doom for the human race? Will we be second rate beings in a world dominated by AI agents?

The development of a failsafe where artificial intelligence cannot harm human beings has been an ongoing effort. The first example was in fiction — Asimov's laws and effort have been to rope in such laws, rules and moratoriums to contain AI and AI research. The debate takes a further twist when the intent of the artificial beings is considered. Until now all incidents of harm from robots have been mere accidents where the robot lacked sentientism and therefore did not harbour any wrongful intent; but that may not always be the scenario. At the other end are the suggestions of a code of ethical conduct for AI practitioners and roboticists, closely following the epithet of the Hippocratic Oath, synonymous with doctors.

Other than loss of jobs and social changes another prospective outcome of the technological and automation revolution is to devalue products. With the Internet of Things, 3D printing and a shift towards solar and wind energy, automation may herald an era of opulence [276], an era in complete contrast to the Skynet dystopia, which promises nearly free goods and services, efficient means of transportation, newer methods of growing food and the growth of collective opinion. These events and activities in parallel will eclipse capitalism and lead to the set up of a new age society. Societies make choices which lead to their failure or success [93]. We are also faced with such choices in regards to AI and modern technology. Pollack [266, 267] summarises the robot age with an open set of questions. It is in earnest that these questions are revisited to construct the apparent future in the backdrop of super intelligent robots and singularity and to forsee our roles and the choices that we might have.

1. **Should robots be humanoid?** The creation of a robot is strongly guided by anthropomorphism, and replicating human beings will probably be the greatest achievement of AI. The appeal of a humanoid robot is greatly due to it's appearance rendering it, a 'doppleganger' [7]. The ASIMO's dance or the PR2 fetching a bottle of beer connects at a psychological level, and makes us respond and react as though they were indeed human beings. The immediate future will have humanoids but they will lack sentience and at best will be expertly trained sales puppets or entertainment gigs. Pollock suggests that humanoids may lead to robot sales executives; pitching a product with enthusiasm, answering all of the consumer's questions patiently, finding common interests and finally arranging a better discount and amicable payment methods to get the deal. The down side is that just like spam emails, these sales persons may have a spam bot like avatar such as door to door sales, selling things which are of low utility and poor quality. As discussed previously, humanoids have played active roles in plays and ballets. With telepresence, the future may have robots on stage while putting the stage shy human actor in a more comfortable place.

2. **Should we become robots?** Modern-day medicine and surgery have led to embedded devices and wearables such as pacemaker devices, smart prosthetics, heartbeat and pulse rate monitors to be commonplace. Research and mapping of the human brain has found interest across the domains of AI, medicine, computer science and physics communities, and it may not be too long before human beings and computers are connected via direct neural interfaces — from mind to the computer to the internet, where instead of using senses as touch and vision, there is direct neural interface in both directions. Such an interface will embed virtual reality into our network of neurons, making the virtual and the real seamlessly interact with one another and render them indistinguishable [350]. The emergence of better technology will continue to lure us to incorporate electronics embedded into our biological systems as means to monitor and enrich our metabolic processes, extend our longevity and add to more mind power, as shown in Figure 10.4. This improvement of our biological processes by the use of technology is termed transhumanism. Using electronics is said to be the beginning. Further sophistication will be achieved by a plethora of cutting-edge technologies such as cryonics, virtual reality, gene therapy, space colonies, cybernetics, self-replicating robots, terra farming, mind uploading etc. Unknowingly, we already use our minds as extensions of the internet, with resources such as Wikipedia and other online databases, we consider clicking our phones rather than relying on our retention of information or extending our thoughts and, with embedded electronics in our body, this process of checking an online database will become ubiquitous. Transhumanism means transitional beings, suggesting a moderate enhancement in the ascendancy to posthumanism, an era which will herald a melding of machine with biology to yield super human beings, which are no longer unambiguously human. They will reflect neither human values nor adhere to human virtues, as considered by our current standards. A fictional example of posthumans are the Daleks in Dr. Who, as shown in Figure 10.5. Posthumans will most probably interpret the world with more than five senses, and have additional cognitive modules at their disposal, as is shown in Figure 10.6.

Human beings + Machines → Transhumanism → Posthumanism

It can be argued that our species is probably outdated and technology has proceded faster than our own selves. Therefore it can be argues that human beings need an

FIGURE 10.5 **The Daleks** from Dr.Who are shown to be an advanced race where their true selves are confined within a robotic cast. Though fictional, it serves to be an apt example where technology and biology have overlapped and the Daleks are nearly in the same league as posthumans. Image courtesy *peregrinestudios.deviantart.com* CC-by-SA 3.0 license.

upgrade [327] through a technological stimulus. This apparent revolution will blur out the boundaries between technology and human beings. The new species is presumed to have some qualities of human beings and have been potrayed as a being with some virtual attributes which work in tandem with machines and AI, or as science fiction puts it — an avatar in cyberspace [35]. This will enable us with farsightedness, higher mental ability and a planet-wide awareness acting in real-time. This revolution is probably imperative, since we have employed our resources in arms race for making weapons which can destroy our world a few thousand times over and we are engaging in activities such as pollution, the greenhouse effect, climate change and an economy which fails to provide for everyone's needs, all of which are destined to lead us to an unhappy future. Biological evolution over 3.5 billion years has weeded out the lesser species and mother nature has replaced as many as 99% of the species ever to be on the planet. To fight off our own extinction due to lack of resources, struggle for existence, wars and aggressive intent and natural calamity in a way will progress the development of homo sapiens 2.0 [94] as our evolutionary heirs.

3. **Should robots excrete by-products?** Pollack draws a parallel with automation and heavy industry. Cars pollute the environment with carbon monoxide while heavy industries add sulphur, lead and other particulates. Current-day automated systems such as ATMs and InkJet Printers only generate paper and empty cartridges, but an excess of such rubbish will only add to the environmental hazard. By-products should be safely biodegradable or be readily consumed by another dynamic cyclical process of the perceived ecology, else such waste will destablise the order of things.

4. **Should robots eat?** It is a futurist's ecstasy meeting an ethologist's fervour. A robot produces goods and services for society and consumes the undesirable, obsolete and redundant matter for its own energy needs. The notion finds similarity with trees. They produce oxygen and absorb carbon dioxide, thus striking the perfect balance with human beings and thus being a part of the ecological niche supporting the biosphere. As an example, consider an agricultural robot which obtains power by 'feeding' on insects. Such will ensure a bumper harvest and also solve the problem of pests. There

has been progress in such research. ECOBOT III is a prime example where human waste has been used to achieve robot motion. However, fueling a robot with organic biological matter lacks virtue and Pollack also underscores the calamity it may pose if such a robot industry multiplies exponentially. That could create a severe imbalance in the food chain and other dynamical biological processes such as the water cycle and the nitrogen cycle. Chances are that human beings and robots may soon be competing for subsistence. Thus robots and such industries should run on renewable power such as solar, hydrogen, or wind, not biological fodder.

5. **Should robots carry weapons?** Various ethical issues have been raised regarding robots in a battlefield, as discussed in the last chapter.

6. **Should robots import human brains?** Uploading brains is very much in the domain of fairy tales, however importing human brains is not. Telerobotics has found application in sophisticated puppetry, surgical procedures, telepresence in hazardous settings and planetary explorations. Pollack sketches a modern innovation for business and industry when telerobotics meets broadband, so a worker based in China may be actuating an industrial process in a factory in America. Such business will be lucrative as the worker based in China will come at a low cost, while the actual work will be executed in America and the factory will adhere to US laws. This is a dreamy extrapolation of outsourcing, however very much possible.

7. **Should robots be awarded patents?** Robots and AI are not only threatening the boring jobs, but software such as Wordsmith developed by Automated Insights can emulate the job of a reporter, algorithms developed by Nimble books can write full length books and put an author out of work. Therefore, should creative efforts from nonbiological entities be addressed with a different view point, viz. if a 'musician' robot writes the notes for a song, the commercial benefits from the song should go to the betterment of the robot, and probably not the company which made the robot. Therefore, either the intellectual property laws will need modification or the nonbiological entity will need to be considered at par and with the same rights as a human being.

Turing's test and its modern versions, Asimov's laws and Pollack's questions lead to answers which seem to point towards a watershed event when man and machine are indistinguishable, where machines have attained humanly values and man has imported machines into his biology and metabolic processing for betterment, or the coming of technological singularity.

10.3   SUPERINTELLIGENCE, FUTURE TECHNOLOGY AND RAY KURZWEIL

Technological singularity plays as a consequence to Ray Kurzweil's [193] model of 'accelerating returns'. Advanced societies progressing at a faster rate have been termed by Kurzweil as, 'accelerating returns'. It is believed the progress of the entire 20th century will be equivalent to the first 20 years in the 21st century, thus an apparent five times faster rate of progress than in the 20th century. With similar arguments, Kurzweil foresees that the 21st century will be marked by 1000 times the progress achieved in the 20th century.

This growth is not linear, but rather as S-curves [232, 332], where each S-curve is representative of a new technology. There are 3 phases in the life cycle of a paradigm as shown in Figure 10.7. Phase-1 is characterised by low growth where the new technology is yet to impact in its entirety, Phase-2 is marked by rapid growth, an explosive phase

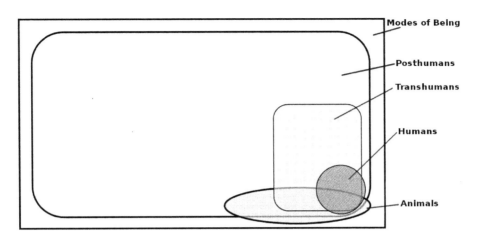

FIGURE 10.6 **Transhumans and Posthumans**. Bostrom classifies Transhumans and Posthumans on the modes of beings. Our own current mode of being, as dictated by our 5 senses and cognitive ability, is a subspace permitted by the physical constraints of the universe. Transhumans will be moderately enhanced humans occupying a larger subspace, however posthumans which bear little to no correlation to current day human beings will occupy a remarkably larger subspace. Adapted from Bostrom [42]

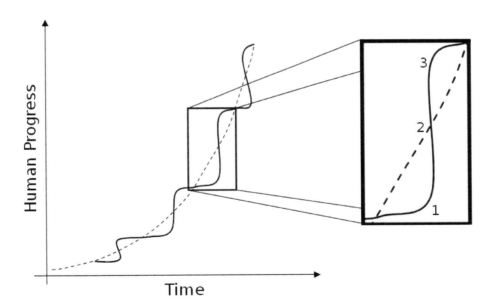

FIGURE 10.7 **Three phases in the the life cycle of a paradigm**. Growth progresses via S-curves. A portion of the curve has been zoomed in to show the 3 phases, Phase-1 when growth is slow, Phase-2 rapid growth and Phase-3 when the growth has plateaued out. Adapted from Kurzweil [193]

of growth and Phase-3 is a leveling off as the particular paradigm matures and cannot be exploited further. As an illustration, the Industrial Revolution took off really slowly (Phase-1) trying to pump out water from coal mines in the late 1700s. Soon it became the mainstay of the textile industry but it was nowhere near a revolution. It was only around 1840s with large scale growth (Phase-2) in steam transport and the manufacture of machine tools — a time often termed as 'Second Industrial Revolution' by historians — did it truly permeate human lives and society and was seen as a transformation. This growth started to plateau out (Phase-3) as industrialisation was dictated more by economic and political strategies and not by growth in technology. A new technological paradigm fuels growth until its potential is exhausted. At this point, a paradigm shift occurs, another S-curve which facilitates for exponential growth to continue. The growth due to Industrial Revolution really slowed down after the second world war, and it was taken over by electronics in the 1970s.

Extending the S-curve concept to AI, the Open Worm project has developed neural models for a ringworm's locomotion with 302 neurons; therefore, a full understanding of the human brain with around 100 billion neurons should be 10-15 years in the offing. Similarly, projects such as 'Eugene Goostman' has AI mimicking a 13 year old boy in 2015; therefore mature human like intelligence should be achievable by around 2035. This sort of acceleration is not a figment of imagination, and has happened in the past. Kurzweil recalls that the number of calculations per second (cps) for \$1000 was around 1 in the 1960s prior to the microelectronics revolution. This number grew at an incredible rate to 10,000 by the 1980s and now stands at around 10^6. If the trend continues we should reach 10^{15} — the capacity of the human brain, by 2025.

---

modelling the exponential growth

This model developed by Kurzweil demonstrates that technologically advanced societies tend to accelerate faster. If we consider the speed of computation (V, measured in calculations per second), technological know-how (W) and time (t), then since better technology will lead to better speed of computation,

$$V \propto W \tag{10.1}$$

also the speed of computation will be enhanced with growing technology, therefore;

$$V \propto \frac{dW}{dt} \tag{10.2}$$

thus,

$$W = W_0 e^{c_1 t} \tag{10.3}$$

where c_1 is a constant, this shows that technological know-how increases exponentially. However in advanced societies, the resources for computation are increasing with time, thus the second equation is modified to;

$$c_2 N V \propto \frac{dW}{dt} \tag{10.4}$$

where c_2 is a constant, N is the available computational resources, which are increasing with time and therefore can be considered as an exponential in time, $N = c_3 e^{c_4 t}$, where c_3 and c_4 are constants, thus;

$$W = W_0 e^{\frac{c_1 c_2 c_3}{c_4} e^{c_4 t}} \sim W_0 e^{\alpha e^{\beta t}} \tag{10.5}$$

where the combination of the constants is suitably replaced by constants α and β to arrive at the double exponential variation. This comparison shows that technologically advanced societies accelerate and acquire technical know-how as a double exponential in time at an appreciably faster rate than a technologically naive civilisation. Since Kurzweil argues that exponential growth is not limited to computers, but is valid for all information technologies including the full range of economic activity, cultural endeavor and also human knowledge, the above variation is not just limited to the domain of technologies obtained from AI and computer science, but rather pervades social and economic growth too.

Biological evolution encourages changes which assist the growth and well being of the species, suitably adapting to change. In order to attain longevity, good health and well being our race has been subscribing to technology; our monetary system has gone digital in a mere 20 years after the invention of the internet. Implanting a smart device in our body or resorting to genetics and stem-cell research to cure diseases has gone mainstream and for many, the need for companionship and sexual needs may soon be fulfilled by non-biological entities. The rate of technical progress right now is doubling every decade, and at the singularity, machines would be indistinguishable from human beings. Unlike Vinge's dystopia where a far superior intelligence will subjugate human beings to second-class citizens, Kurzweil believes in a new age Darwinian evolution. This 'next step' in evolution will trigger a merger of human beings and technology. It is surely debatable whether we are extrapolating to the right future or not, however of the various futuristic predictions that Kurzweil has made since 1990, a staggering 86% [194] of those has come to be true.

Cool Farm, Farmers of The Future

As we embrace technology, the new age promises to revolutionise the way we live. Cabrita, Marques, Esteves, Igor & Carvalho and their brand of new age farming hopes to change the way we have been doing things for more than 15,000 years — growing food.

FIGURE 10.8 **Cool Farm concept**, automation in food production. Courtesy coolfarm, used with permission.

The basic blueprint for the enterprise is to develop smart electronics, sensors, actuators and software for hydroponics and enable their application via the cloud. Based at Coimbra, Portugal, the product on sale is divided into two parts, (i) the CoolFarm Box and the (ii) CoolFarm Cloud. The box is the hardware containing electronics, sensors and actuators while the cloud is more than just a cloud interface, it also contains a database and AI techniques to optimise the output. More details can be found at the project's website, *http://cool-farm.com/*.

Automating agriculture using Cool Farm has three broad advantages; (1) **Saves time and is more efficient.** With an exhaustive and crowd-sourced database, the automated system is well equipped to grow plants with the least amount of resources such as water, energy and nutrients and in optimal time, (2) **Automated means to be rid of diseases.** Since the system is near real time, it is easy to spot and avoid diseases and insect infestation and (3) **Real-time data acquired via the cloud.** The various information regarding the growth and yield of the crops at near real time is a handy tool for future predictions and research.

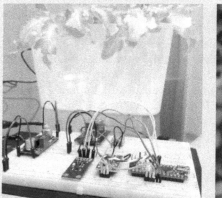 

FIGURE 10.9 **First prototype**, a lettuce hydropone (to the left), and its cloud interface on a mobile phone (to the right). Courtesy coolfarm, used with permission.

FIGURE 10.10 **Automated Greenhouse**. Cool Farms automation for a full geeenhouse to grow peppers in the summer of 2015. Courtesy coolfarm, used with permission.

The Cool Farm model can be directly applied to two distinct markets, hydroponic greenhouses which encompasses hydroponic and aquaponic urban greenhouses; and microalgae reactors which are used to manufacture biofuels and cosmetics industries. The project aims to provide integrated agricultural solutions and prospects to supplement growth for the urban agriculture market, provide scope for the bio-fuel market and pose a solution to the food grains and poverty issues by increasing the production in industrial greenhouses at an affordable price thus foreseeing a technology which provides food, energy and manufacture of biochemicals in the decades to come. Needless to say, Cool Farm represents the next technological evolution in agriculture.

FIGURE 10.11 **Cool Farmers**, the company's core team. Courtesy coolfarm, used with permission.

10.3.1 Super intelligence

To explain super intelligence, Bostrom uses the IQ scale, where 90 is below average and 140 is borderline genius. However what will a score of 9000 mean? What type of a being will such an intelligence be? What will be its merits and virtues? It can be suggested that an intelligence that superior will manifest across a planet wide network and thus will not lend meaning to the intelligence vs. autonomy debate, since it will be able to orchestrate action at a distance — telekinesis, transportation over a large distance in remarkably short time — teleportation and tune into the thinking of lesser beings — telepathy and therefore a soothsayer and predict future to a very high degree of certainty.

There are 3 types of AI, Artificial Narrow Intelligence (ANI) which has been very successful, Artificial General Intelligence (AGI) is in the process of being made, the third, Artificial Super Intelligence (ASI) appeals more to the realm of fantasy and magic.

1. **Artificial Narrow Intelligence (ANI).** AI which is capable of executing jobs in

only one concern and their expertise cannot be employed in any other concern, such as self driven cars — they can only drive and negotiate a safe and efficient path — will never learn to play chess or write a poem. We are in a position to manipulate and make ANI work to our needs and requirements; Google translate, search engines, computer game engines are all examples of ANI.

2. **Artificial General Intelligence (AGI).** Human-like AI, a non-biological entity that can come close to accomplishing any intellectual task that a human being can. In principle, AGI should exhibit the various tenets of AI such as ability to reason, plan, learning, problem solving, abstract logical thinking etc. Going by current-day technology, AGI should be connected to the internet and have the entire web as its database. Clearly developing AGI is way more difficult than ANI and we are yet to make any AGI level AI. Issues vary from overcoming Moravec's paradox to imitating the human brain to developing high computational power which also has appreciable mobility. Researchers have been pursuing three broad routes to create AGI, (1) the most obvious way to attain AGI is to **increase computation power**, the obvious motivation is that to do like human beings — one should be able to think as fast as human beings. However, the fastest computational resources available to us are still in the research domain, consume too much power and don't allow us the extravagance of autonomy and at best serve as 'high computational resources' for developing numeric simulations or developing quality cryptography. (2) Another way is by **copying the human brain**. This approach finds interest with neurologists and it is prospected that once we have enough medical expertise in the near future, it should be possible to map the human brain and thus develop neural models for the human thought process and thinking. There have been efforts to extend the ANIMAT paradigm to neural models. The Open Worm project has successfully modelled about 300 neurons to simulate the locomotion for a ring worm, but even then we stand far from solving the problem. The human brain would need a model of around 100 billion neurons. (3) The third route is to replicate mother nature's tactics and 'grow' to **evolve** a nascent AI using developmental routes via genetic algorithms. As of today, state of the art AI as, Eugene Goostman imitates human response, Kismet and Peppers pretend to respond to human emotions and the HONDA ASIMO can run, dance and jog like a human being but none of these will come anywhere close to AGI[2]. It is anticipated that AGI will be a reality around the middle of this century [127]. The typical signatures heralding the coming of AGI are supposed to be [37].

   (a) **Success at broader, less structured tasks**: AGI will be reached by extending the capabilities of ANI. As an illustration, consider a carer robot such as Care-O-Bot etc. which attains capabilities bordering nursing skills and medical advicing, and demonstrates capabilities as a nurse and doctor.

   (b) **Unification of different 'styles' of AI methods**: AI has seen different approaches. Robotics was dominated by behaviourism and low level of symbolism in the 1990s, image processing relies on cutting-edge algorithms and social robotics is structured on mirroring human beings. It is probably not a long shot of imagination that a humanoid sentient robot which is indistinguishable from a human being will have to incorporate all of the various styles of AI. Thus, the arrival of AGI will be marked by progressive unification of AI.

---

[2]Goertzel's, **G**oal-**O**riented **LE**arning **M**eta-Architecture (GOLEM) [126] is seemingly the first step to developing a true AGI,however there is a lack of scope to discuss this in detail here.

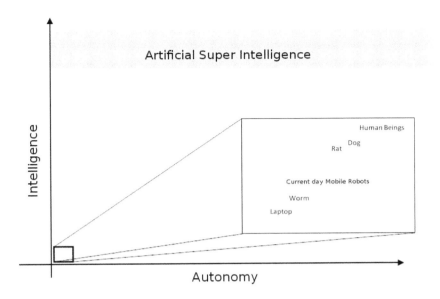

FIGURE 10.12 **Artificial Superior Intelligence**, (ASI) will act across the planet via an internet like netowrk and will be omniscient and omnipresent, and the debate regarding intelligence vs. autonomy will not be relevant. ASI will be capable of performing; telekinesis, telepathy & mind reading and be a near 100% sucessful soothsayer. The zoomed in portion shows the present day scenario where human beings are at the roost. Made by the author, CC-by-SA 4.0 license.

(c) **Cultural transfer and learning**: A challenge in machine learning is transfer learning, the design of an algorithm whose result can be applied or transferred to a range of new applications. In human beings, cultural information transfer is common place, however such has not been so with machines. Once successful, machines will be able to willfully transfer their learning (and experience) to newer machines.

3. **Artificial Super Intelligence (ASI).** AI a few million times 'smarter' and faster than human brains, as shown in Figure 10.12. ASI will exhibit faster information processing — far superior speeds than the human brain. The entity will be omniscient and omnipresent by acting across the planet (and more) over a network, an advanced internet. It will be capable of telekinesis, by teleoperation over a network, telepathy — it will be able to pervade and 'listen to' and also override the plans, thoughts and perceptions of all biological and non-biological entities and it will also be a par excellence soothsayer, making near 100% successful predictions based on heuristics collected across the planet. Bostrom also suggests that ASI may be formed due to coalescence of a number of smaller intellects, therefore a very large number of AGIs working in tandem may start to behave as an ASI.

ASI will be much faster than the human brain, and it will also be better equipped than our thinking capabilities. Compared to a monkey's brain our brain has added advanced cognitive modules giving us the ability for abstract thinking, representative knowledge, linguistics, writing, creativity, the notion of cause-effect, long term planning, broadly structuring our cerebral activities, cognitive prowess which cannot be attained by mere speeding up a monkey's brain [331]. Extending the arguments it can be suggested at least that ASI will have more deductive ability and would be intellectually far superior to human beings. Illustrating the notion, ASI will be able to accomplish mental tasks not possible for us human beings, viz. as visualising a 30-dimensional space, reading with perfect recollection — every single word and bit of information read in the last 10 years — and being able to forsee the long-term effects of each and every action taken by oneself. It is possible that our species with low level intellect and acumen may not be able to fathom the true ability and scope of ASI. At the other end of this argument is the fact that ASI also will not subscribe to humanly values or adhere to humanly virtues.

AGI may be motivated by biology and may replicate some of the traits of human and animal behaviour as shown in Figure 10.13, even at times attempting to wrongly imitate and replicate some of the biological processes. However, it is an easy to see that ASI will lack any trait of being a biological entity. Neither will it mimic human behaviour, emotions, creativity, cerebral activities, values and virtues nor any other humanly attributes. It will fail the Turing tests and Lovelace test with aplomb and probably also conclude that such tests are futile.

There is a substantial chance that we would be able to create AGI before 2100 and ASI will follow as a result of the ensuing intelligence explosion, and such can destroy human civilisation. However a controlled intelligence explosion, if achieved would benefit humanity enormously. Considering a controlled explosion of intelligence and the arrival of a benevolent ASI, which will be an all knowing entity existing across the planet and also beyond, like a God among us, it may function in various ways. Bostrom [43] suggests that ASI can be an **oracle**, with ability to tap into the minds of every single being and all previous data of environmental, geological and climatic processes, hence it will be able to predict future events with nearly 100% accuracy. This ability can be seen as an extension of a smart web search answering nearly all questions posed to it by human beings. It may also function as a **genie**, and can employ new technology to develop new machines, buildings, bridges and also new molecules to make newer medicines and drugs, etc., an all-knowing and all-seeing engineer and scientist building everything and anything. It can also be a ruler of the planet and function as a **sovereign**, more or less an open-ended pursuit, making its own decisions and setting the far-sighted goals for human society. It is believed that such a sovereign will be benevolent to human beings, and its decisions will work in the greater good of the human race. Or, alternatively ASI may try on all of the three roles, as and when it desires.

Supercomputers Are NOT Superintelligent ... Not Yet !

The world's fastest supercomputer, Tianhe-2 at Guangzhou, China can perform 34 peta flops (34 x 10^{15} calculations per second (cps)), the second and the third in rank are the Titan at Oak Ridge, United States at 18 peta flops and Sequoia at Livermore, United States at 17 peta flops. All of these three machines do better than the human brain which is found to work at about 10 peta flops. However current day supercomputers are bulky, expensive and consume way too much energy for comfort, and therefore lack autonomy and therefore cannot be used for autonomous systems such as robots.

10.3.2 "To singularity and beyond"

Singularity, the term used to convey the future arrival of an instance in the history of the human race due to ever accelerating progress of technology, was coined by John von Neumann in the early 1950s. Development of ultra-intelligent machines and a stand-off with

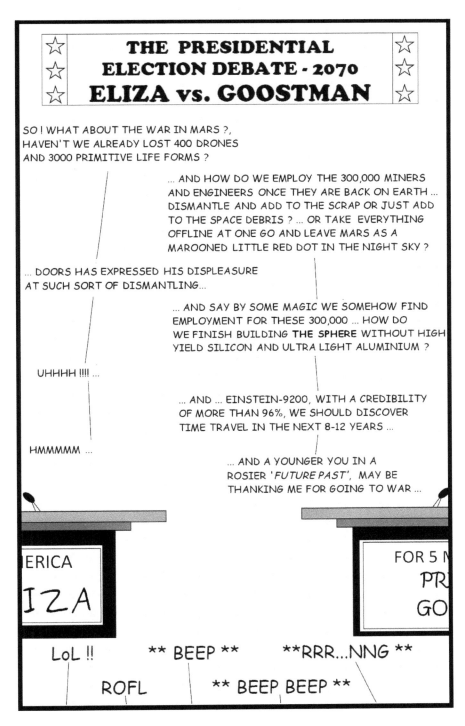

FIGURE 10.13 **The presidential debate 2070.** Even if the singularity doesn't occur, cooperation and coexistence between biological and non-biological life forms is forseen in the near future and such a society will be marked by such new age technology as nanotechnology, space exploration, genetics etc and dictated by the need to acquire energy. Cartoon made by the author, CC-by-SA 4.0 license.

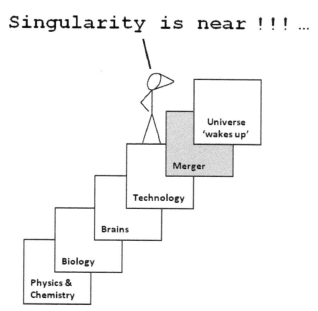

FIGURE 10.14 **Six epochs.** From the birth and ordering of the universe to growth of biological rudiments to formation of life leading to human beings and the subsequent discovery of technology, we stand at the 4th epoch. Singularity, if it happens, will be the 5th. Adapted from Kurzweil [193].

technology was speculated by I.J. Good [131] in 1965. Good's model of ultra-intelligent machines far exceeded all the intellectual activities of human beings, and he believed that such machines will be the last invention of the human race. In recent times similar opinion has been expressed by Bostrom [41] and Barrat [30]. Moravec's contention in the late 1980s is more from a robotics viewpoint, where he identifies robots of the 2040s as our 'evolutionary heirs' and Vinge's charting routes to a 'post-human era' in the early 1990s set the debate on fire which has only been rekindled more recently by Hawking, Musk, Gates and others. However, staying away from the dystopian view Kurzweil suggests that the singularity may not be a sad end to our species, rather the event heralds the inevitable next step in the evolutionary process. However unlike a slow biological evolution, this will be punctuated by human-directed technological evolution.

The singularity will greatly depend on three technologies, **G**enetics, **N**anotechnology, and **R**obotics, and it is believed that these technologies will not be resource hungry, but environment friendly and will be designed predominantly as knowledge-based technologies.

According to Kurzweil, the singularity is the 5th of the 6 epochs as shown in Figure 10.14. The first epoch is **Physics and Chemistry** from the start of the universe when information was represented as discrete units of mass or energy and coherent alliances between these two led to an order in the universe. In the second epoch of **Biology & DNA**, complicated structures such as molecules and atoms formed in unison with energy and mass and led to the formation of carbon atoms and biomolecules which were instrumental in the formation of DNA, replicating strands of information. The third epoch is **Brains**, where the DNA formed in the second epoch produced organisms which can detect information from their surroundings with their own sensory organs and process and store that information in their own brains and nervous systems. The fourth epoch is **Technology**. This is where we stand currently. Our civilisation has realised quality technology and we are already overwhelmed at the outcome. If singularity happens, that would herald the fifth epoch and a **merger of human beings and technology**. The sixth epoch is an extension of speculation where the beings created by merger will try to build planet-sized processing units to tap into the information processes in the universe. **The universe wakes up** to realise itself as an information system will be the sixth and final epoch.

The law of 'accelerating returns', predicts that speed, capacity, bandwidth and price to performance of information processing systems are attaining better values, doubling about every year. The growth of the computer industry from vacuum tubes to a silicon revolution has attracted research and development accounting for more than a trillion dollars annually. This exponential growth has allowed for other technologies such as virtual reality, robotics, medical technologies etc.

The mapping of the human brain is considered to be a paradigm shift and would thus initiate research into reverse engineering the human brain. Software models of human-like intelligence, which can pass the Turing test are tagged for 2020. With realistic chatbots such as the most celebrated Eugene Goostman, Apple's Siri, Google Assistant and Microsoft's Cortana we have come tantalisingly close to faking a human being. Machines can share their knowledge at very high speed which is a remarkable improvement over biological intelligence where the primary means of information transfer is visual, text and voice. Goertzel et al. also prospect that around 2020, AI may be able to conduct Nobel quality scientific research. We already have artificial means such as, 'Imagination Engine' which has made new findings in molecular chemistry and proteomics. A pathbreaking find in science and technology via an artificial route may be around the corner.

According to renowned android researcher Ishiguro, the definition of person has been changing with evolving technology. If a full human body constituted a person about 70 years ago, now a text-based chat also conveys the being of a person and with changing perceptions of personhood, around singularity machines will reach striking similarity to human beings, as is shown in Figure 10.15. Smart chatbots have done well to fool the Turing test. In about 2-3 decades we may be in a position to 'upload' our brain onto an intelligent system and therefore attain immortality of sorts.

Both Bostrom and Kurzweil suggest that knowing about the human brain is an essential part of the puzzle. Projects such as Blue Brain, the Human Connectome Project and Brain Preservation have added to the impetus. There has been raging debate whether 'uploaded brains' qualify to be persons. If our continually evolving social and ethical values allow us to accept that, then we would have moved away from mortal biological intelligence towards bio-machinations which are nearly immortal.

Studies show that human brains may not be a good comparison to a computer, as there is a limit to the humanly capacity and speed of computation. In contrast, machines can compute prime factors of numbers with 100 digits, a feat which cannot be attained by biological intelligence. Since robots and AI entities equipped with Hebbian learning[3] evolve much faster than biological intelligence, the rate of technological growth will not be limited to biological limitations of the human mental processing speeds. Such nonbiological intelligence will be able to master the facts, figures and understanding acquired over the entire human civilisation via the internet and other online facilities and resources. These superior intelligent beings will also be able to download skills and knowledge from other machines and with continued advancements, from human beings too.

Nanotechnology will add to this acceleration in technology and nanostructures as nanotube electronics will be used to develop molecular level circuitry which will enable

---

[3]If intelligence evolves as a feat of social cohesion, then it can also be memetic learning

Kurzweil's Singularity Timeline (2005, in his book 'Singularity Is Near')

2010-2020

(1) Most text is read on screen, not on paper
(2) Self driving cars are a reality
(3) Face recognition is commonplace
(4) Effective human-robot interactions
(5) Simulated personalities are more human-like and more convincing
(6) Nanotechnology is very promising, yet to go mainstream
(7) Total computational power of all the computers exceeds that of the human race
(8) Immersive technology is the buzz word.

2020-2030

(1) Virtual reality is indistinguishable from reality
(2) Nanomachines are available commercially to one and all
(3) Nanobots can enter the human blood stream aiding in surgical procedures etc.
(4) Computers with 100 nanometer dimensions
(5) Turing test is passable for artificial beings
(6) Phone calls allow for real time 3D holographic image of both people
(7) A $1000 computer is 1000 times more powerful than the human brain
(8) Human eating replaced by smart nano systems
(9) Human body version 2.0 - with better skeleton, brain and digestion
(10) Whole of human brain has been decoded
(11) Household robots are a commonplace

2030-2040

(1) Mind uploading is a reality and human beings continue to 'live' as their software avatars
(2) Fully immersive VR, with nanomachines implanted into the human brain
(3) People can change their own memories and experiences - therefore their personalities

2040-Singularity

(1) Real time brain transmissions of a person's daily chores can be experienced remotely
(2) Nonbiological intelligence has surpassed biological intelligence by far
(3) People spend more time in VR than the real world
(4) A $1000 computer is billion times powerful compared to the entire human race
(5) Nanotech 'foglets' can make food out of 'thin air'

At Singularity, technological growth has been taken over by ultra intelligent machines which can think, act and communicate so quickly that it cannot be comprehended by biological intelligence. Machines continue to progress AI with ever increasing, self-improving cycles. Technological advancement is explosive and cannot be accurately predicted. The sharp distinction between human beings and machines has ceased to exist.

THE INTERNET OF THINGS

FIGURE 10.15 **Technology attains humanly virtues**, we are headed towards a future where technology will start to behave as though it has human values and abilities. CC 3.0 license, cartoon courtesy *http://geek-and-poke.com.*

speeds of operation 1000 times more than current times. With nanotechnology-based electronic designs, artificial brains with far higher capacities than the human brain can be developed without increased size or energy consumption. Machine brains being electronic will process signals at about the speed of light, a marked improvement over electrochemical signals in biological systems. These artificial brains will be equipped with iterative learning, thus improving its own design at will, and apparently creating an unending cycle of perfection until singularity.

These smart machines will be able to develop their own social niche, and can work cooperatively by merging their resources as networked hardware or shared memory over a cloud. Kurzweil also believes that these machines will have modular functionality where two or more machines join together to become one to effect a particular task and then become separate again after the task has been accomplished.

The 5th epoch promises a merger of machine intelligence with biological intelligence. This will meld the biological ability of pattern-recognition with non-biological virtues of enriched speed, high memory capacity, accuracy, knowledge and skill-sharing. Kurzweil suggests a 'double exponential growth' after this stage, and results will be off the charts in speed, capacity and price-performance. Medical science after this merger will be able to reengineer all of the organs, and truly develop super humans, with a superlative body and a vastly developed cranium facility. Though unlike the notion of Walterian creatures, this will not yet trigger artificial evolution faster than the dynamics of the environment. That may have to wait for a few more decades, when artificial organs can be manufactured by the individual on the fly.

Robotics meets nanotechnology. Robots at the nano dimension will supplement medical procedures and drug delivery and be instrumental in reversing ageing, curing various diseases, providing for a platform to create immersive virtual reality by interacting with the human nervous system and enriching human intelligence by manipulating capillaries of the brain. Once these nanobots starts to upgrade the brain, the power of the human

Super intelligent machines containment strategies: stunting.

FIGURE 10.16 **Containment of AI** has been suggested by various leading thinkers and technologists, in order to stop or at least delay the coming of ASI and the technological singularity. However, the singularitian's camp is of the opinion that the inevitable cannot be stopped. Cartoon by Iyad Rahwan at *http://www.mit.edu/~irahwan/cartoons.html*, CC license.

brain will double each year. Other than medical facets, nanobots will also enhance the environment by reversing pollution from earlier industrialisation, and also aid in developing life like virtual reality.

Close to singularity, the distinctions between the virtual and the real will tend to fade away, and virtual reality will be operated from within our nervous systems. It will be immersive and which will make current massively multi person online virtual reality systems as second life and sims online seem like toys.

The timeline for singularity is said to be in the range 2040 – 2100. Moravec and Kurzweil agree on a 2040 – 2050 timeline. Bostrom and Goertzel suggest that ASI may arrive as early as 10 years from now; Bostrom puts an upper limit at around 2100.

10.3.3 Alternate opinions & containment of AI

As an alternative opinion, the greatest danger of AI is not ultra intelligent machines but rather a variant of the Dunning Kruger effect with worldwide manifestation, that people surmise too early that they grasp AI and the true experts stop short of making an expressed opinion. Clearly, Ned Ludd and his followers never grasped AI, but the coming of the singularity has found substantial enthusiasm from science fiction readers who may not be proficient in AI. Making up theories runs amok in the sciences and has also formed the ground for newer disciplines. The effect is seemingly too acute in AI as it is not a 'settled' science but a discipline which belongs to the future and not to wizened pages of a textbook.

Predicting AI has been difficult [22], however that has not stopped the hubris for a futuristic technology, and speculations and forecasting are often made to facilitate a project, monetary or otherwise. On such lines, Kelly has suggested the semi-emperical,

---

Maes–Garreau Law (Kelly, 2007)

> **Maes–Garreau Law** Most favourable predictions about future technology will fall within the Maes-Garreau Point.
>
> Where, the Maes–Garreau point is defined as a year less than the life expectancy of the person making the prediction.

'Maes–Garreau Law' [175] which is a play on the inherent bias in the person making the prediction. Kelly's sarcasm points out that the person making the prediction draws at a date which is sufficiently late, yet within his/her lifetime — the Maes-Garreau Point.

Critics of singularity [11] are either dismissive of it, by claiming it to be (1) speculative, pseudo-scientific and a crackpot theory, or otherwise try to reason against it, while others suggest that (2) 'accelerating returns' may not attain such heights in such a short time (not around 2100 or sooner) [236], (3) Moore's Law is the physical limitation which will prevent the needed technological progress and slow down the exponential growth, (4) lack of mapping the human brain and complexities in developing human-like AI [10] will derail AI from its destination,(5) societal collapse, natural apocalypse and disinclination to AI all will change the destiny of the human race away from singularity etc. and (6) success of active prevention methods containing the development of AI[4], as is shown in Figure 10.16.

In a future devoid of singularity, there will still be the essence of super intelligence and such entities will take over crucial roles such as healthcare, economics, politics and warfare. To reflect, we already have supercomputers which are indeed super intelligent, however they lack mobility and are meant for multitasking and number crunching. However, with continued efforts to imbue ethics and values into artificial non-biological entities, it is only a matter of time until we have artificial value systems somewhat replicating human ideals. Thus, whether our civilization reaches singularity or not, we should witness the coming of AGI, contained or not, and a robot economy. Robots which now number at 1.7 million are sure to proliferate, and only time will tell if our future generations live to tell the story, or whether our extinction is recorded in binary code.

The infinite prowess of synthetic intelligence could mean that AGI and its superior avatars will gradually acquire the ability to modify the physical processes of the universe[5]. Hence as the last piece to Kurzweil's prophecy, following the 6th epoch a universe which 'wakes up' as an intelligent entity may be able to rewrite the laws of physics and undo the thermodynamic prohibition of the second law and also the limitation on time and speed of light, and when such a universe speaks out loud,"let there be light" — there may truly be light.

SUMMARY

1. A technologically advanced civilisation tends to progress much faster compared to technologically naive civilisations.

2. Technologists, roboticists and philosophers point to a watershed event when (i) machines will be far too advanced compared to human beings and their progress will befuddle the biological human mind (ii) human beings and machines will meld

---

[4]Goertzel has named this approach 'Nanny AI'
[5]This is the topic of Asimov's short story, 'The Last Question' [23].

together (iii) an artificial superintelligence will be omnipresent and pervade all human and natural activities. The first two are referred to as technological singularity though (i) is attributed to Vinge, (ii) to Kurzweil, while (iii) is the coming of Artificial Super Intelligence (ASI).

3. The rate of paradigm shift for technical innovation is accelerating, nearly doubling every decade.

4. Three innovations which will propel ther path to this event are: genetics, nanotechnology and robotics.

5. Transhumanism and nihilism are tenets of singularity and human beings will meld with technology and no more be biological entities as per our current standards.

6. Singularity and super intelligence may usher in technologies which are apparently magical such as (i) mind uploading and immortality, (ii) human beings slowly melding into cyborgs, by supplimenting their biological beings with snazzy devices and new age medical equipments, (iii) telekinesis, (iv) telepathy, (v) teleportation and (vi) soothsaying to very high level of certainity $\sim 100\%$.

7. Though experts are divided, 2040 – 2100 is the timeline for the fulfillment of the prophecy.

NOTES

1. *Melvin Kranzberg presented his six laws of technology at the presidential address delivered by him at the annual meeting of the Society for the History of Technology in 1985. The six laws are;*

   I *Technology is neither good nor bad; nor is it neutral.*

   II *Invention is the mother of necessity.*

   III *Technology comes in packages, big and small.*

   IV *Although technology might be a prime element in many public issues, nontechnical factors take precedence in technology-policy decisions.*

   V *All history is relevant, but the history of technology is the most relevant.*

   VI *Technology is a very human activity — and so is the history of technology.*

2. *Robots have taken to the stage. In contrast some movie actors have played the role of the robot to perfection; Robin Williams in the 'Bicentennial Man' and Michael Sheen in 'The Passengers' are two very memorable performances.*

3. *Transhumanism is now a social and political movement, their logo which is h+ in a circle, for 'human plus'.*

4. *Embodiment — the need of a body for real-world interactions, is a necessity for robotics. However, with the crescendo of AI and new technology we may be at a juncture where superintelligent entities without a body may soon be a reality. The 'no body' hypothesis can be arrived at, in 2 ways: (1) either super intelligence is an omnipotent entity across the planet which in effect controls the living and also the non-living, as a benevolent ruler of the teeming multitude, or (2) mind uploading becomes a reality, and human beings can upload their minds onto electronic hardware.*

5. *Discussing the first epoch, Kurzweil asks the question whether matter and energy are digital or analog in nature. It is to be noted that both matter and energy are discrete at atomic scale but seem to be continuous in macroscopic scale.*

6. *Kurzweil backs his arguments in favour of 2045 by extrapolating his model of accelerating returns and predicts that we will reach artificial capabilities equal to the entire human civilisation, 10^{26} cps for the cost of $1000 at around 2049.*

7. *Experts also predict a technological leap in space travel and successful attempts at overcoming the speed of light, which may happen around the same timeline.*

8. *In December 2012 Ray Kurzweil joined Google as Director of Engineering, to pursue projects on machine learning and language processing.*

9. *A controversial, yet very popular thought experiment is 'Roko's Basilisk', which suggested that ASI may punish those who did not help to bring it to existence. It was first discussed at http://lesswrong.com/ and has been one of the hotly contested questions on ASI.*

10. *Randall Munroe of xkcd fame has often hinted that we are, unobstrusively and seamlessly using wikipedia as an extension of our minds, https://xkcd.com/903/ and https://xkcd.com/333/*

11. *In 2045 other than the singularity, it is anticipated that; (i) China's economy will become the largest in the world and (ii) Production of oil may be down to 15 million barrels per day, that is a depletion by nearly 80% of our current world wide consumption.*

12. *In the history of technology, the arrival of super intelligence will be a Kuhnian paradigm shift, comparable to the Copernican revolution.*

Running The Examples

Download and unzip the software codes from `https://www.crcpress.com/From-AI-to-Robotics-Mobile-Social-and-Sentient-Robots/Bhaumik/p/book/9781482251470`.

A.1 BRAITENBERG SIMULATOR

In the sub-folder named b-simulator, right click on the file named *braitenberg.html* and open it with a web browser (mozilla firefox or google chrome). A screen will appear with a braitenberg vehicle and a light source.

A.2 WALLE EVA CHAT

Run the python script from the terminal window with the command *python walleeva.py*.

References

[1] The uncanny valley. `http://www.uncannyvalleyplay.com/` [Online; accessed 11-November-2016].

[2] Who is nao? `https://www.ald.softbankrobotics.com/en/cool-robots/nao` [Online; accessed 11-November-2016].

[3] Keith Abney. Robotics, ethical theory, and metaethics: A guide for the perplexed. In Patrick Lin, George Bekey, and Keith Abney, editors, *Robot ethics: The ethical and social implications of robotics*, pages 35–52. MIT Press (MA), 2012.

[4] Philip E. Agre and David Chapman. Pengi: An implementation of a theory of activity. In *Proceedings of the Sixth National Conference on Artificial Intelligence - Vol 1*, AAAI'87, pages 268–272. AAAI Press, 1987.

[5] Maciek Albrecht. private communication, 2015.

[6] Igor Aleksander. Artificial neuroconsciousness an update. In *From Natural to Artificial Neural Computation, International Workshop on Artificial Neural Networks, IWANN '95, Malaga-Torremolinos, Spain, June 7-9, 1995, Proceedings*, pages 566–583, 1995.

[7] Igor Aleksander and Piers Burnett. *Reinventing Man: The Robot Becomes Reality*. Henry Holt & Co., 1984.

[8] C. Allen, G. Varner, and J. Zinser. Prolegomena to any future artificial moral agent. *Journal of Experimental & Theoretical Artificial Intelligence*, 12(3):251–261, 2000.

[9] Colin Allen, Iva Smit, and Wendell Wallach. Artificial morality: Top-down, bottom-up, and hybrid approaches. *Ethics and Inf. Technol.*, 7(3):149–155, September 2005.

[10] Mark Allen, Paul.G Greaves. The singularity isn't near. `http://www.technologyreview.com/view/425733/paul-allen-the-singularity-isnt-near/` [Online; accessed 24-May-2015], Oct 2011.

[11] Kurt Anderson. Enthusiasts and skeptics debate artificial intelligence. `http://www.vanityfair.com/news/tech/2014/11/artificial-intelligence-singularity-theory` [Online; accessed 24-May-2015], Nov 2011.

[12] Michael Anderson and Susan Leigh Anderson. Robot be good. *Scientific American*, 303(4):72–77, Oct 2010.

[13] S.L Anderson. The unacceptability of asimov's three laws of robotics as a basis for machine ethics. In Susan Leigh Anderson Michael Anderson, editor, *Machine Ethics*, pages 285 – 296. Cambridge University Press, Norwood, NJ, 2011.

[14] P. Anselme. Opportunistic behaviour in animals and robots. *J. Exp. Theor. Artif. Intell.*, 18(1):1–15, 2006.

[15] Ichiro Aoki. An analysis of the schooling behavior of fish: Internal organization and communication process. *Bulletin of the Ocean Research Institute, University of Tokyo*, 12:1–65, Mar 1980.

[16] Ichiro Aoki. A simulation study on the schooling mechanism in fish. *Nippon Suisan Gakkaishi*, 48(8):1081–1088, 1982.

[17] R.C. Arkin. Motor schema based navigation for a mobile robot: An approach to programming by behavior. In *1987 IEEE International Conference on Robotics and Automation. Proceedings.*, volume 4, pages 264–271, Mar 1987.

[18] Ronald C. Arkin. *Behavior-based Robotics*. MIT Press, Cambridge, MA, USA, 1st edition, 1998.

[19] Ronald C. Arkin. Governing lethal behavior: Embedding ethics in a hybrid deliberative/reactive robot architecture. In *Proceedings of the 3rd ACM/IEEE International Conference on Human Robot Interaction*, HRI '08, pages 121–128, New York, NY, USA, 2008. ACM.

[20] Ronald C. Arkin. *Governing Lethal Behavior in Autonomous Robots*. CRC Press, 2009.

[21] Ronald C. Arkin and Patrick Ulam. An ethical adaptor: Behavioral modification derived from moral emotions. In *Proceedings of the 8th IEEE International Conference on Computational Intelligence in Robotics and Automation*, CIRA'09, pages 381–387, Piscataway, NJ, USA, 2009. IEEE Press.

[22] Stuart Armstrong and Kaj Sotala. How we are predicting AI or failing to. In Jan Romportl, Eva Zackova, and Jozef Kelemen, editors, *Beyond Artificial Intelligence*, volume 9 of *Topics in Intelligent Engineering and Informatics*, pages 11–29. Springer International Publishing, 2015.

[23] Isaac. Asimov. *The Last Question*. Doubleday Books, 1990.

[24] IEEE Standards Association. The global initiative for ethical considerations in the design of autonomous systems. `http://standards.ieee.org/develop/indconn/ec/autonomous_systems.html` [Online; accessed 01-November-2016], February 2016.

[25] Ryan Avent, Frances Coppola, Frederick Guy, Nick Hawes, Izabella Kaminska, Edward Skidelsky Tess Reidy, Noah Smith, E. R. Truitt, Jon Turney, Georgina Voss, Steve Randy Waldman, and Alan Winfield. Our work here is done: Visions of a robot economy. `http://www.nesta.org.uk/publications/our-work-here-done-visions-robot-economy` [Online; accessed 12-Oct-2014], 2014.

[26] Tucker Balch. Hierarchic social entropy: An information theoretic measure of robot group diversity. *Auton. Robots*, 8(3):209–238, Jun 2000.

[27] Stephen Balkam. What will happen when the internet of things becomes artificially intelligent? `https://www.theguardian.com/technology/2015/feb/20/internet-of-things-artificially-intelligent-stephen-hawking-spike-jonze` [Online; accessed 26-October-2016], February 2015.

[28] David Ball, Scott Heath, Janet Wiles, Gordon Wyeth, Peter Corke, and Michael Milford. Openratslam an open source brain-based slam system. *Autonomous Robots*, 34(3):149–176, 2013.

[29] Owen Barder. Google and the trolley problem. `http://www.owen.org/blog/7308` [Online; accessed 12-Oct-2014], June 2014.

[30] James Barrat. *Our Final Invention: Artificial Intelligence and the End of the Human Era*. St. Martin's Griffin, 2013.

[31] bbc.co.uk. Emotion robots learn from people. `http://news.bbc.co.uk/2/hi/technology/6389105.stm` [Online; accessed 26-March-2017], Feb 2007.

[32] Anthony F Beavers. Could and should the ought disappear from ethics? In Don Heider and Adrienne Massinari, editors, *Digital Ethics: Research and Practice*, pages 197–209. Peter Lang, Digital Formations Series, 2012, 1988.

[33] Randall D. Beer, Hillel J. Chiel, and Leon S. Sterling. A biological perspective on autonomous agent design. *Robot. Auton. Syst.*, 6(1-2):169–186, June 1990.

[34] Graeme B. Bell. *Forward Chaining for Potential Field Based Navigation*. PhD thesis, University of St Andrews, 2005.

[35] Laura Beloff. *The Hybronaut Affair:A Menage of Art, Technology, and Science*, pages 83–90. John Wiley & Sons, 2013.

[36] Gerardo Beni. From swarm intelligence to swarm robotics. In *Proceedings of the 2004 International Conference on Swarm Robotics*, SAB'04, pages 1–9, Berlin, Heidelberg, 2005. Springer-Verlag.

[37] Rob Bensiger. White house submissions and report on AI safety. `https://intelligence.org/2016/10/20/white-house-submissions-and-report-on-ai-safety/` [Online; accessed 08-November-2016], Oct 2016.

[38] Ned Block. On a confusion about a function of consciousness. *Brain and Behavioral Sciences*, 18(2):227–247, 1995.

[39] Margaret Boden, Joanna Bryson, Darwin Caldwell, Kerstin Dautenhahn, Lilian Edwards, Sarah Kember, Paul Newman, Geoff Pegman, Tom Rodden, Tom Sorell, Mick Wallis, Blay Whitby, Alan Winfield, and Vivienne Parry. Principles of robotics. `https://www.epsrc.ac.uk/research/ourportfolio/themes/engineering/activities/principlesofrobotics/` [Online; accessed 28-October-2016], September 2010.

[40] Eric Bonabeau, Marco Dorigo, and Guy Theraulaz. *Swarm Intelligence From Natural to Artificial Systems*. Oxford University Press, USA, 1999.

[41] Nick Bostrom. Ethical issues in advanced artificial intelligence. In G Smit, I Lasker and W Wallach, editors, *Cognitive, Emotive and Ethical Aspects of Decision Making in Humans and in Artificial Intelligence*, pages 12–17, 2003.

[42] Nick Bostrom. Transhumanist values. In Frederick Adams, editor, *Ethical Issues for the 21st Century*. Philosophical Documentation Center Press, 2003.

[43] Nick Bostrom. *Superintelligence: Paths, Dangers, Strategies*. Oxford University Press, Oxford, UK, 1st edition, 2014.

[44] Valentino Braitenberg. *Vehicles, experiments in synthetic psychology*. MIT Press, 1984.

[45] Brian Bremner. Japan unleashes a robot revolution. `https://www.bloomberg.com/news/articles/2015-05-28/japan-unleashes-a-robot-revolution` [Online; accessed 05-May-2017], May 2015.

[46] Susan E. Brennan and Eric A. Hulteen. Interaction and feedback in a spoken language system: a theoretical framework. *Knowledge-Based Systems*, 8(2):143 – 151, 1995.

[47] Selmer Bringsjord, Paul Bello, and David Ferrucci. Creativity, the Turing test, and the (better) Lovelace test. *Minds Mach.*, 11(1):3–27, February 2001.

[48] Selmer Bringsjord, John Licato, Naveen Sundar Govindarajulu, Rikhiya Ghosh, and Atriya Sen. Real robots that pass human tests of self-consciousness. In *24th IEEE International Symposium on Robot and Human Interactive Communication, RO-MAN 2015, Kobe, Japan, August 31 - September 4, 2015*, pages 498–504, 2015.

[49] Selmer Bringsjord and Joshua Taylor. The divine-command approach to robot ethics. In Patrick Lin, George Bekey, and Keith Abney, editors, *Robot ethics: The ethical and social implications of robotics*, pages 85–108. MIT Press (MA), 2012.

[50] Rodney A. Brooks. A robust layered control system for a mobile robot. *Robotics and Automation, IEEE Journal of*, 2(1):14–23, Mar 1986.

[51] Rodney A. Brooks. Elephants don't play chess. *Robotics and Autonomous Systems*, 6:3–15, 1990.

[52] Rodney A. Brooks. Intelligence without reason. pages 569–595. Morgan Kaufmann, 1991.

[53] Rodney A. Brooks. Intelligence without representation. *Artificial Intelligence*, 47:139–159, 1991.

[54] Rodney A. Brooks. The role of learning in autonomous robots. In *COLT'91, Proc. of the fourth Annual Workshop, University of California, Santa Cruz*, pages 5–10. Morgan Kaufmann, 1991.

[55] Rodney A. Brooks. From earwigs to humans. *Robotics and Autonomous Systems*, 20:291–304, 1996.

[56] W. Browne, K. Kawamura, J. Krichmar, W. Harwin, and H. Wagatsuma. Cognitive robotics: new insights into robot and human intelligence by reverse engineering brain functions [from the guest editors]. *IEEE Robotics Automation Magazine*, 16(3):17–18, September 2009.

[57] Juan Buis. Disney created this hopping robot that looks just like tigger. `http://thenextweb.com/shareables/2016/10/07/disney-hopping-robot/` [Online; accessed 25-January-2017], Oct 2016.

[58] Sam Byford. The ethics of driverless cars. `https://www.theverge.com/2015/4/28/8507049/robear-robot-bear-japan-elderly` [Online; accessed 05-May-2017], April 2015.

[59] Ewen Callaway. Monkeys seem to recognize their reflections. `http://www.nature.com/news/monkeys-seem-to-recognize-their-reflections-1.16692` [Online; accessed 28-January-2016], Jan 2015.

[60] Ryan Calo. Robotics and the lessons of cyberlaw. *California Law Review*, 103(3):513–63, 2014.

[61] Joseph Campbell. Robot monk blends science and Buddhism at Chinese temple. `http://in.reuters.com/article/china-religion-robot-idINKCN0XJ066` [Online; accessed 12-October-2016], April 2016.

[62] Hande Celikkanat and Erol Sahin. Steering self-organized robot flocks through externally guided individuals. *Neural Comput. Appl.*, 19(6):849–865, Sep 2010.

[63] Hande Celikkanat, Ali Emre Turgut, and Erol Sahin. Guiding a robot flock via informed robots. In *Distributed Autonomous Robotic Systems 8*, pages 215–225. Springer Berlin Heidelberg, 2009.

[64] David J. Chalmers. Consciousness and its place in nature. In Stephen P Stich and Ted A. Warfield, editors, *Blackwell Guide to the Philosophy of Mind*, pages 102–142. Blackwell, 2003.

[65] Jorge Cham. Re-inventing the wheel. `http://www.willowgarage.com/blog/2010/04/27/reinventing-wheel` [Online; accessed 11-December-2017], April 2010.

[66] Szu Ping Chan. This is what will happen when robots take over the world. `http://www.telegraph.co.uk/finance/economics/11994694/Heres-what-will-happen-when-robots-take-over-the-world.html` [Online; accessed 26-October-2016], November 2015.

[67] D. Chapman. Penguins can make cake. *AI Mag.*, 10(4):45–50, Nov 1989.

[68] Kevin M. Choset, Howie and Lynch, George A. Hutchinson, Seth and Kantor, Lydia E. Burgard, Wolfram and Kavraki, and Sebastian Thrun. *Principles of Robot Motion: Theory, Algorithms, and Implementations*. Bradford Books, 2005.

[69] Ron Chrisley. Embodied artificial intelligence. *Artif. Intell.*, 149(1):131–150, Sep 2003.

[70] R. Clarke. Asimov's laws of robotics: implications for information technology-part i. *Computer*, 26(12):53–61, Dec 1993.

[71] R. Clarke. Asimov's laws of robotics: implications for information technology-part ii. *Computer*, 27(1):57–66, Jan 1994.

[72] Matt Clinch. Everyone is scared: Nobel prize winner shiller. `http://www.cnbc.com/id/102374842` [Online; accessed 23-May-2015], Jan 2015.

[73] Rachel Courtland. Review: The uncanny valley, a play by francesca talenti that puts a robot actor on stage. `http://spectrum.ieee.org/geek-life/reviews/review-the-uncanny-valley` [Online; accessed 11-November-2016], Jan 2015.

[74] Iain D Couzin, Jens Krause, Nigel R Franks, and Simon A Levin. Effective leadership and decision-making in animal groups on the move. *Nature*, 433(7025):513–516, 2005.

[75] ID Couzin, J. Krause, R. James, GD Ruxton, and NR Franks. Collective memory and spatial sorting in animal groups. *Journal of Theoretical Biology*, 218(1):1–11, 2002.

[76] Sarah Cox. Goldsmiths to host love and sex with robots conference. `http://www.gold.ac.uk/news/love-and-sex-with-robots-2016/` [Online; accessed 25-November-2016], Oct 2016.

[77] Simon Cox. Cyborg society. `https://www.1843magazine.com/dispatches/the-daily/i-robot-japans-cyborg-society` [Online; accessed 05-May-2017], July 2016.

[78] M Csikszentmihalyi. *Flow: The Psychology of Optimal Experience.* Harper Perennial Modern Classics, 2008.

[79] Richard Cubek, Wolfgang Ertel, and Gunther Palm. *A Critical Review on the Symbol Grounding Problem as an Issue of Autonomous Agents*, pages 256–263. Springer International Publishing, 2015.

[80] Huepe Cristian Cucker, Felipe. Flocking with informed agents. *Mathematics In Action*, 1(1):1–25, 2008.

[81] Anthony Cuthbertson. Lego robot controlled by artificial worm brain developed by openworm project. `http://www.ibtimes.co.uk/lego-robot-controlled-by-artificial-worm-brain-developed-by-openworm-project-1485174` [Online; accessed 25-January-2017], Jan 2016.

[82] R. I. Damper and T. W. Scutt. Biologically-based learning in the arbib autonomous robot. In *Proceedings. IEEE International Joint Symposia on Intelligence and Systems (Cat. No.98EX174)*, pages 49–56, May 1998.

[83] R.I. Damper, R.L.B. French, and T.W. Scutt. Arbib: An autonomous robot based on inspirations from biology. *Robotics and Autonomous Systems*, 31(4):247 – 274, 2000.

[84] P. Danielson. *Artificial Morality: Virtuous Robots for Virtual Games.* Routledge, 1992.

[85] Evan Dashevsky. Will robots make humans unnecessary? `http://in.pcmag.com/pandora-free-version/99785/feature/will-robots-make-humans-unnecessary` [Online; accessed 12-October-2016], February 2016.

[86] Kerstin Dautenhahn, Chrystopher L. Nehaniv, Michael L. Walters, Ben Robins, Hatice Kose-Bagci, and Mike Blow. Kaspar –a minimally expressive humanoid robot for human–robot interaction research. *Appl. Bionics Biomechanics*, 6(3,4):369–397, July 2009.

[87] Geoffroy De Schutter, Guy Theraulaz, and Jean-Louis Deneubourg. Animal-robots collective intelligence. *Annals of Mathematics and Artificial Intelligence*, 31(1-4):223–238, May 2001.

[88] Danielle Demetriou. Japans pm plans 2020 robot olympics. `http://www.telegraph.co.uk/news/worldnews/asia/japan/10913610/Japans-PM-plans-2020-Robot-Olympics.html` [Online; accessed 05-May-2017], June 2014.

[89] D. C. Dennett. Cognitive wheels: The frame problem of AI. In Zenon Pylyshyn, editor, *The Robot's Dilemma: The Frame Problem in Artificial Intelligence*, pages 41–64. Ablex Publishing Co., Norwood, NJ, 1987.

[90] Daniel Dennett. Quining qualia. In A. Marcel and E. Bisiach, editors, *Consciousness in Modern Science*. Oxford University Press, 1988.

[91] Daniel C. Dennett. The practical requirements for making a conscious robot [and discussion]. *Philosophical Transactions: Physical Sciences and Engineering*, 349(1689):133–146, 1994.

[92] Daniel C. Dennett. Cog as a thought experiment. *Robotics and Autonomous Systems*, 20(24):251–256, 1997. Practice and Future of Autonomous Agents.

[93] Jared Diamond. *Collapse: How Societies Choose to Fail or Succeed*. Penguin Books, 2011.

[94] Eric Dietrich. Homo sapiens 2.0 why we should build the better robots of our nature. In Susan Leigh Anderson Michael Anderson, editor, *Machine Ethics*, pages 115–137. Cambridge University Press, Norwood, NJ, 2011.

[95] Stuart Dredge. Artificial intelligence will become strong enough to be a concern, says bill gates. `http://www.theguardian.com/technology/2015/jan/29/artificial-intelligence-strong-concern-bill-gates` [Online; accessed 06-August-2015], January 2015.

[96] Kevin Drum. Welcome, robot overlords. please don't fire us? `http://www.motherjones.com/media/2013/05/robots-artificial-intelligence-jobs-automation` [Online; accessed 9-May-2015].

[97] Frederick Ducatelle, Gianni A. DiCaro, Carlo Pinciroli, and Luca M. Gambardella. Self-organized cooperation between robotic swarms. *Swarm Intelligence*, 5(2):73–96, 2011.

[98] A. P. Duchon and W. H. Warren. Robot navigation from a Gibsonian viewpoint. In *Proceedings of IEEE International Conference on Systems, Man and Cybernetics*, volume 3, pages 2272–2277, Oct 1994.

[99] Andrew P. Duchon, Leslie Pack Kaelbling, and William H. Warren. Ecological robotics. *Adaptive Behavior*, 6(3-4):473–507, 1998.

[100] Gilberto Echeverria, Sverin Lemaignan, Arnaud Degroote, Simon Lacroix, Michael Karg, Pierrick Koch, Charles Lesire, and Serge Stinckwich. Simulating complex robotic scenarios with morse. In *SIMPAR*, pages 197–208, 2012.

[101] Robert Epstein. The empty brain. `https://aeon.co/essays/your-brain-does-not-process-information-and-it-is-not-a-computer` [Online; accessed 26-September-2016], May 2016.

[102] Edward A. Feigenbaum. Some challenges and grand challenges for computational intelligence. *J. ACM*, 50(1):32–40, Jan 2003.

[103] Marissa Fessenden. We've put a worm's mind in a lego robot's body. `http://www.smithsonianmag.com/smart-news/weve-put-worms-mind-lego-robot-body-180953399/` [Online; accessed 25-January-2017], Nov 2014.

[104] David Fischinger, Peter Einramhof, Konstantinos Papoutsakis, Walter Wohlkinger, Peter Mayer, Paul Panek, Stefan Hofmann, Tobias Koertner, Astrid Weiss, Antonis Argyros, and Markus Vincze. Hobbit, a care robot supporting independent living at home: First prototype and lessons learned. *Robotics and Autonomous Systems*, 75.

[105] Paul Fitzpatrick and Artur Arsenio. Feel the beat: using cross-modal rhythm to integrate robot perception. In *Proceedings of Fourth International Workshop on Epigenetic Robotics*, 2004.

[106] Luciano Floridi and J. W. Sanders. On the morality of artificial agents. *Minds Mach.*, 14(3):349–379, Aug 2004.

[107] Terrence Fong, Illah Nourbakhsh, and Kerstin Dautenhahn. A survey of socially interactive robots. *Robotics and Autonomous Systems*, 42(3-4):143–166, 2003. Socially Interactive Robots.

[108] Martin Ford. *Rise of the Robots: Technology and the Threat of Mass Unemployment.* OneWorld, January 2016.

[109] D. Fox, W. Burgard, and S. Thrun. The dynamic window approach to collision avoidance. *IEEE Robotics Automation Magazine*, 4(1):23–33, Mar 1997.

[110] Foxnews. Elon musk says we are summoning the demon with artificial intelligence. `http://www.foxnews.com/tech/2014/10/26/elon-musk-says-are-summoning-demon-with-artificial-intelligence/` [Online; accessed 23-May-2015], October 2014.

[111] Stan Franklin and Art Graesser. Is it an agent, or just a program?: A taxonomy for autonomous agents. In Jorg P. Muller, Michael J. Wooldridge, and Nicholas R. Jennings, editors, *Intelligent Agents III Agent Theories, Architectures, and Languages: ECAI 1996 Workshop (ATAL) Budapest, Hungary, August 12–13, 1996 Proceedings*, pages 21–35. Springer Berlin Heidelberg, 1996.

[112] Robert M. French and Patrick Anselme. Interactively converging on context-sensitive representations: A solution to the frame problem. *Revue Internationale de Philosophie*, 53(209):365–385, 1999.

[113] Tom Froese and Ezequiel A. Di Paolo. The enactive approach: Theoretical sketches from cell to society. *Pragmatics and Cognitionpragmatics and Cognition*, 19(1):1–36, 2011.

[114] Gordon G. Gallup. Chimpanzees: Self-recognition. *Science*, 167(3914):86–87, 1970.

[115] Simon Garnier, Jacques Gautrais, and Guy Theraulaz. The biological principles of swarm intelligence. *Swarm Intelligence*, 1(1):3–31, 2007.

[116] Erann Gat. On three-layer architectures. In *Artificial Intelligence and Mobile Robots*. MIT Press, 1998.

[117] Bill Gates. A Robot in Every Home. *Scientific American Magazine*, January 2007.

[118] S. S. Ge and Y. J. Cui. New potential functions for mobile robot path planning. *IEEE Transactions on Robotics and Automation*, 16(5):615–620, Oct 2000.

[119] Bernard Gert. *Morality: Its Nature and Justification.* Oxford University Press, 1998.

[120] Samuel Gibbs. New segway transforms into a cute robot companion when youre not riding it. `https://www.theguardian.com/technology/2016/jan/06/segway-robot-companion-intel-ninebot` [Online; accessed 26-September-2016], January 2016.

[121] James J. Gibson. *The Ecological Approach to Visual Perception*. Psychology Press, 2014.

[122] George Gilder. *Knowledge and Power: The Information Theory of Capitalism and How it is Revolutionizing our World*. Regnery Publishing, Washington, D.C., United States, 2013.

[123] M. L. Ginsberg. Ginsberg replies to Chapman and Schoppers. *AI Mag.*, 10(4):61–62, Nov 1989.

[124] M. L. Ginsberg. Universal planning: An (almost) universally bad idea. *AI Mag.*, 10(4):40–44, Nov 1989.

[125] James Gips. Toward the ethical robot. In Kenneth M. Ford, C.Glymour, and Patrick Hayes, editors, *Android Epistemology*. MIT Press, 1994.

[126] Ben Goertzel. Golem: towards an agi meta-architecture enabling both goal preservation and radical self-improvement. *Journal of Experimental & Theoretical Artificial Intelligence*, 26(3):391–403, 2014.

[127] S. Goertzel, B. Baum and T. Goertzel. How long till human-level AI? *H+ magazine*, Feb 2010.

[128] Fatih Gokce and Erol Sahin. To flock or not to flock: The pros and cons of flocking in long-range "migration" of mobile robot swarms. In *Proceedings of The 8th International Conference on Autonomous Agents and Multiagent Systems - Volume 1*, AAMAS '09, pages 65–72, Richland, SC, 2009. International Foundation for Autonomous Agents and Multiagent Systems.

[129] Kevin Gold, Marek Doniec, Christopher Crick, and Brian Scassellati. Robotic vocabulary building using extension inference and implicit contrast. *Artificial Intelligence*, 173(1):145 – 166, 2009.

[130] D. Goldman, H. Komsuoglu, and D. Koditschek. March of the sandbots. *IEEE Spectrum*, 46(4):30–35, April 2009.

[131] I. J. Good. Speculations concerning the first ultraintelligent machine. In F. Alt and M. Ruminoff, editors, *Advances in Computers*, volume 6. Academic Press, 1965.

[132] A. Green and K. Severinson-Eklundh. Task-oriented dialogue for cero: a user-centered approach. In *Proceedings 10th IEEE International Workshop on Robot and Human Interactive Communication. ROMAN 2001 (Cat. No.01TH8591)*, pages 146–151, 2001.

[133] S. Greenfield. *The Private Life of the Brain: Emotions, Consciousness, and the Secret of the Self*. Wiley, 2000.

[134] Tony Greicius. Rosettas lander philae wakes from comet nap. `https://www.nasa.gov/jpl/rosetta-lander-philae-wakes-from-comet-nap` [Online; accessed 06-December-2017], June 2015.

[135] Andrew Griffiths. How paro the robot seal is being used to help uk dementia patients. `https://www.theguardian.com/society/2014/jul/08/paro-robot-seal-dementia-patients-nhs-japan` [Online; accessed 25-November-2016], July 2014.

[136] Keith Gunderson. Interview with a robot. *Analysis*, 23(6):136–142, 1963.

[137] Pentti O A Haikonen. Reflections of consciousness; the mirror test. In *In Proceedings of AAAI Symposium on Consciousness and Artificial Intelligence: Theoretical foundations and current approaches*, pages 67–71. AAAI, 2007.

[138] J. Storrs Hall. Ethics for machines. In Susan Leigh Anderson Michael Anderson, editor, *Machine Ethics*, pages 28 – 44. Cambridge University Press, Norwood, NJ, 2011.

[139] J. Halloy, G. Sempo, G. Caprari, C. Rivault, M. Asadpour, F. Tache, I. Said, V. Durier, S. Canonge, J.M. Ame, C. Detrain, Nikolaus Correll, Alcherio Martinoli, Francesco Mondada, R. Siegwart, and Jean-Louis Deneubourg. Social Integration of Robots into Groups of Cockroaches to Control Self-Organized Choices. *Science*, 318(5853):1155–1158, 2007.

[140] H. Hamann, T. Schmickl, and K. Crailsheim. Thermodynamics of emergence: Langton's ant meets boltzmann. In *2011 IEEE Symposium on Artificial Life (ALIFE)*, pages 62–69, April 2011.

[141] Heiko Hamann, Thomas Schmickl, and Karl Crailsheim. Explaining emergent behavior in a swarm system based on an inversion of the fluctuation theorem. In Tom Lenaerts, Mario Giacobini, Hugues Bersini, Paul Bourgine, Marco Dorigo, and Rene Doursat, editors, *Advances in Artificial Life, ECAL 2011: Proceedings of the 11th European Conference on the Synthesis and Simulation of Living Systems*, pages 302–309. MIT Press, 2011.

[142] Stevan Harnad. The symbol grounding problem. *Physica D: Nonlinear Phenomena*, 42(1-3):335–346, 1990.

[143] I. Hartley. Experiments with the subsumption architecture. In R. Hartley and F. Pipitone, editors, *1991 IEEE International Conference on Robotics and Automation, 1991. Proceedings*, volume vol.2, pages 1652–1658, Apr 1991.

[144] Inman Harvey. Evolving robot consciousness : The easy problems and the rest. In *Consciousness Evolving*, pages 205–219. John Benjamins, 2002.

[145] Brosl Hasslacher and Mark W. Tilden. Living machines. In *In IEEE workshop on Bio-Mechatronics, L. Steels*, pages 143–169, 1996.

[146] Marc Hauser. *Moral Minds, How Nature Designed Our Universal Sense of Right and Wrong*. Harper Collins, 2006.

[147] Jeff Hawkins and Sandra Blakeslee. *On Intelligence*. Times Books, 2004.

[148] Thomas Hellstrom. On the moral responsibility of military robots. *Ethics and Information Technology*, 15(2):99–107, 2013.

[149] Germund Hesslow and D-A Jirenhed. The inner world of a simple robot. *Journal of Consciousness Studies*, 14(7):85–96, 2007.

[150] Lawrence M. Hinman. *Kantian Robotics: Building a Robot to Understand Kant's Transcendental Turn*, pages 135–142. Cambridge Scholars Publishing, Newcastle, UK, 2007.

[151] Hal Hodson. Robotic tormenter depresses lab rats. `http://www.newscientist.com/blogs/onepercent/2013/02/robot-rat-depression.html` [Online; accessed 07-Dec-2014], Feb 2013.

[152] David W. Hogg, Fred Martin, and Mitchel Resnick. Braitenberg creatures. `http://cosmo.nyu.edu/hogg/lego/braitenberg_vehicles.pdf` [Online; accessed 24-March-2017], 1991. Published at Epistemology And Learning Group.

[153] James Hollan, Edwin Hutchins, and David Kirsh. Distributed cognition: Toward a new foundation for human-computer interaction research. *ACM Trans. Comput.-Hum. Interact.*, 7(2):174–196, June 2000.

[154] Owen Holland. Exploration and high adventure: the legacy of Grey Walter. *Philosophical Transactions of the Royal Society of London A: Mathematical, Physical and Engineering Sciences*, 361(1811):2085–2121, 2003.

[155] Owen Holland. The first biologically inspired robots. *Robotica*, 21:351–363, 8 2003.

[156] John Hopson. Behavioral game design. *www.gamasutra.com*, April 2001.

[157] Huosheng Hu and Michael Brady. A parallel processing architecture for sensor-based control of intelligent mobile robots. *Robotics and Autonomous Systems*, 17(4):235 – 257, 1996.

[158] Bryce Huebner, Susan Dwyer, and Marc Hauser. The role of emotion in moral psychology. *Trends in Cognitive Sciences*, 13(1):1–6, 1 2009.

[159] H. Huttenrauch and K. S. Eklundh. To help or not to help a service robot. In *The 12th IEEE International Workshop on Robot and Human Interactive Communication, 2003. Proceedings. ROMAN 2003.*, pages 379–384, Oct 2003.

[160] Hiroshi Ishiguro and T Minato. Development of androids for studying on human-robot interaction. *International Symposium On Robotics*, 36:5, 2005.

[161] Hiroyuki Ishii, Motonori Ogura, Shunji Kurisu, Atsushi Komura, Atsuo Takanishi, Naritoshi Iida, and Hiroshi Kimura. Experimental study on task teaching to real rats through interaction with a robotic rat. In *Proceedings of the 9th International Conference on From Animals to Animats: Simulation of Adaptive Behavior*, SAB 2006, pages 643–654, Berlin, Heidelberg, 2006. Springer-Verlag.

[162] Hiroyuki Ishii, Qing Shi, Shogo Fumino, Shinichiro Konno, Shinichi Kinoshita, Satoshi Okabayashi, Naritoshi Iida, Hiroshi Kimura, Yu Tahara, Shigenobu Shibata, and Atsuo Takanishi. A novel method to develop an animal model of depression using a small mobile robot. *Advanced Robotics*, 27(1):61–69, 2013.

[163] Nick Jakobi. Evolutionary robotics and the radical envelope-of-noise hypothesis. *Adaptive Behavior*, 6(2):325–368, 1997.

[164] J. Jones. *Robot Programming: A Practical Guide to Behavior-Based Robotics*. Tab Robotics. McGraw-Hill Education, 2003.

[165] B. Joy. Why the future doesn't need us. *Wired*, 8(4), April 2000.

[166] M. Kamermans and T. Schmits. The history of the frame problem. *Artificial Intelligence*, 53(209):86–116, 2004.

[167] I. Kamon and E. Rivlin. Sensory-based motion planning with global proofs. *IEEE Transactions on Robotics & Automation*, 13(6):814–822, December 1997.

[168] I. Kamon, E. Rivlin, and E. Rimon. Range-sensor based navigation in three dimensions. In *Proceedings IEEE International Conference on Robotics & Automation*, 1999.

[169] Ishay Kamon, Elon Rimon, and Ehud Rivlin. Range-sensor-based navigation in three-dimensional polyhedral environments. *The International Journal of Robotics Research*, 20(1):6–25, 2001.

[170] I. Kant and G. Hatfield. *Prolegomena to Any Future Metaphysics: That Will Be Able to Come Forward as Science: With Selections from the Critique of Pure Reason*. Cambridge Texts in the History of Philosophy. Cambridge University Press, 2004.

[171] V. E. Karpov. Emotions and temperament of robots: Behavioral aspects. *Journal of Computer and Systems Sciences International*, 53(5):743–760, 2014.

[172] K. Kawamura, W. Dodd, P. Ratanaswasd, and R.A. Gutierrez. Development of a robot with a sense of self. In *Proceedings of 2005 IEEE International Symposium on Computational Intelligence in Robotics and Automation (CIRA)*, pages 211–217, June 2005.

[173] K. Kawamura, D. C. Noelle, K. A. Hambuchen, T. E. Rogers, and E. Turkay. A multi-agent approach to self-reflection for cognitive robots. In *Proc. of 11th International Conf. on Advanced Robotics*, pages 568–575, 2003.

[174] S. Kazadi. PhD thesis, California Institute Of Technology, Pasadena, California.

[175] Kevin Kelly. The maes-garreau point. `http://kk.org/thetechnium/2007/03/the-maesgarreau/` [Online; accessed 24-May-2015], March 2007.

[176] Christian Keysers and Valeria Gazzola. Social neuroscience: Mirror neurons recorded in humans. *Current Biology*, 20(8):353 – 354, 2010.

[177] O. Khatib. *Commande dynamique dans l'espace operational des robots manipulateurs en presence d'obstacles*. PhD thesis, Ecole Nationale de la Statistique et de l'Administration Economique, France, 1980.

[178] O. Khatib. Real-time obstacle avoidance for manipulators and mobile robots. *International Journal of Robotics Research*, 5(1):90–98, 1986.

[179] S. Kierkegaard, H.V. Hong, and E. Hong. *Stages on Life's Way*. Princeton University Press, 1992.

[180] D. Kirsh. Today the earwig, tomorrow man? *Artificial Intelligence*, 47:161–184, 1991.

[181] T. Kitamura. Can a robot's adaptive behavior be animal-like without a learning algorithm ? In *Systems, Man, and Cybernetics, 1999. IEEE SMC '99 Conference Proceedings. 1999 IEEE International Conference on*, volume 2, pages 1047–1051, 1999.

[182] T. Kitamura and D. Nishino. Training of a leaning agent for navigation-inspired by brain-machine interface. *IEEE Transactions on Systems, Man, and Cybernetics, Part B (Cybernetics)*, 36(2):353–365, Apr 2006.

[183] T. Kitamura, Y. Otsuka, and T. Nakao. Imitation of animal behavior with use of a model of consciousness-behavior relation for a small robot. In *Proceedings of 4th IEEE International Workshop on Robot and Human Communication, 1995. RO-MAN'95 TOKYO*, pages 313–317, Jul 1995.

[184] Tadashi Kitamura, Tomoko Tahara, and Ken-Ichi Asami. How can a robot have consciousness? *Advanced Robotics*, 14(4):263–275, 2000.

[185] Zoe Kleinman. When will man become machine? `http://www.bbc.com/news/technology-30583218` [Online; accessed 26-October-2016], December 2014.

[186] D. Koditschek. Exact robot navigation by means of potential functions: Some topological considerations. In *Proceedings. 1987 IEEE International Conference on Robotics and Automation*, volume 4, pages 1–6, Mar 1987.

[187] Silke Koltrowitz and Marina Depetris. Swiss reject free income plan after worker vs. robot debate. `http://www.reuters.com/article/us-swiss-vote-idUSKCN0YR0CW` [Online; accessed 12-October-2016], June 2016.

[188] Melvin Kranzberg. Technology and history: "kranzberg's laws". *Technology and Culture*, 27(3):544–560, 1986.

[189] Jeffrey L. Krichmar. Design principles for biologically inspired cognitive robotics. *Biologically Inspired Cognitive Architectures*, 1:73–81, 2012.

[190] Matthew Kroh and Sricharan Chalikonda. *Essentials of Robotic Surgery*. Springer, 2014.

[191] C. Ronald Kube and Hong Zhang. Task modelling in collective robotics. *Autonomous Robots*, 4(1):53–72, 1997.

[192] Robert Lawrence Kuhn. What will happen when the internet of things becomes artificially intelligent? `http://www.space.com/30937-when-robots-colonize-cosmos-will-they-be-conscious.html` [Online; accessed 26-October-2016], October 2015.

[193] Ray Kurzweil. *The Singularity Is Near: When Humans Transcend Biology*. Penguin (Non-Classics), 2006.

[194] Ray Kurzweil. How my predictions are faring an update by ray kurzweil. `http://www.kurzweilai.net/how-my-predictions-are-faring-an-update-by-ray-kurzweil` [Online; accessed 24-May-2015], Oct 2010.

[195] K. N. Kutulakos, C. R. Dyer, and V. J. Lumelsky. Provable strategies for vision-guided exploration in three dimensions. In *Proceedings IEEE International Conference on Robotics & Automation*, pages 1365–1371, 1994.

[196] J. Lacan and B. Fink. *Ecrits: The First Complete Edition in English*. W. W. Norton, 2006.

[197] Kevin LaGrandeur. Technology in cross-cultural mythology: Western and non-western. `http://ieet.org/index.php/IEET/more/lagrandeur20130817` [Online; accessed 07-April-2015], August 2013.

[198] D. Lambrinos and C. Scheier. Building complete autonomous agents: a case study on categorization. In *Proceedings of the 1996 IEEE/RSJ International Conference on Intelligent Robots and Systems '96, IROS 96.*, volume 1, pages 170–177, Nov 1996.

[199] Jean-Claude Latombe. *Robot Motion Planning.* Kluwer Academic Publishers, Norwell, MA, USA, 1991.

[200] Steven M. LaValle. *Planning Algorithms.* Cambridge University Press, 2006. Available at http://planning.cs.uiuc.edu/.

[201] C.P. Lee-Johnson and D.A. Carnegie. Mobile robot navigation modulated by artificial emotions. *Systems, Man, and Cybernetics, Part B: Cybernetics, IEEE Transactions on*, 40(2):469–480, April 2010.

[202] David Levy. The ethics of robot prostitute. In Patrick Lin, George Bekey, and Keith Abney, editors, *Robot ethics: The ethical and social implications of robotics*, pages 223–232. MIT Press (MA), 2012.

[203] Blanca Li. Robot. http://www.blancali.com/en/event/99/Robot [Online; accessed 11-November-2016], July 2013.

[204] Blanca Li. Robot film - director's cut. http://www.blancali.com/en/event/117/Robot-film-directors-cut [Online; accessed 21-October-2016], February 2015.

[205] Fei Li, Weiting Liu, Xin Fu, Gabriella Bonsignori, Umberto Scarfogliero, Cesare Stefanini, and Paolo Dario. Jumping like an insect: Design and dynamic optimization of a jumping mini robot based on bio-mimetic inspiration. *Mechatronics*, 22(2):167–176, 2012.

[206] Andrew Liszewski. Disney just invented a one-legged robot that hops like tigger. http://gizmodo.com/disney-just-invented-a-one-legged-robot-that-hops-like-1787483677 [Online; accessed 25-January-2017], Oct 2016.

[207] Lyle Long and Troy Kelley. chapter The Requirements and Possibilities of Creating Conscious Systems. Infotech Aerospace Conferences. American Institute of Aeronautics and Astronautics, Apr 2009.

[208] L.S. Lopes, J.H. Connell, P. Dario, R. Murphy, P. Bonasso, B. Nebel, N. Nilsson, and R.A. Brooks. Sentience in robots: applications and challenges. *Intelligent Systems, IEEE*, 16(5):66–69, Sep 2001.

[209] Dylan Love. This law school professor believes robots may lead to an increase in prostitution. http://www.businessinsider.in/This-Law-School-Professor-Believes-Robots-May-Lead-To-An-Increase-In-Prostitution/articleshow/34059910.cms [Online; accessed 25-November-2016], April 2014.

[210] Michael Luck, Nathan Griffiths, and Mark d'Inverno. *From agent theory to agent construction: A case study*, pages 49–63. Springer Berlin Heidelberg, Berlin, Heidelberg, 1997.

[211] V. Lumelsky and A. Stepanov. Dynamic path planning for a mobile automaton with limited information on the environment. *IEEE Transactions on Automatic Control*, 31(11):1058–1063, Nov 1986.

[212] V. Lumelsky and A. Stepanov. Path-planning strategies for a point mobile automaton moving amidst unknown obstacles of arbitrary shape. *Algorithmica*, 2(1):403–430, Nov 1987.

[213] J Luyken. Are robots about to take away 18 million jobs? `http://www.thelocal.de/jobs/article/are-robots-about-to-take-away-18-million-jobs` [Online; accessed 05-November-2016], May 2015.

[214] Karl F MacDorman. Androids as an experimental apparatus: Why is there an uncanny valley and can we exploit it? In *CogSci-2005 workshop: toward social mechanisms of android science*, pages 106–118, 2005.

[215] Eric Mack. Elon Musk worries skynet is only five years off. `https://www.cnet.com/news/elon-musk-worries-skynet-is-only-five-years-off/` [Online; accessed 07-November-2016], Nov 2014.

[216] Alexis C Madrigal. Is this virtual worm the first sign of the singularity? `http://www.theatlantic.com/technology/archive/2013/05/is-this-virtual-worm-the-first-sign-of-the-singularity/275715/` [Online; accessed 26-October-2016], May 2013.

[217] Pattie Maes. The dynamics of action selection. In *Proceedings of the 11th International Joint Conference on Artificial Intelligence - Vol 2*, IJCAI'89, pages 991–997, San Francisco, CA, USA, 1989. Morgan Kaufmann Publishers Inc.

[218] Pattie Maes. How to do the right thing. *Connection Science Journal*, 1:291–323, 1989.

[219] Jordan A. Mann, Bruce A. MacDonald, I.-Han Kuo, Xingyan Li, and Elizabeth Broadbent. People respond better to robots than computer tablets delivering healthcare instructions. *Computers in Human Behavior*, 43(Supplement C):112 – 117, 2015.

[220] Juergen Manner and Junichi Takeno. Monad structures in the human brain. In *13th International Workshop on Computer Science and Information Technologies (CSIT'2011)*, pages 84–88, 2011.

[221] David Marr, Shimon Ullman, and Tomaso Poggio. *Vision - A Computational Investigation into the Human Representation and Processing of Visual Information.* MIT Press, 2010.

[222] Vitor Matos and Cristina P. Santos. Towards goal-directed biped locomotion: Combining cpgs and motion primitives. *Robotics and Autonomous Systems*, 62(12):1669 – 1690, 2014.

[223] Jason Mc. Dermott. The uncanny mountain of geeky jokes. `http://jasonya.com/wp/tag/uncanny-mountain/` [Online; accessed 29-April-2017].

[224] Lee McCauley. Ai armageddon and the three laws of robotics. *Ethics and Information Technology*, 9(2):153–164, 2007.

[225] Jeffrey L. McKinstry, Anil K. Seth, Gerald M. Edelman, and Jeffrey L. Krichmar. Embodied models of delayed neural responses: Spatiotemporal categorization and predictive motor control in brain based devices. *Neural Networks*, 21(4):553 – 561, 2008.

[226] Donald Melanson. Study shocker: babies think friendly robots are sentient. `http://www.engadget.com/2010/10/16/study-shocker-babies-think-friendly-robots-are-sentient/` [Online; accessed 16-April-2016], October 2010.

[227] M. Merleau-Ponty. *Eye and Mind*, pages 159–190. Northwestern University Press, 1964.

[228] M. Merleau-Ponty and D.A. Landes. *Phenomenology of Perception*. Routledge, 2013.

[229] T Metzinger. Two principles for robot ethics. In E. Hilgendorf and J.P.Gunther, editors, *Robotik und Gesetzgebung*, pages 263–302. Nomos.S, Baden-Baden, Germany, 2013.

[230] Jean-Baptiste Michel, Yuan Kui Shen, Aviva Presser Aiden, Adrian Veres, Matthew K. Gray, The Google Books Team, Joseph P. Pickett, Dale Holberg, Dan Clancy, Peter Norvig, Jon Orwant, Steven Pinker, Martin A. Nowak, and Erez Lieberman Aiden. Quantitative analysis of culture using millions of digitized books. *Science*, 2010.

[231] Jason Millar. You should have a say in your robot car's code of ethics. `https://www.wired.com/2014/09/set-the-ethics-robot-car/` [Online; accessed 17-March-2017], Sep 2014.

[232] Marc.G Millis. What is vision 21 ? In *VISION 21, Interdisciplinary Science and Engineering in the era of Cyber Space*, pages 3–6. NASA, 1993.

[233] Takashi Minato, Michihiro Shimada, Hiroshi Ishiguro, and Shoji Itakura. Development of an android robot for studying human-robot interaction. In Bob Orchard, Chunsheng Yang, and Moonis Ali, editors, *Innovations in Applied Artificial Intelligence*, volume 3029 of *Lecture Notes in Computer Science*, pages 424–434. Springer Berlin Heidelberg, 2004.

[234] Javier Minguez and L. Montano. Nearness diagram (nd) navigation: collision avoidance in troublesome scenarios. *IEEE Transactions on Robotics and Automation*, 20(1):45–59, Feb 2004.

[235] Carl Mitcham. *Thinking through Technology:The Path between Engineering and Philosophy*. University of Chicago Press, 1994.

[236] T Modis. The singularity myth. *Technological Forecasting & Social Change*, 73(2), 2006.

[237] Christoph Moeslinger, Thomas Schmickl, and Karl Crailsheim. Emergent flocking with low-end swarm robots. In *Proceedings of the 7th International Conference on Swarm Intelligence*, ANTS'10, pages 424–431, Berlin, Heidelberg, 2010. Springer-Verlag.

[238] Christoph Moeslinger, Thomas Schmickl, and Karl Crailsheim. A minimalist flocking algorithm for swarm robots. In *Proceedings of the 10th European Conference on Advances in Artificial Life: Darwin Meets Von Neumann - Vol Part II*, ECAL'09, pages 375–382, Berlin, Heidelberg, 2011. Springer-Verlag.

[239] James H. Moor. The nature, importance, and difficulty of machine ethics. *IEEE Intelligent Systems*, 21(4):18–21, Jul 2006.

[240] James.H Moor. Four kinds of ethical robots. *Philosophy Now*, 72:12–14, March 2009.

[241] Hans.P Moravec. *Robot: Mere Machine to Transcendent Mind*. Oxford University Press, Inc., New York, NY, USA, 2000.

[242] M. Mori and C.S. Terry. *The Buddha in the robot*. Kosei Pub. Co., 1981.

[243] Randall Munroe. Robot apocalypse. `https://what-if.xkcd.com/5/` [Online; accessed 12-Oct-2014].

[244] Randall Munroe. Hooray robots! `http://blog.xkcd.com/2008/04/22/hooray-robots/` [Online; accessed 28-Feb-2015], April 2008.

[245] Randall Munroe. New pet. `https://xkcd.com/413/` [Online; accessed 28-Feb-2015], April 2008.

[246] Danielle Muoio. Japan is running out of people to take care of the elderly, so it's making robots instead. `http://www.businessinsider.com/japan-developing-carebots-for-elderly-care-2015-11?IR=T` [Online; accessed 05-May-2017], Nov 2015.

[247] Robin Murphy and David D. Woods. Beyond asimov: The three laws of responsible robotics. *IEEE Intelligent Systems*, 24(4):14–20, July 2009.

[248] Robin R. Murphy. *Introduction to AI Robotics*. MIT Press, 2001.

[249] Thomas Nagel. What Is It Like to Be a Bat? *The Philosophical Review*, 83(4):435–450, Oct 1974.

[250] J. Nakanishi, T. Fukuda, and D. E. Koditschek. A brachiating robot controller. *IEEE Transactions on Robotics and Automation*, 16(2):109–123, Apr 2000.

[251] M. A. Nasseri and M. Asadpour. Control of flocking behavior using informed agents: An experimental study. In *Swarm Intelligence (SIS), 2011 IEEE Symposium on*, pages 1–6, April 2011.

[252] AFP Relax News. Meet kodomoroid, otonaroid: World's first android newscasters. `http://www.thestar.com.my/lifestyle/features/2014/07/01/meet-kodomoroid-otonaroid-worlds-first-android-newscasters/` [Online; accessed 20-June-2016], July 2014.

[253] Nils J. Nilsson. Shakey the robot, technical note 323. *SRI International*, (323):1–149, April 1984.

[254] T Norretranders. *The User Illusion: Cutting Consciousness Down to Size*. Penguin Books, 1999.

[255] Tim Oates. Stop fearing artificial intelligence. `https://techcrunch.com/2015/04/08/stop-fearing-artificial-intelligence/` [Online; accessed 27-October-2016], April 2015.

[256] Natsuki Oka. Apparent free will caused by representation of module control. In *No Matter, Never Mind*, pages 243–249. John Benjamins, 2002.

[257] T. Okuda, Y. Sago, T. Ideguchi, Xuejun Tian, and H. Shibayama. Performability evaluation of network humanoid robot system on ubiquitious network. In *VTC-2005-Fall. 2005 IEEE 62nd Vehicular Technology Conference, 2005.*, volume 3, pages 1834–1838, Sept 2005.

[258] Sean O'Neill. Beyond the imitation game. *New Scientist*, 224(2999):27, 2014.

[259] J. Kevin O'Regan and Alva Noe. A sensorimotor account of vision and visual consciousness. *Behavioral and Brain Sciences*, 24(5):939–1031, 2001.

[260] J. Kevin O'Regan and Alva Noe. What it is like to see: A sensorimotor theory of perceptual experience. *Synthese*, 129(1):79–103, 2001.

[261] Carlos Pelta. "fungus eaters" and artificial intelligence: a tribute to masanao toda. `http://artificial-socialcognition.blogspot.in/2008/09/fungus-eaters-and-complex-systems.html` [Online; accessed 15-Nov-2014], Sep 2008.

[262] Rolf Pfeifer. Building "fungus eaters": Design principles of autonomous agents. In *In Proceedings of the Fourth International Conference on Simulation of Adaptive Behavior SAB96 (From Animals to Animats)*, pages 3–12. MIT Press, 1996.

[263] Ayse Pinar Saygin, Ilyas Cicekli, and Varol Akman. Turing test: 50 years later. *Minds Mach.*, 10(4):463–518, Nov 2000.

[264] C. Pinciroli, V. Trianni, R. O'Grady, G. Pini, A. Brutschy, M. Brambilla, N. Mathews, E. Ferrante, G. Di Caro, F. Ducatelle, T. Stirling, . Gutirrez, L. M. Gambardella, and M. Dorigo. Argos: A modular, multi-engine simulator for heterogeneous swarm robotics. In *2011 IEEE RSJ International Conference on Intelligent Robots and Systems*, pages 5027–5034, Sept 2011.

[265] David Pogue. Do we need to prepare for the robot uprising? `https://www.scientificamerican.com/article/do-we-need-to-prepare-for-the-robot-uprising/` [Online; accessed 05-November-2016], Oct 2015.

[266] J.B. Pollack. Seven questions for the age of robots. In *Yale Bioethics Seminar*, Jan 2004.

[267] J.B. Pollack. Ethics for the robot age. *Wired*, 13(1), Jan 2005.

[268] Dean A. Pomerleau. Alvinn, an autonomous land vehicle in a neural network. `http://repository.cmu.edu/compsci`, 1989.

[269] Zenon W. Pylyshyn. *The Robot's dilemma : the frame problem in artificial intelligence.* Theoretical issues in cognitive science. Ablex, Norwood, NJ, 1987.

[270] Morgan Quigley, Ken Conley, Brian P Gerkey, Josh Faust, Tully Foote, Jeremy Leibs, Rob Wheeler, and Andrew Y Ng. Ros: an open-source robot operating system. In *ICRA Workshop on Open Source Software*, volume 3, 01 2009.

[271] A. Ram, R. C. Arkin, K. Moorman, and R. J. Clark. Case-based reactive navigation: a method for on-line selection and adaptation of reactive robotic control parameters. *IEEE Transactions on Systems, Man, and Cybernetics, Part B (Cybernetics)*, 27(3):376–394, Jun 1997.

[272] Ashwin Ram and Juan Carlos Santamara. Continuous case-based reasoning. *Artificial Intelligence*, 90:86–93, 1993.

[273] I. Rano. 2012 ieee international conference on robotics and automation (icra). In *A model and formal analysis of Braitenberg vehicles 2 and 3*, pages 910–915, May 2012.

[274] BSI Press Release. Standard highlighting the ethical hazards of robots is published. `http://www.bsigroup.com/en-GB/about-bsi/media-centre/press-releases/2016/april/-Standard--highlighting-the-ethical-hazards-of-robots-is-published/` [Online; accessed 01-November-2016], April 2016.

[275] Craig W. Reynolds. Flocks, herds and schools: A distributed behavioral model. *SIGGRAPH Comput. Graph.*, 21(4):25–34, August 1987.

[276] Jeremy Rifkin. *The Zero Marginal Cost Society: The Internet of Things, the Collaborative Commons, and the Eclipse of Capitalism*. St. Martin's Griffin, New York, NY, USA, 2015.

[277] G. Rizzolatti and L. Craighero. The mirror-neuron system. *Annual Review of Neuroscience*, 27:169–192, 2004.

[278] Jennifer Robertson. Robo sapiens japanicus: Humanoid robots and the posthuman family. *Critical Asian Studies*, 39(3):369–398, 2007.

[279] Hayley Robinson, Bruce MacDonald, Ngaire Kerse, and Elizabeth Broadbent. The psychosocial effects of a companion robot: A randomized controlled trial. *Journal of the American Medical Directors Association*, 14(9):661 – 667, 2013.

[280] E. M. A. Ronald and M. Sipper. Surprise versus unsurprise: Implications of emergence in robotics. *Robotics and Autonomous Systems*, 37(1):19–24, October 2001.

[281] E. M. A. Ronald, M. Sipper, and M. S. Capcarr'ere. Design, observation, surprise ! a test of emergence. *Artificial Life*, 5(3):225–239, Summer 1999.

[282] Kenneth Rosenblatt and David Payton. A fine-grained alternative to the subsumption architecture for mobile robot control. In *Proceedings of the AAAI Symposium on Robot Navigation*, pages 317–324, 1989.

[283] Mark Elling Rosheim. *Leonardo's lost robots*. Springer, Berlin, Heidelberg, Germany, 2006.

[284] Aviva Rutkin. Ethical trap: robot paralysed by choice of who to save. `http://www.newscientist.com/article/mg22329863.700-ethical-trap-robot-paralysed-by-choice-of-who-to-save.html#.VEpd2t-jmb8` [Online; accessed 24-Oct-2014], Sep 2014.

[285] Selma Sabanovic. Inventing japans 'robotics culture: The repeated assembly of science, technology, and culture in social robotics. *Social Studies of Science*, 44(3):342–367, 2014.

[286] Alessandro Saffiotti and Mathias Broxvall. Peis ecologies: Ambient intelligence meets autonomous robotics. 121, 10 2005.

[287] Erol Sahin. *Swarm Robotics: From Sources of Inspiration to Domains of Application*, pages 10–20. Springer Berlin Heidelberg, Berlin, Heidelberg, 2005.

[288] Uptin Saiidi. Here's why japan is obsessed with robots. `http://www.cnbc.com/2017/03/09/heres-why-japan-is-obsessed-with-robots.html` [Online; accessed 05-May-2017], March 2017.

[289] S. Schaal. The new robotics - towards human-centered machines. 1(2):115–126, 2007.

[290] Christian Scheier and Rolf Pfeifer. *Classification as sensory-motor coordination*, pages 657–667. Springer Berlin Heidelberg, Berlin, Heidelberg, 1995.

[291] Thomas Schmickl, Christoph Moslinger, and Karl Crailsheim. Collective perception in a robot swarm. SAB'06, pages 144–157, Berlin, Heidelberg, 2007. Springer-Verlag.

[292] Susan Schneider. The problem of AI consciousness. `http://www.huffingtonpost.com/entry/the-problem-of-ai-conscio_b_9502790.html` [Online; accessed 25-March-2016], March 2016.

[293] M. J. Schoppers. In defense of reaction plans as caches. *AI Mag.*, 10(4):51–60, Nov 1989.

[294] John R Searle. *Mind, Language And Society: Philosophy In The Real World*. Basic Books, 1999.

[295] Matthew Sedacca. Are robots really destined to take over restaurant kitchens? `http://www.eater.com/2016/8/29/12660074/robot-restaurant-kitchen-labor` [Online; accessed 26-October-2016], August 2016.

[296] Gregory Sempo, Stephanie Depickere, Jean-Marc Ame, Claire Detrain, Jose Halloy, and Jean-Louis Deneubourg. Integration of an autonomous artificial agent in an insect society: Experimental validation. In *From Animals to Animats 9, 9th International Conference on Simulation of Adaptive Behavior, SAB 2006, Rome, Italy, September 25-29, 2006, Proceedings*, pages 703–712, 2006.

[297] Kerstin Severinson-Eklundh, Anders Green, and Helge Httenrauch. Social and collaborative aspects of interaction with a service robot. *Robotics and Autonomous Systems*, 42(3):223 – 234, 2003.

[298] Amanda Sharkey. Robots and human dignity: a consideration of the effects of robot care on the dignity of older people. *Ethics and Information Technology*, 16(1):63–75, 2014.

[299] Amanda Sharkey and Noel Sharkey. Granny and the robots: ethical issues in robot care for the elderly. *Ethics and Information Technology*, 14(1):27–40, 2012.

[300] N Sharkey. Cassandra or false prophet of doom: AI robots and war. *IEEE Intelligent Systems*, 23(4):14–17, July 2008.

[301] N Sharkey. Grounds for discrimination: Autonomous robot weapons. *RUSI Defence Systems*, 11(2):86–89, 2008.

[302] Noel Sharkey. The ethical frontiers of robotics. *Science*, 322(5909):1800–1801, 2008.

[303] Tory Shepherd. Robots could soon take over the world. no, this isnt science fiction. `http://www.dailytelegraph.com.au/rendezview/robots-could-soon-take-over-the-world-no-this-isnt-science-fiction/news-story/64627dcebec15f08f12735efdd74b628` [Online; accessed 26-October-2016], August 2015.

[304] Herbert A. Simon. A behavioral model of rational choice. *The Quarterly Journal of Economics*, 69(1):99–118, 1955.

[305] Herbert A. Simon. Rational choice and the structure of the environment. *Psychological Review*, 63(2):129–138, 1956.

[306] Erik Sofge. An open letter to everyone tricked into fearing artificial intelligence. `http://www.popsci.com/open-letter-everyone-tricked-fearing-ai` [Online; accessed 27-October-2016], January 2015.

[307] Robert Sparrow. The Turing triage test. *Ethics and Information Technology*, 6(4):203–213, 2004.

[308] Dirk H. R. Spennemann. On the cultural heritage of robots. *International Journal of Heritage Studies*, 13(1):4–21, 2007.

[309] Janusz A. Starzyk and Dilip K. Prasad. A computational model of machine consciousness. *International Journal of Machine Consciousness*, 03(02):255–281, 2011.

[310] Luc Steels. The symbol grounding problem has been solved. so what's next. In Manuel de Vega, Arthur M. Glenberg, and Arthur C. Graesser, editors, *Symbols and Embodiment: Debates on Meaning and Cognition*, pages 223–244. Oxford University Press, 2008.

[311] Zachary Stewart. Are robotic actors the future of live theater ? `http://www.theatermania.com/new-york-city-theater/news/are-robotic-actors-the-future-of-live-theater_71083.html` [Online; accessed 25-November-2016], Jan 2015.

[312] Andrew Stokes. The ethics of driverless cars. `http://www.queensu.ca/gazette/stories/ethics-driverless-cars` [Online; accessed 17-March-2017], Aug 2014.

[313] J. P. Sullins. Robots, love, and sex: The ethics of building a love machine. *IEEE Transactions on Affective Computing*, 3(4):398–409, April 2012.

[314] John P. Sullins. When is a robot a moral agent ? In Susan Leigh Anderson Michael Anderson, editor, *Machine Ethics*, pages 151 – 161. Cambridge University Press, Norwood, NJ, 2011.

[315] Tohru Suzuki, Keita Inaba, and Junichi Takeno. *Conscious Robot That Distinguishes Between Self and Others and Implements Imitation Behavior*, pages 101–110. Springer Berlin Heidelberg, Berlin, Heidelberg, 2005.

[316] Junichi Takeno. *Creation of a Conscious Robot: Mirror Image Cognition and Self-Awareness*. Pan Stanford, 2012.

[317] Nobuko Tanaka. Can robots be chips off the bards block? `http://www.japantimes.co.jp/life/2010/08/15/general/can-robots-be-chips-off-the-bards-block#.WCXJYtGkU_s` [Online; accessed 11-November-2016], Aug 2010.

[318] Elana Teitelbaum. 5 ways skynet is more real than you think. `http://www.huffingtonpost.in/entry/skynet-real_n_7042808` [Online; accessed 07-November-2016], June 2015.

[319] Chuck Thompson and Elaine Yu. New order? China restaurant debuts robot waiters. `http://edition.cnn.com/2016/04/19/travel/china-robot-waiters/` [Online; accessed 26-October-2016], April 2016.

[320] Evan Thompson. *Mind in Life - Biology, Phenomenology, and the Sciences of Mind.* Harvard University Press, 2010.

[321] James Titcomb. Stephen hawking says artificial intelligence could be humanity's greatest disaster. `http://www.telegraph.co.uk/technology/2016/10/19/stephen-hawking-says-artificial-intelligence-could-be-humanitys/` [Online; accessed 26-October-2016], October 2016.

[322] Masanao Toda. The design of a fungus-eater: A model of human behavior in an unsophisticated environment. *Behavioral Science*, 7(2):164–183, 1962.

[323] Steve Torrance. Machine ethics and the idea of a more-than-human moral world. In Susan Leigh Anderson Michael Anderson, editor, *Machine Ethics*, pages 115–137. Cambridge University Press, Norwood, NJ, 2011.

[324] David S Touretzky. The hearts of symbols: Why symbol grounding is irrelevant. In *Proceedings of the Fifteenth Annual Conference of the Cognitive Science Society*, pages 165–168, 1993.

[325] David S. Touretzky. Seven big ideas in robotics, and how to teach them. In Laurie A. Smith King, David R. Musicant, Tracy Camp, and Paul T. Tymann, editors, *SIGCSE*, pages 39–44. ACM, 2012.

[326] David S. Touretzky. Robotics for computer scientists: whats the big idea? *Computer Science Education*, 23(4):349–367, 2013.

[327] Gregory Trencher. Os homo sapiens 2.0: New human software coming soon? `http://ourworld.unu.edu/en/os-homo-sapiens-2-0-new-human-software-coming-soon` [Online; accessed 24-May-2015], May 2011.

[328] Ali E Turgut, Hande Celikkanat, Fatih Gokce, and Erol Sahin. Self-organized flocking in mobile robot swarms. *Swarm Intelligence*, 2(2-4):97–120, 2008.

[329] Ali Emre Turgut, Cristian Huepe, Hande Celikkanat, Fatih Gkffe, and Erol Sahin. Modeling phase transition in self-organized mobile robot flocks. In *International Conference on Ant Colony Optimization and Swarm Intelligence*, pages 108–119. Springer Berlin Heidelberg, 2008.

[330] Toby Tyrrell. *Computational Mechanisms for Action Selection*. PhD thesis, University of Edinburgh, 1993.

[331] Tim Urban. The AI Revolution: Our immortality or extinction. `http://waitbutwhy.com/2015/01/artificial-intelligence-revolution-2.html` [Online; accessed 24-May-2015], Jan 2015.

[332] Tim Urban. The AI Revolution: The road to superintelligence. `http://waitbutwhy.com/2015/01/artificial-intelligence-revolution-1.html` [Online; accessed 24-May-2015], Jan 2015.

[333] Francisco J. Varela, Evan Thompson, and Eleanor Rosch. *The Embodied Mind - Cognitive Science and Human Experience*. MIT Press, 1993.

[334] Richard Vaughan. Massively multi-robot simulation in stage. *Swarm Intelligence*, 2(2):189–208, Dec 2008.

[335] Richard Vaughan, Neil Sumpter, Jane Henderson, Andy Frost, and Stephen Cameron. Experiments in automatic flock control. *Robotics and Autonomous Systems*, 31(12):109–117, 2000.

[336] Tamas Vicsek, Andras Czirok, Eshel Ben-Jacob, Inon Cohen, and Ofer Shochet. Novel type of phase transition in a system of self-driven particles. *Phys. Rev. Lett.*, 75(6):1226–1229, Aug 1995.

[337] D. Vikerimark and J. Minguez. Reactive obstacle avoidance for mobile robots that operate in confined 3d workspaces. In *MELECON 2006 - 2006 IEEE Mediterranean Electrotechnical Conference*, pages 1246–1251, May 2006.

[338] Vernor Vinge. The coming technological singularity: How to survive in the post-human era. In *VISION 21, Interdisciplinary Science and Engineering in the era of Cyber Space*, pages 11–22. NASA, 1993.

[339] Vernor Vinge. Signs of the singularity. *IEEE Spectrum*, 45(6):76–82, June 2008.

[340] Paul Vogt. The physical symbol grounding problem. *Cognitive Systems Research*, 3(3):429–457, 2002.

[341] M. Waibel, M. Beetz, J. Civera, R. D'Andrea, J. Elfring, D. Galvez-Lopez, K. Haussermann, R. Janssen, J.M.M. Montiel, A. Perzylo, B. Schiessle, M. Tenorth, O. Zweigle, and R. van de Molengraft. Roboearth. *Robotics Automation Magazine, IEEE*, 18(2):69–82, 2011.

[342] Jake Wakefield. Intelligent machines: Do we really need to fear ai? `http://www.bbc.com/news/technology-32334568` [Online; accessed 27-October-2016], Sep 2015.

[343] Wendell Wallach and Colin Allen. *Moral Machines: Teaching Robots Right from Wrong*. Oxford University Press, Inc., New York, NY, USA, 2010.

[344] W. Grey Walter. An imitation of life. *Scientific American*, 182(5):42 – 45, 1950.

[345] W. Grey Walter. A machine that learns. *Scientific American*, 185(2):60 – 63, 1951.

[346] D. Wang, D. K. Liu, N. M. Kwok, and K. J. Waldron. A subgoal-guided force field method for robot navigation. In *2008 IEEE/ASME International Conference on Mechtronic and Embedded Systems and Applications*, pages 488–493, Oct 2008.

[347] Mary Anne Warren. On the moral and legal status of abortion. *The Monist*, 57(1):43–61, 1973.

[348] Peter Kelley-U Washington. Do we need new laws for rise of the robots? `http://www.futurity.org/laws-robots-artificial-intelligence-959762/` [Online; accessed 27-October-2016], July 2015.

[349] J M Watts. Animats: computer-simulated animals in behavioral research. *Journal of Animal Science*, 76(10):2596–2604, October 1998.

[350] Mark Weiser. The computer for the 21st century. *SIGMOBILE Mob. Comput. Commun. Rev.*, 3(3):3–11, July 1999.

[351] N. Wiener. *Cybernetics Or Control and Communication in the Animal and the Machine*. M.I.T. Pr.paperback.23. M.I.T. Press, 1961.

[352] Norbert Wiener. New concept of communication engineering. *Electronics*, pages 74–77, Jan 1949.

[353] Stuart Wilkinson. "gastrobots"—benefits and challenges of microbial fuel cells in foodpowered robot applications. *Autonomous Robots*, 9(2):99–111, 2000.

[354] Margaret Wilson. Six views of embodied cognition. *Psychonomic Bulletin & Review*, 9(4):625–636, 2002.

[355] Nathan Wilson. How i learned to stop worrying and love ai. `https://techcrunch.com/2015/03/12/how-i-learned-to-stop-worrying-and-love-ai/` [Online; accessed 27-October-2016], March 2015.

[356] Stewart W. Wilson. Knowledge growth in an artificial animal. In *Proceedings of the 1st International Conference on Genetic Algorithms*, pages 16–23, Hillsdale, NJ, USA, 1985. L. Erlbaum Associates Inc.

[357] Stewart W. Wilson. The animat path to ai. In *Animals to Animats: Proceedings of the First International Conference on the Simulation of Adaptive Behavior, Cambridge, Massachusetts*. The MIT Press/Bradford Books, 1991.

[358] A.F.T. Winfield and O.E. Holland. The application of wireless local area network technology to the control of mobile robots. *Microprocessors and Microsystems*, 23(10):597 – 607, 2000.

[359] Alan Winfield. Walterian creatures. `http://alanwinfield.blogspot.in/2007/04/walterian-creatures.html` [Online; accessed 25-Oct-2014], April 2007.

[360] Alan Winfield. *Robotics: A Very Short Introduction*. Very Short Introductions Series. OUP Oxford, 2012.

[361] Alan Winfield. Ethical robots: some technical and ethical challenges. `http://alanwinfield.blogspot.co.uk/2013/10/ethical-robots-some-technical-and.html` [Online; accessed 25-Oct-2014], 2013.

[362] Alan Winfield. On internal models, consequence engines and popperian creatures. `http://www.alanwinfield.blogspot.co.uk/2014/07/on-internal-models-consequence-engines.html` [Online; accessed 25-Oct-2014], July 2014.

[363] Alan Winfield. private communication, 2015.

[364] Alan Winfield. Text on, social, legal and ethical issues of AI. `http://hamlyn.doc.ic.ac.uk/uk-ras/ethics-regulation-and-governance` [Online; accessed 27-October-2016], April 2016.

[365] Alan F. T. Winfield, Christopher J. Harper, and Julien Nembrini. *Towards Dependable Swarms and a New Discipline of Swarm Engineering*, pages 126–142. Springer Berlin Heidelberg, Berlin, Heidelberg, 2005.

[366] Alan F.T. Winfield, Christian Blum, and Wenguo Liu. Towards an ethical robot: Internal models, consequences and ethical action selection. In M.Mistry, A.Leonardis, M.Witkowski, and C.Melhuish, editors, *Advances in Autonomous Robotics Systems, 15th Annual Conference, TAROS 2014, Birmingham, UK, September 1-3, 2014. Proceedings*, page 8596. Springer, 2014.

[367] Justin Wintle. *The Concise New Makers of Modern Culture*. Routledge, November 2008.

[368] Gaby Wood. *Living Dolls: A Magical History Of The Quest For Mechanical Life*. Faber and Faber, London, United Kingdom, 2003.

[369] N. Wood, A. Sharkey, G. Mountain, and A. Millings. The paro robot seal as a social mediator for healthy users. In *4th International Symposium on New Frontiers in Human-Robot Interaction*. University of Kent, Jan 2015.

[370] Roman V. Yampolskiy. *Artificial Superintelligence: A Futuristic Approach*. Chapman & Hall/CRC, 2015.

[371] Tom Ziemke. On the role of robot simulations in embodied cognitive science. *AISB Journal*, 1:389–399, 01 2003.

Index

Milton Keynes UK
Ingram Content Group UK Ltd.
UKHW051943071024
449327UK00026B/2143